Digital Audio Fundamentals
in MATLAB

Mohammad Nuruzzaman

Electrical Engineering Department
King Fahd University of Petroleum & Minerals
Dhahran, Saudi Arabia

C R E A T E S P A C E
100 Enterprise Way, Suite A200
Scotts Valley, California 95066
www.createspace.com

Dr. Mohammad Nuruzzaman
Electrical Engineering Department
King Fahd University of Petroleum and Minerals
KFUPM BOX 1286
Dhahran 31261, Saudi Arabia
Email: nzaman@kfupm.edu.sa
Web Link: http://faculty.kfupm.edu.sa/EE/NZAMAN/

ISBN-10: 1-4536-2501-1
ISBN-13: 978-1453625019

Printed in the United States of America

This book is printed on acid-free paper.

To my parents
Mohammad Shamsul Haque & Nurbanu Begum

Other titles by the author:

1. M. Nuruzzaman, *"Modern Approach to Solving Electromagnetics in MATLAB"*, January, 2009, BookSurge Publishing, Charleston, South Carolina.
2. M. Nuruzzaman, *"Signal and System Fundamentals in MATLAB and SIMULINK"*, July, 2008, BookSurge Publishing, Charleston, South Carolina.
3. M. Nuruzzaman, *"Electric Circuit Fundamentals in MATLAB and SIMULINK"*, October, 2007, BookSurge Publishing, Charleston, South Carolina.
4. M. Nuruzzaman, *"Technical Computation and Visualization in MATLAB for Engineers and Scientists"*, February, 2007, AuthorHouse, Bloomington, Indiana.
5. M. Nuruzzaman, *"Digital Image Fundamentals in MATLAB"*, September, 2005, AuthorHouse, Bloomington, Indiana.
6. M. Nuruzzaman, *"Modeling and Simulation in SIMULINK for Engineers and Scientists"*, January, 2005, AuthorHouse, Bloomington, Indiana.
7. M. Nuruzzaman, *"Tutorials on Mathematics to MATLAB"*, March, 2003, AuthorHouse, Bloomington, Indiana.

Preface

The text emphasizes the fundamental principles and applications of digital audio processing realization in the platform of MATLAB, a scientific and a technical computing software. Tremendous development of digital signal processing (DSP) chips and fast computers have shifted the trend in the study and research of audio engineering field toward digital form rather than the analog. By no means we will ever get rid of analog concepts and terminologies as far as true nature of an audio is concerned. A realistic and unprocessed audio signal is always continuous in time or analog no matter how split resolution signal processing device we develop. For this reason every so often we have drawn parallelism between the continuous and the discrete theory subject matters in the book.

The aim of this text is to render implementional and accessible MATLAB tools and material connected to the fundamental theory and application of digital audio signals. But the goal is ambitious since digital audio processing covers diverse spectrum of topics. Certainly several volumes would be required to include the entire subject area. We have tried to present the very basics of the digital audio theory implementation in an easeful way so that average and interested readers develop a useful and a hand-on understanding of modern digital audio engineering. Almost every chapter introduces the terms and concepts first considering a prototype discrete signal, computer execution of which for a digital audio follows latter. As a branch of signal processing we believe this approach of explaining concept-to-implementation for digital audio study is neoteric furthermore it does not require extra hardware or DSP chip.

Audio engineering is an ever evolving field. Importance of the field lies in the fact that video or audio devices attract human attention more than any other consumer electronics do. Scientists, researchers, and government organizations have been continuously working in the field to improve the digital audio related algorithms or devices and the end of the field is no where soon. The resolution whether in audio amplitude or time scale has advanced significantly especially in the last two decades e.g. from micro to nano and probably in future from nano to optical. However some contemporary use of the digital audio related devices is in the following:

✦ ✦ Audio or music files downloaded from the internet are basically digital audio
✦ ✦ Indoor audio components such as CD and computer music use digital audio
✦ ✦ In detection fields such as speech and music recognitions
✦ ✦ Usage of speech coding-decoding in digital cellular telephony
✦ ✦ Most communication and navigation systems use digital audio
✦ ✦ Text to speech (TTS) systems use digital audio
✦ ✦ Voice operated systems such as automatic door use digital audio
✦ ✦ Voice over internet, PC to Phone, etc use digital audio
        …   and many more.

"What is the implication of practicing audio study and research in MATLAB while other dedicated software and hardware are available?" – this question might be asked by the reader first. Twenty first century themes need twenty first century solution. Literally every one is having mobile and PC or laptop these days. Study and research materials need to be reached to individual beside the personnel of corporate audio laboratories. An individual may also contribute in the field of research if proper material access is given. In this era of globalization our way of solving problems demands change. Incidentally the text introduces evolutionary approach to present the text material contrarily the traditional style. MATLAB (abbreviation for matrix laboratory) as a computing software is gaining more and more attentions from scientific and engineering communities. Built-in functions embedded in MATLAB do not require reprogramming an already solved problem, nor does clumsy compiling often encountered in base language such as in FORTRAN or C. Facilitating introduction of the study tools for digital audio makes this text perfect for self-learners. The readers themselves can verify audio theory terms through MATLAB means rather than heavily following the instructor note.

Chapter 1 presents a brief introduction to MATLAB's getting-started features. Chapter 2 addresses digital audio fundamentals. Chapter 3 merely demonstrates the elementary operations on a digital audio but initiating focus on a discrete time system since a digital audio is one kind of discrete time system. Issues related to the discrete Fourier transform (DFT) linking to digital audio both in forward and inverse domains cover the chapter 4. If audio signal spectral characteristic is sought against time, then the discrete short time Fourier transform (STFT) is the correct tool for the study i.e. STFT is a time-frequency based analysis-synthesis approach whose discourse is placed in chapter 5. The Z transform and its associated digital audio problems are handled in chapter 6. Understanding features of an audio by matrix algebra is an essential reading of digital audio which we concentrate on in chapter 7. Chapter 8 demonstrates digital audio statistics related problem implementation. Linear prediction coding (LPC) has widespread applications on digital audio that is why chapter 9 is devised with examples. The spectrum of audio enhancement is huge truly speaking another title is necessary to focus the noise and enhancement issues completely. Nevertheless some flavor of the issue is illustrated in chapter 10. To make the text less voluminous yet covering all relevant topics, some necessary fundamental audio processing terms are addressed in chapter 11. Lastly chapter 12 randomly presents the exercise problems under the heading mini project ranging from the easiest to hardest along with the solutions.

I wish to express my acknowledgement to the King Fahd University of Petroleum and Minerals (KFUPM). I sincerely appreciate the library and laboratory facilities that I received from the King Fahd University. All audio processing problems covered in the text have been conducted by a Pentium IV Laptop on Microsoft Windows operating system.

Mohammad Nuruzzaman

Table of Contents

Chapter 5
Discrete Short Time Fourier Transform and Digital Audio

Chapter 6
Z Transform and Digital Audio

Chapter 7
Matrix Algebra and Digital Audio

Chapter 8
Statistics and Digital Audio

Chapter 9
Linear Prediction and Digital Audio

Chapter 10
Noise and Elementary Speech Enhancement

Chapter 11
Miscellaneous Topics

Chapter 12
Mini Problems on Discrete Signal-System and Digital Audio

Appendices

Chapter 1

 Introduction to MATLAB

MATLAB is a computing software, which provides the quickest and the easiest way to compute scientific and technical problems and visualize the solutions. As worldly standard for simulation and analysis, engineers, scientists, and researchers are becoming more and more affiliated with MATLAB. The general questionnaires about MATLAB platform before one gets started with are the contents of this chapter. Much of the MATLAB computational style presupposes that the element to be handled is a vector or matrix. Our explanation highlights the following:

✦ ✦ MATLAB features found in its opened command window
✦ ✦ The easiest and the quickest way to get started in MATLAB beginning from scratch
✦ ✦ Frequently encountered questions while working in MATLAB environment
✦ ✦ Relevant introductory topics and forms of assistance about MATLAB

1.1 What is MATLAB?

MATLAB is mainly a scientific and technical computing software whose elaboration is matrix laboratory. The command prompt of MATLAB (>>) provides an interactive system. In the workspace of MATLAB, most data element is dealt as a matrix without dimensioning. The package is incredibly advantageous for the matrix-oriented computations. MATLAB's easy-to-use platform enables us to compute and manipulate matrices, perform numerical analyses, and visualize different variety of one/two/three dimensional graphics in a matter of second or seconds without conventional programming as done in FORTRAN, PASCAL, or C.

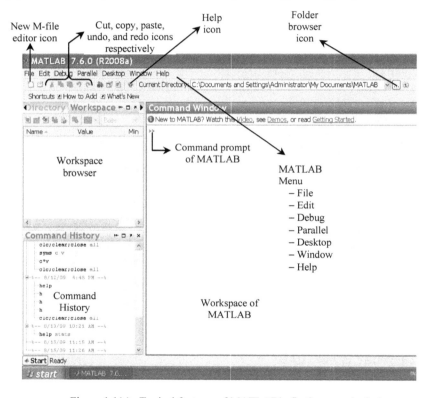

1.2 MATLAB's opening window features

If you do not have MATLAB installed in your personal computer, contact MathWorks (owner and developer, www.mathworks.com) for the installation CD. If you know how to get in MATLAB and its basics, you can skip the chapter. Assuming the package is installed in your system, run MATLAB from the Start of the Microsoft Windows. Let us get familiarized with MATLAB's opening window features. Figure 1.1(a) shows a typical firstly opened MATLAB window. Depending on the desktop setting or MATLAB version, your MATLAB window may not look like the figure 1.1(a) but the descriptions of the features by and large are appropriate.

Figure 1.1(a)   Typical features of MATLAB's firstly opened window

♦ ♦ **Command prompt of MATLAB**

Command prompt means that you tell MATLAB to do something from here. As an interactive system, MATLAB responds to user through this prompt. MATLAB cursor will be blinking after >> prompt once you open MATLAB that says MATLAB is ready to take your commands. To enter any command, type executable MATLAB statements from keyboard and to execute that, press the Enter key (the symbol ↵ is for the 'Hit the Enter Key' operation).

♣ ♣ **MATLAB Menu**

MATLAB is supplied with seven submenus namely File, Edit, Debug, Parallel, Desktop, Window, and Help. Each submenu has its own features. Use the mouse to click different submenus and their brief descriptions are as follows:

Figure 1.1(b) Submenu File

Figure 1.1(c) Submenu Edit

Submenu File: It (figure 1.1(b)) opens a new M-file, figure, model, or Graphical User Interface (GUI) layout maker, opens a file which was saved before, loads a saved workspace, imports data from a file, saves the workspace variables, sets the required path to execute a file, prints the workspace, and keeps the provision for changing the command window property.

Figure 1.1(d) Submenu Debug

Submenu Edit: The second submenu Edit (figure 1.1(c)) includes cutting, copying, pasting, undoing, and clearing operations. These operations are useful when you frequently work at the command prompt.

Figure 1.1(e) Submenu Parallel

Submenu Debug: The submenu Debug (figure 1.1(d)) is mainly related with the text mode or M-file programming.

Submenu Parallel: This submenu (figure 1.1(e)) provides necessary links or tools for parallel computing.

Submenu Desktop: The fifth submenu Desktop (figure 1.1(f)) is equipped with control on command window or its subwindow opening options for example workspace browser of figure 1.1(a) will remain opened or not – this sort of option.

Submenu Window: You may open some graphics window from MATLAB command prompt or running some M-files. From the sixth submenu Window (figure 1.1(g)), one can see how many graphics window under MATLAB are open and can switch from one window to other by clicking the mouse to the required window.

Figure 1.1(f) Submenu Desktop

Submenu Help: MATLAB holds abundant help facilities. The last submenu shows Help (figure 1.1(h)) in different ways. Latter in this chapter, we mention how one gets specific help. The submenu also provides the easiness to get connected with the MathWorks Website provided that your system is connected to internet.

✦ ✦ Icons

Available icons are shown in the icon bar (down the menu bar) of the figure 1.1(a). Frequently used operations such as opening a new file, opening an existing file, getting help, etc are found in the icon bar so that the user does not have to go through the menu bar over and over.

✦ ✦ MATLAB workspace

Workspace (figure 1.1(a)) is the platform of MATLAB where one executes MATLAB commands. During execution of commands, one may have to deal with some input and output variables. These variables can be one-dimensional array, multi-dimensional array,

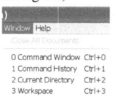

Figure 1.1(g) Submenu Window

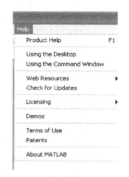

Figure 1.1(h) Submenu Help

characters, symbolic objects, etc. Again to deal with graphics window, we have texts, graphics, or object handles. Workspace holds all those variables or handles for you. As a subwindow of the figure 1.1(a), the browser exhibits the types or properties of those variables or handles. If the browser is not seen in the opening window of MATLAB, click the Desktop down Workspace in the menu bar to bring the subwindow (figure 1.1(f)).

✦ ✦ MATLAB command history

There is a subwindow in the figure 1.1(a) called Command History which holds all previously used commands at the command prompt. Depending on the desktop setting, it may or may not appear during the opening of MATLAB. If it is not seen, click the Command History pulldown menu of the figure 1.1(f) under the Desktop.

1.3 How to get started in MATLAB?

New MATLAB users face a common question how one gets started in MATLAB. This tutorial is for the beginners in MATLAB. Here we address the terms under the following bold headings.

✦ ✦ How one can enter a vector/matrix

The first step is the user has to be in the command window of MATLAB. Look for the command prompt >> in the command window. One can type anything from the keyboard at the command prompt. Row or column matrices are termed as vectors. We intend to enter the row matrix R=[2   3   4   −2   0] into the workspace of MATLAB. Type the following from the keyboard at the command prompt:

>>R=[2 3 4 -2 0]   ← Arial font set is used for executable commands in the text

i.e. R⇔R

There is one space gap between two elements of the matrix R but no space gap at the edge elements. All elements are placed under the [ ]. Press Enter key after the third brace] from the keyboard and we see

R =

    2   3   4   -2   0

>>   ← command prompt is ready again

It means we assigned the row matrix to the workspace variable R. Whenever we call R, MATLAB understands the whole row matrix. Matrix R is having five elements. Even if the R had 100 elements, it would understand the whole matrix that is one of many appreciative features of MATLAB. Next we wish

to enter the column matrix $C=\begin{bmatrix} 7 \\ 8 \\ 10 \\ -11 \end{bmatrix}$. Again type the following from the

keyboard at the blinking cursor:

>>C=[7;8;10;-11] ↵   you will see (↵ means 'Press the Enter Key'),

C =

    7
    8
    10
    −11

>>   ← command prompt is ready again

This time we also assigned the column matrix to the workspace variable C. For the column matrix, there is one semicolon ; between two consecutive elements of the matrix C but no space gap is necessary. As another option, the matrix C could have been entered by writing C=[7 8 10 -11]'. The operator ' of the keyboard is the matrix transposition operator in MATLAB. As if you entered a row matrix but at the end just the transpose operator ' is attached. After that the rectangular matrix $A=\begin{bmatrix} 20 & 6 & 7 \\ 5 & 12 & -3 \\ 1 & -1 & 0 \\ 19 & 3 & 2 \end{bmatrix}$ is to be entered:

    >>A=[20 6 7;5 12 -3;1 -1 0;19 3 2] ↵   you will see,

    A =
            20   6   7
             5 12   -3
             1 -1   0
            19   3   2

Two consecutive rows of A are separated by a semicolon ; and consecutive elements in a row are separated by one space gap. Instead of typing all elements in a row, one can type the first row, press Enter key, the cursor blinks in the next line, type the second row, and so on.

❖ ❖ How one can use the colon and the semicolon operators

The operators semicolon ; and colon : have special significance in MATLAB. Most MATLAB statements and M-file programming use these two operators almost in every line. Vector generation is easily performed by the colon operator no matter how many elements we need. Let us carry out the following at the command prompt to see the importance of the colon operator:

    >>A=1:4 ↵   you will see,

    A =
         1   2   3   4   ← We created a vector A or row matrix
                                    where A=[1   2   3   4]

Interact with MATLAB for an incremental vector by the following:

    >>R=1:3:10 ↵   you will see,

    R =
         1   4   7   10   ← We created a vector or row matrix R whose elements
                              form an arithmetic progression with first element 1,
                              last element 10, and common difference or
                              increment 3

Vector with decrement can also be generated:

    >>C=[0:-2:-10]' ↵   you will see,

    C =
            0
           -2
           -4   ← We created a vector or column matrix C whose
           -6   consecutive elements have the decrement 2 with the
           -8   first element 0 and the last element −10
          -10

MATLAB also generates vectors whose elements are decimal numbers. Let us form a row matrix R whose first element is 3, last element is 6, and increment is 0.5 and which we accomplish as follows:

>>R=3:0.5:6 ↵   you will see,

R =
        3.0000   3.5000   4.0000   4.5000   5.0000   5.5000   6.0000

Then what is the use of the semicolon operator? Append a semicolon at the end in the last command and execute that:

>>R=3:0.5:6; ↵   you will see,

>>   ← Assignment is not shown

Type R at the command prompt and press Enter:

>>R ↵

R =
        3.0000   3.5000   4.0000   4.5000   5.0000   5.5000   6.0000

It indicates that the semicolon operator prevents MATLAB from displaying the contents of the workspace variable R (called suppression command).

✦ ✦ How one can call a built-in MATLAB function

In MATLAB thousands of M-files or built-in function files are embedded. Knowing the descriptions of the function, the numbers of input and output arguments, and the class or type of the arguments is mandatory in order to execute a built-in function at the command prompt. Let us start with a simplest example. We intend to find $\sin x$ for $x = \frac{3\pi}{2}$ which should be −1. The MATLAB counterpart (appendix A) of $\sin x$ is **sin(x)** where **x** can be any real or complex number in radians and can be a matrix too. The angle $\frac{3\pi}{2}$ is written as **3*pi/2**(π is coded by **pi**) and let us perform it as follows:

>>sin(3*pi/2) ↵

ans =
        −1

By default the return from any function is assigned to workspace **ans**. If you wish to assign the return to **S**, you would write **S=sin(3*pi/2);**.

As another example, let us factorize the integer 84 (84=2×2×3×7). The MATLAB built-in function **factor** finds the factors of an integer and the implementation is as follows:

>>f=factor(84) ↵

f =
        2   2   3   7

The output of the **factor** is a row matrix which we assigned to workspace **f** in fact the **f** can be any user-supplied variable. Thus you can call any other built-in function from the command prompt provided that you have the knowledge about the calling of inputs to and outputs from the function.

✦✦ How one can open and execute an M-file

This is the most important start up for the beginners. An M-file can be regarded as a text or script file. A collection of executable MATLAB statements are the contents of an M-file. Ongoing discussion made you familiarize with entering a matrix, computing a sine value, and factorizing

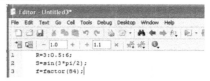

Figure 1.2(a) Last three executed statements are typed in the M-file editor of MATLAB

an integer. These three executions took place at the command prompt. They can be executed from an M-file as well. This necessitates opening the M-file editor. Referring to the figure 1.1(b), you find the link for the M-file editor as File → New → M-file and click it to see the new untitled M-file editor. Another option is click the New M-file editor icon of the figure 1.1(a). However after opening the new M-file editor, we typed the last three executable statements in the untitled file as shown in the figure 1.2(a). The next step is to save the untitled file by clicking the Save icon or from the File Menu of the M-file editor window. Figure 1.2(b) presents the File Save dialog window. We typed the File name of the

Figure 1.2(b) Save dialog window for naming the M-file

figure 1.2(b) as test (type after deleting Untitled, can be any name of your choice) in the window. The M-file has the file extension .m but we do not type .m only the file name is enough. After saving the file, let us move on to the MATLAB command prompt and conduct the following:

```
>>test  ↵
>>         ← command prompt is ready again
```

It indicates that MATLAB executed the M-file by the name test and is ready for the next command. We can check calling the assignees whether the previously performed executions occurred exactly as follows:

```
>>R ↵

R =
      3.0000   3.5000   4.0000   4.5000   5.0000   5.5000   6.0000
>>S ↵                               >>f ↵

S =                               f =
      -1                               2    2    3    7
```

The last returns are the same ones what we found before. Thus one can run any executable MATLAB statements in the M-file.

The reader might ask in which folder or path the file **test** was saved. Figure 1.1(a) shows one slot for the **Current Directory** in the upper middle portion of the window for example the shown one is **C:\Documents and Settings\Administrator\My Documents\MATLAB**. That is the location of your saved file. If you want to save the M-file in other folder or directory, change your path by clicking the folder browser icon (figure 1.1(a)) before saving the file. When you call the **test** or any other file from the command prompt of MATLAB, the command prompt must be in the same directory where the file is in or its path must be defined to MATLAB.

✦✦ Input and output arguments of a function file

MATLAB is supplied with numerous M-files. Some files are executed without any return and some return results which are called function files (appendix C). You have seen the use of function **sin(x)** which has one input argument **x**. The statement **test(x,y)** means that the **test** is a function file which has two input arguments – **x** and **y**. Again the **test(x,y,z)** means the **test** is a function file which needs three input arguments – **x**, **y**, and **z**. Similar style also follows for the return but under the third bracket. The **[a,b]=test(x,y)** means there are two output arguments from the **test** which are **a** and **b** and the **[a,b,c]=test(x,y)** means three returns from the **test** which are **a**, **b**, and **c**.

✦✦ How one can plot a graph

MATLAB is very convenient for plotting different sorts of graphs. The graphs are plotted either from the mathematical expression or from the data.

Let us plot the function $y = -2\sin 2x$. The MATLAB function **ezplot** plots y versus x type graph taking the expression as its input argument. The MATLAB code (appendix A) for the function $-2\sin 2x$ is - **2*sin(2*x)**. The functional code is input argumented by using quote hence we conduct

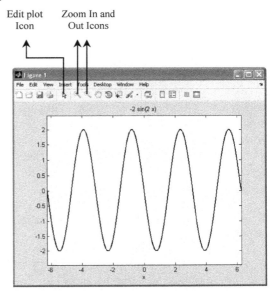

Figure 1.2(c) Graph of $-2\sin 2x$ versus x

the following at the command prompt:
>>ezplot('-2*sin(2*x)') ↵

Figure 1.2(c) presents the outcome from above execution. The window in which the graph is plotted is called the MATLAB figure window. Any graphics is plotted in the figure window, which has its own menu (such as File, Edit, etc) as seen in the figure 1.2(c).

1.4 Some queries about MATLAB environment

Users need to know the answers of some questions when they start working in MATLAB. Some MATLAB environment related queries are presented in the following.

⊟ How to change the numeric format?

When you perform any computation at the command prompt, the output is returned up to the four decimal display due to the short numeric format which is the default one. There are other numeric formats also. To reach the numeric format dialog box, the clicking operation sequence is MATLAB command window ⇒ File ⇒ Preferences ⇒ Command Window ⇒ Text Display ⇒ Numeric Format (select from the popup menu e.g. long).

⊟ How to change the font or background color settings?

One might be interested to change the background color or font color while working in the command window. The clicking sequence is MATLAB command window ⇒ File ⇒ Preferences ⇒ Colors.

⊟ How to delete some/all variables from the workspace?

In order to delete all variables present in the workspace, the clicking sequence is MATLAB command window ⇒ Edit ⇒ Clear Workspace (figure 1.1(c)). If you want to delete a particular workspace variable, select the concern variable by using the mouse pointer in the workspace browser (assuming that it is open like the figure 1.2(d)) and then rightclick ⇒ delete. See section 1.2 for bringing the workspace browser in front in case it is not opened.

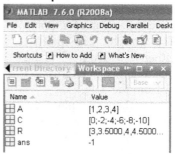

Figure 1.2(d) Workspace browser displays variable information

⊟ How to clear workspace but not the variables?

Once you conduct some sessions at the command prompt, monitor screen keeps all interactive sessions. You can clear the screen contents without removing the variables present in the workspace by the command clc or performing the clicking operation MATLAB command window ⇒ Edit ⇒ Clear Command Window (figure 1.1(c)).

⊟ How to know the current path?

In the upper portion of the figure 1.1(a), the **Current Directory** bar is located that indicates in which path the command prompt is or execute **cd** (abbreviation for the current directory) at the command prompt.

⊟ How to see different variables in the workspace?

There are two ways of viewing this – either use the command **who** or look at the workspace browser (like figure 1.2(d)) which displays information about workspace variables for example **R** is the name of the variable holding some values and their data class is **double** precision. One can view, change, or edit the contents of a variable by doubleclicking the concern variable situated in the workspace browser as conducted in Microsoft Excel.

⊟ How to enter a long command line?

MATLAB statements can be too long to fit in one line. Giving a break in the middle of a statement is accomplished by the ellipsis (three dots are called ellipsis). We show that considering the entering of the vector **x=[1:3:10]** as follows:

```
>>x=[1:3: . . . ↵
        10] ↵
x =
        1   4   7   10
```

Typing takes place in two lines and there is one space gap before the ellipsis.

⊟ Editing at the command prompt

This is advantageous specially for them who work frequently in the command window without opening an M-file. Keyboard has different arrow keys marked by ← ↑ → ↓. One may type a misspelled command at the command prompt causing error message to appear. Instead of retyping the entire line, press uparrow (for previous line) or downarrow (for next line) to edit the MATLAB statement. Or you can reexecute any past statement this way.

For example we generate a row vector 1 to 10 with increment 2 and assign the vector to **x**. The necessary command is **x=1:2:10**. Mistakenly you typed **x+1:2:10**. The response is as follows:

```
>>x+1:2:10 ↵
??? Undefined function or variable 'x'.
```

You discovered the mistake and want to correct that. Press ↑ key to see,

```
>>x+1:2:10
```

Edit the command by going to the **+** sign by using the left arrow key or mouse pointer. At the prompt, if you type **x** and press ↑ again and again, you see used commands that start with **x**.

⊟ Saving and loading data

User can save workspace variables or data in a binary file having the extension **.mat**. Suppose you have the matrix $A = \begin{bmatrix} 3 & 4 & 8 \\ 0 & 2 & 1 \end{bmatrix}$ and wish to save the A in a file by the name **data.mat**. Let us carry out the following:

>>A=[3 4 8;0 2 1]; ↵   ← Assigning the A to A

Now move on to the workspace browser (figure 1.2(d)) and you see the variable A including its information located in the subwindow. Bring the mouse pointer on the A, rightclick the mouse, and click the **Save As**. The Save dialog window appears and type only **data** (not the **data.mat**) in the slot of **File name**. If it is necessary, you can save all workspace variables by using the same action but clicking **File** ⇒ **Save Workspace As** (figure 1.1(b)). One retrieves the **data** file by clicking the menu **File** ⇒ **Import Data** (figure 1.1(b)). Another option is use the command **load data** at the command prompt.

🖹 **How to delete a file from the command prompt?**

Let us delete just mentioned **data.mat** by executing the command **delete data.mat** at the command prompt.

🖹 **How to see the data held in a variable?**

Figure 1.2(d) presents some variable information in which you find R. Doubleclick the R or your variable in the workspace browser and find the matrix contents of R in a data sheet.

1.5 How to get help?

Help facilities in MATLAB are plentiful. One can access to the information about a MATLAB function in a variety of ways. Command **help** finds the help of a particular function file. You are familiar with the function **sin(x)** from earlier discussion and can have the command prompt help regarding the **sin(x)** as follows:

>>help sin ↵   ← Function name without the argument
SIN   Sine of argument in radians.
    SIN(X) is the sine of the elements of X.

    See also asin, sind.

    Overloaded methods:
      distributed/sin
      sym/sin

    Reference page in Help browser
      doc sin

One disadvantage of this method is that the user has to know the exact file name of a function. For a novice, this facility may not be appreciative.

Casually we know the partial name of a function or try to check whether a function exists by that name. Suppose we intend to see whether any function by the name **audio** exists. We execute the following by the intermediacy of the command **lookfor** (no space gap between **look** and **for**) to see all possible functions bearing the file name **audio** or having the file name **audio** partly:

```
>>lookfor audio ⏎
```

audioplayer	- Audio player object.
audiorecorder	- Audio recorder object.
audiodevinfo	- Audio device information.
audioplayerreg	- audiorecorderreg - AUDIOPLAYERREG
⋮	

The return shows all possible matches of functions containing the word **audio**. Now the command **help** can be conducted to go through a particular one for example the first one is **audioplayer** and we execute **help audioplayer** to see its description at the command prompt.

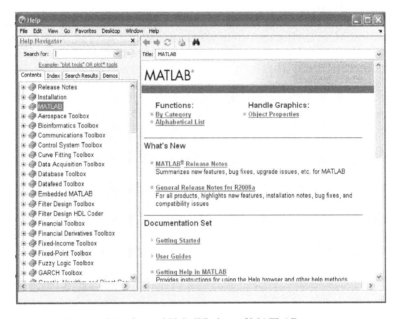

Figure 1.2(e) General Help Window of MATLAB

In order to have window form help, click the Help icon of the figure 1.1(a) and MATLAB responds with the opening Help window of the figure 1.2(e). As the figure shows, help is available by content or index. If you have some search word for MATLAB, you search that through the **Search for** of the figure 1.2(e). This help form is better when one navigates MATLAB's capability not looking for a particular function.

Execute just **help** at the command prompt, you find scores of libraries embedded in MATLAB as follows:

```
>>help ⏎
```

HELP topics:

matlab\general   - General purpose commands.
matlab\ops   - Operators and special characters.
matlab\lang   - Programming language constructs.
matlab\elmat   - Elementary matrices and matrix manipulation.

⋮

MATLAB displays a long list of libraries or demonstrating functions in which we find some by the name **signal** and which is related to the signal processing of audio signals. In order to know about a specific library or demonstration for example the **signal**, we execute **help signal** at the command prompt.

Hidden algorithm or mathematical expression is often necessary in conducting MATLAB executions whose assistance we can have through the search option from MathWorks Website provided that the PC is connected to the Internet (figure 1.1(h)).

However we close the introductory discussion on MATLAB with this.

Chapter 2

 ## Digital Audio
Fundamentals

The subject matter in this chapter is to address the very basics of a digital audio to the context of MATLAB. Our illustration starts with a modular audio or softcopy file, which is understood by an average reader. A digital audio assumes the form of a large size row or column matrix depending on the audio signal duration. A distinct link is explained how one obtains the discrete audio signal from its continuous counterpart. However the know-how details of the chapter highlight the following:

- ✦ ✦ Concept of a digital audio as well as its mathematical model
- ✦ ✦ Tools for obtaining sampling information and amplitude details from a digital audio
- ✦ ✦ Audio playing options in accordance with the theory
- ✦ ✦ Different ways to obtain a test audio either from PC System or outside MATLAB

2.1 Digital audio

Can you imagine our life without music or sound these days? When you hear your mobile is ringing, that ring tone is digital audio. If you turn on your PC or laptop, at the beginning the PC plays an introductory music – that is a digital audio. If you buy latest CD of some super popstar, that CD contains digital music. Very often teenagers buy or download audio through internet, that audio is also a digital one. In a nutshell almost all aspects of the twenty first century education, business, transaction, and entertainment use a digital audio in some form.

As such a digital audio mathematically is nothing but a row or column matrix of large size. In a simplistic way the row or column matrix data holds your favorite music. Suppose ten thousand data as a row matrix represents your favorite music which you listen after playing in a MP3 or Windows media player for 20secs. Then the first five thousand data in the row matrix will let you listen the music for 10secs – the explanation of a digital audio is as simple as that.

2.2 Mathematical representation of a digital audio

In order to understand the mathematical model of a digital audio, we first need to understand the conversion of a continuous time signal to discrete time signal. A realistic audio signal is always continuous never discrete. For example when we listen some news on TV, that sound signal is continuous.

Audio loudness
proportional to $v(t)$

Figure 2.1(a) A typical audio signal

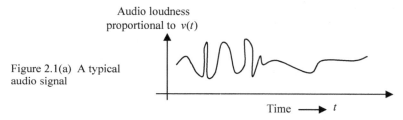

Time ⟶ t

In audio engineering every signal handling happens in terms of voltage or current. Let us say the audio signal is expressed in terms of voltage $v(t)$. If an audio recipient gets more loudness, more voltage is assigned for the signal or less loudness-less voltage. Figure 2.1(a) shows a typical audio signal.

Now if we wish to record, store, or manipulate an audio signal, our sound device should be ready for all instant of time for example at time $t = 0$sec, $t = 0.00001$sec, or $t = 1$sec. At every time point we have an

Discrete audio
proportional to $v[n]$

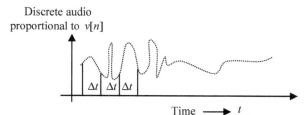

Time ⟶ t

Figure 2.1(b) Digitization of the audio signal in figure 2.1(a) in time domain

audio loudness or voltage value. Thus a realistic audio signal needs infinite time points and infinite voltage values for perfect manipulation, storage, or recoding which is solely impractical. This sort of signal, infinite watch points and infinitely watched values, is called continuous (mathematics term) or analog (engineering term).

In practical sense we consider the voltage values every after certain time Δt rather than at all instants of an interval. This Δt has serious implication in audio engineering, it is called sampling period T_s or resolution. Sampling frequency then becomes $f_s = \dfrac{1}{T_s} = \dfrac{1}{\Delta t}$. Now the signal $v(t)$ is having discrete values as regard to the figure 2.1(b) therefore the $v(t)$ has discrete-time representation and we label that as $v[n]$. The term discrete is in mathematics whose synonym is digital in engineering. Mathematically we write discrete $v[n] = v(nT_s) = v(n\Delta t)$ where n is purely positive integer including 0. For a particular digital audio system, the T_s, f_s, or Δt is predecided.

In figure 2.1(b) we showed the discreteness in time, the same happens in the $v(t)$ amplitudes as well. Instead of taking a true amplitude value we take the amplitude value closest to the predecided amplitude. For example in figure 2.1(c) the Amp represents a particular amplitude level. Suppose we assign the sound intensity of an audio system as follows: no sound meaning 0 and full intensity sound within the limit of the device meaning 1 or 100%. We wish to confine only to 5 voltage levels. Therefore the predecided $v[n]$ values are 0%, 25%, 50%, 75%, and 100%. Let us say any value of $v[n]$ is 20%, then we assume that value as 25% because the 25% level is the closest.

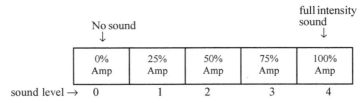

Figure 2.1(c) Illustration of sounud levels in a digital audio

We chose five levels for the amplitudes in above example – this is just for the sake of illustration. In a practical audio literally there are dozens of levels (let us say integer M) sometimes even hundreds. Usually these level numbers are as the power of 2 for example 128, 256, 512, ... etc. Earlier we mentioned the Δt as the resolution. Appropriately that should have been stated as the time resolution. Figure 2.1(c) cited level assignment is related with the amplitude resolution.

Now for different n, the $v[n]$ becomes a row or column matrix which is also known as a vector. From all these discussions, what do we come up with? Any practical audio signal is basically a row or column matrix data subject to sampling frequency f_s and amplitude level number M in digital domain – that is schematically shown below:

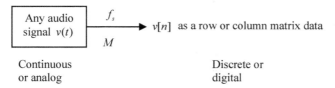

In a computer when we study or analyze some digital audio, we basically work on the row or column matrix $v[n]$ knowing the f_s and M. Prior to working on any softcopy digital audio for example **test.wav**, the first concern of the reader should be how to get the $v[n]$, f_s, and M from the **test.wav** – will be discussed in the following.

2.3 How to listen a digital audio in MATLAB?

The first step of listening a digital audio is to obtain the softcopy of the digital audio (section 2.7) and place the audio file with the help of the Windows Explorer to your working path of MATLAB. Several audio file names are mentioned in section 2.7. Obtain a softcopy say **bird.wav** of section 2.7 and place it by the Windows Explorer in your working directory of MATLAB. The command **dir** shows all softcopy files existing in the working path of MATLAB so execute it:

```
>>dir ↵
.    ..   Digital.doc  bird.wav    chap2.doc
                          ↑
```
Meaning the file is in your working path

First the machine has to read the audio **bird.wav** and then you can play it. The built-in function **wavread** reads a digital audio with the syntax [user-supplied variable for the discrete audio data, user-supplied variable for the sampling frequency]=**wavread**(file name under quote) therefore we exercise the following:

```
>>[v,fs]=wavread('bird.wav'); ↵
```
where the **v** and **fs** are our-chosen variables for the discrete audio data and sampling frequency respectively. There is also a built-in function called **wavplay** which plays the digital audio with the syntax **wavplay**(discrete audio data, sampling frequency). We assume that your PC system sound card and speaker are working fine so let us execute the following:

```
>>wavplay(v,fs) ↵
```
Upon the execution of the last command, you must be listening the sound of a crow.

2.4 How to know a digital audio sample information?

By digital audio sample information we mean the sample and sampling frequency information of the audio signal. Make no confusion that a computer processed audio signal is always a discrete time one never continuous (section 2.2). First we have to select some audio softcopy file for example the **bird.wav** of section 2.7. Then place the digital audio file with the help of the Windows Explorer in your working path and check that the file is in your working path by using the command **dir** (section 2.3).

Now the built-in command **wavfinfo** displays some information of a digital audio file with the syntax [user-supplied two output arguments]=**wavfinfo**(file name under quote) so we exercise the following for the **bird.wav**:

```
>>[a,b]=wavfinfo('bird.wav') ↵

a =

Sound (WAV) file

b =

Sound (WAV) file containing: 19456 samples in 1 channel(s)
```

The **a,b** in the last execution are our chosen variables. The **a** return indicates that the **bird.wav** file is a sound wave file with format **.wav** and the **b** return says that there are 19456 samples or discrete data (i.e. $M = 19456$ of section 2.2) in the digital audio **bird.wav**.

How can we relate this audio data with the mathematical model of an audio signal as discussed in section 2.2? Referring to the figure 2.1(b), this discrete data is simply the functional values of the $v[n]$, not of $v(t)$ of coarse in machine format. Is it a voltage or current data? Certainly we have no idea about that.

Next query is about the integer n. Where does the n start? Well that is user-defined. What is the variation of n? It is completely sample number related – from earlier execution the **bird.wav** has 19456 samples. If we start the n from 0, the last sample number will be 19455. Again if we start the n from 1, the last sample number will be 19456. Mathematically for the **bird.wav** we may write $0 \le n \le 19455$ assuming 0 start.

We explained about the sample, what about the sampling frequency? In section 2.3 we have explained how the machine reads an audio file by applying the **wavread** function and which we exercised on the **bird.wav**. We also mentioned in the same section that the return numbers from the **wavread** are two – **v** and **fs**. This **v** holds the discrete data of $v[n]$ and the **fs** holds the sampling frequency as regard to section 2.2 so execute the following:

```
>>[v,fs]=wavread('bird.wav'); ↵     ← Repetition of section 2.3 command
>>fs ↵                               ← Calling the fs to see its content
```

fs =
　　　22050

Above execution says that to produce the **bird.wav** file, a sampling frequency f_s =22050 Hz was applied (assume that 1 Hz is tantamount to 1 sample/sec). With the sampling frequency, the sampling period is $T_s = \Delta t = \frac{1}{22050}$ =4.5351×10^{-5}sec. The duration of the **bird.wav** is Δt ×(total sample number−1)=4.5351×10^{-5}×19455=0.8823sec. We can say that in the continuous sound source the $v(t)$ (of the section 2.2) interval was $0 \le t \le 0.8823$ sec assuming the t counting started at t =0sec.

2.5 How to know a digital audio amplitude information?

Audio amplitude is important in the sense that most digital analyses use the amplitude data. Amplitude means the value of the signal $v[n]$ or $v(t)$ as cited in section 2.2.

Let us consider the ongoing softcopy **bird.wav**. In section 2.4 we mentioned that the workspace variable **v** holds the samples of $v(t)$ or discrete $v[n]$. Prior to writing the data to **bird.wav**, physical sound data was converted to some pre-selected range which is from −1 to 1. The convention is followed as a unifying approach or standard. That is the range of $v[n]$ is always from −1 to 1 or mathematically $-1 \le v[n] \le 1$ (also written as [−1,1]).

The next question is how do we relate these amplitude values with the time? The workspace variable **v** is a column matrix and holds the amplitude sample values. If you say I need to see those amplitude values, just call the variable:

```
>>v ↵
     ⋮
     0
     0
     0
>>                    ← Prompt is ready again
```

Since there are 19456 amplitudes, so many data can not be placed in MATLAB command window. Above return says that the last three amplitudes have the value 0.

There is another option, look at the workspace browser (section 1.4) under the **Name**, find the **v**, doubleclick the **v**, and find the amplitude data in a workspace like Microsoft Excel. In the same workspace browser under the **Value**, you find **<19456×1 double>**. The **19456×1** means matrix size, which is a column matrix for the second dimension being 1. The **double** means every amplitude is a double precision number (simply a decimal number). Obviously every element in the v is in $[-1,1]$.

From the last section we know that in continuous form $v(t)$ exists over $0 \le t \le 0.8823$ sec and $\Delta t = 4.5351 \times 10^{-5}$ sec. In MATLAB the array indexing starts from the 1 so v(1), v(2), v(3), etc indicate the first, second, third, etc amplitudes respectively. With this,

value of $v(t)$ when $t = 0$ is equivalent to v(1) or $v[n]$ with $n = 0$

value of $v(t)$ when $t = \Delta t$ is equivalent to v(2) or $v[n]$ with $n = 1$

value of $v(t)$ when $t = 2 \Delta t$ is equivalent to v(3) or $v[n]$ with $n = 2$

value of $v(t)$ when $t = 3 \Delta t$ is equivalent to v(4) or $v[n]$ with $n = 3$

$$\vdots$$

and value of $v(t)$ when $t = 0.8823$ sec is equivalent to v(19456) or $v[n]$ with $n = 19455$ where we assumed the t and n start from 0.

Because of discreteness we can not get the t which is fractional of Δt unless interpolation is used for example $2.5 \Delta t$. If you need to see any audio amplitude for example $v(t)$ when $t = 900 \Delta t$, just execute the following:

>>v(901) ↵

ans =

    -0.0156   ← Meaning audio amplitude $v(t) = -0.0156$ at

                              $t = 900 \Delta t = 900 \times 4.5351 \times 10^{-5} = 0.0408$ sec

What if a range of audio amplitudes are needed for example over $900 \Delta t \le t \le 906 \Delta t$ sec or $0.0409 \le t \le 0.0411$ sec? We use the colon operator to pick the amplitudes from v as follows:

>>r=v(901:907) ↵

r =

    -0.0156   ← Meaning audio amplitude $v(t)$ at $t = 900 \Delta t$

    -0.0156

    -0.0078

        0

     0.0078

     0.0078

     0.0078   ← Meaning audio amplitude $v(t)$ at $t = 906 \Delta t$

The r is a user-chosen variable which keeps the required audio amplitudes and the return to which is also a column matrix as the v is.

At this point if we wish to know the maximum amplitude in the audio, we apply the appendix B.5 cited max as follows:

>>max(v) ↵

ans =

     0.8906

Above output indicates that the maximum amplitude value in the audio is 0.8906 i.e. mathematically $v(t)|_{max} = v[n]|_{max} = 0.8906$ where the amplitude varies from −1 to 1. Similarly minimum amplitude value we see by the **min** as:

>>min(v) ↵

ans =

-1   ← Meaning $v(t)|_{min} = v[n]|_{min} = -1$

If you wish to see the audio wave as a graph in continuous sense, you may employ the **plot** function of appendix D. Simply execution of the command

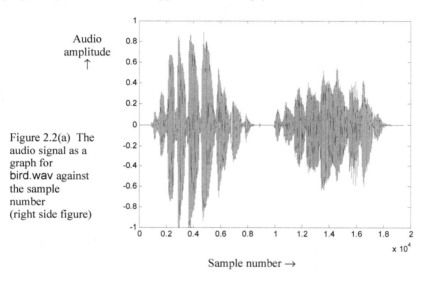

Audio
amplitude
↑

Figure 2.2(a) The audio signal as a graph for bird.wav against the sample number (right side figure)

Sample number →

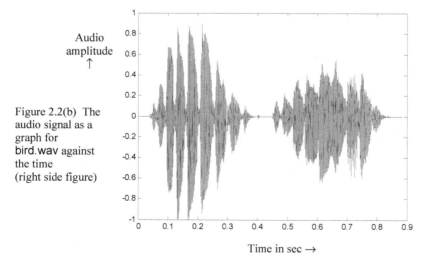

Audio
amplitude
↑

Figure 2.2(b) The audio signal as a graph for bird.wav against the time (right side figure)

Time in sec →

plot(v) returns the graph of figure 2.2(a). The vertical and horizontal axes of the figure 2.2(a) refer to the amplitude (which is between −1 and 0.8906) and sample number (which is between 1 and 19456) respectively.

If you wish to see the horizontal axis over the wave duration time $0 \le t \le 0.8823 \sec$, you need to generate a time vector (section 1.3) with the step size Δt by:

```
>>t=0:1/fs:19455/fs; ↵          ← 1/fs means Δt
>>plot(t,v) ↵          ← Calling the plot with the two input arguments
```

This time the **plot** returns the figure 2.2(b) in which the horizontal axis is changed to time from the sample number.

♣ ♣ Readymade tool of MATLAB

There is a built-in function called **soundview** which is convenient for viewing an audio waveshape without too many details. The function works with the syntax **soundview**(audio file name under quote). Considering the ongoing **bird.wav**, we call the function as follows:

```
>>soundview('bird.wav') ↵
```

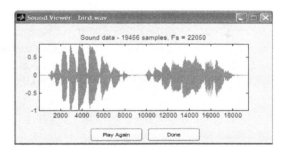

Figure 2.2(c) Audio waveshape viewer of MATLAB displays
the waveshape of bird.wav

Figure 2.2(c) shows the response from the last execution. The waveshape, total audio sample number (which is 19456), and sampling frequency (which is 22050) are also shown in the figure. By inspection the waveshapes of the figures 2.2(a) and 2.2(c) are identical.

2.6 Basic statistics of a digital audio

In research and development (R & D) of digital audio, sometimes we need the basic statistical property of a digital audio. Basic statistical property means mean, standard deviation, range, etc of the audio. There can be two instances – viewing the properties or getting hand on the properties, both of which are addressed in this section.

❖ ❖ Maximum or minimum in an audio

How one obtains the maximum or the minimum audio amplitude value from a digital audio is addressed in the last section.

❖ ❖ Audio data range

Range of a digital audio is the difference between the maximum and the minimum of audio amplitude values. The function **range** finds the range of a digital audio whose syntax is **range**(audio samples as a row or column matrix). In the last section we obtained the audio **bird.wav** maximum and minimum amplitudes as $v[n]|_{max} = 0.8906$ and $v[n]|_{min} = -1$ respectively based on that we should get the range of the digital audio as 1.8906. That just needs us to execute the following:

>>range(v) ⏎

ans =
            1.8906

❖ ❖ Audio data mean or average

Mean is the average of all amplitude values in a digital audio $v[n]$ which is given as $\frac{1}{N}\sum_{n=1}^{N} v[n]$ where N is the number of samples in the audio $v[n]$ and assuming that the n starts at 1. To compute the mean of an audio, we use the function **mean** with the syntax **mean**(audio samples as a row or column matrix). Considering the **bird.wav** of last section (execute first **[v,fs]=wavread('bird.wav');**), the mean we find as follows:

>>mean(v) ⏎

ans =
            -5.3888e-004

Above return indicates that the mean amplitude value of the audio **bird.wav** is -5.3888×10^{-4} obviously over the interval $0 \le t \le 0.8823$ sec (section 2.4).

❖ ❖ Audio data variance and standard deviation

If the mean of a digital audio $v[n]$ is m, the variance of $v[n]$ is defined as $V = \frac{1}{N-1}\sum_{i=1}^{N}(v[i]-m)^2$ where $m = \frac{1}{N}\sum_{i=1}^{N} v[i]$ and N is the number of amplitudes in the audio $v[n]$ and assuming that the n starts at 1. Standard deviation of the audio is defined as the positive square root of the variance which is given by $\sigma = \sqrt{V}$. Suppose four audio amplitudes are given as $\begin{bmatrix} -2 \\ -8 \\ 3 \\ -14 \end{bmatrix}$

and which has the $m = -5.25$ and $V = \frac{(-2+5.25)^2+(-8+5.25)^2+(3+5.25)^2+(-14+5.25)^2}{4-1} = 54.25$ thereby providing $\sigma = 7.3655$. MATLAB function **std** computes the standard

-24-

deviation of an audio with the syntax **std(** $v[n]$ as a row or column matrix) therefore the computation takes place by:
```
>>v=[-2 -8 3 -14]; ↵
>>std(v) ↵
```

ans =
   7.3655

In statistics literature the V is defined as $V = \dfrac{1}{N}\sum\limits_{i=1}^{N}(v[i]-m)^2$ whose standard

deviation we obtain by **std(v,1)** in which the second input argument **1** is reserved to MATLAB.

Next what the standard deviation will be for the audio amplitudes of ongoing **bird.wav** and which we obtain very simply (execute first **[v,fs]= wavread('bird.wav');**) by:
```
>>std(v) ↵
```

ans =
   0.2040

✦ ✦ Audio data median

In ascending order, the middle element of the amplitudes of a digital audio is called the median. If there are N sorted audio amplitudes in a digital

file, the median is the $\left(\dfrac{N+1}{2}\right)^{th}$ amplitude if the number of amplitudes is odd

or the average of the $\left(\dfrac{N}{2}\right)^{th}$ and $\left(\dfrac{N}{2}+1\right)^{th}$ amplitudes if the number of

amplitudes is even. To obtain the median from a digital audio, we use the command **median(** $v[n]$ as a row or column matrix). For example the modular audio data $v[n] = [4 \quad 7 \quad 5 \quad -5 \quad 0 \quad 1 \quad 9]$ has the median 4 and which we obtain as follows:
```
>>v=[4 7 5 -5 0 1 9]; ↵
>>median(v) ↵
```

ans =
   4

For the ongoing audio **bird.wav**, the median amplitude is 0 which we obtain by (execute first **[v,fs]=wavread('bird.wav');**):
```
>>median(v) ↵
```

ans =
   0

Figure 2.2(d) Basic statistics of the digital audio bird.wav

✦ ✦ A comprehensive way to know all these basic statistics

There is a built-in function called **datastats** which provides all these basic digital audio statistics when the audio as a row or column matrix is its input argument. For the ongoing **bird.wav** we obtain that as follows:

```
>>[v,fs]=wavread('bird.wav'); ↵
>>datastats(v) ↵

ans =
      num: 19456          ← Number of audio amplitudes
      max: 0.8906         ← Maximum value of the audio amplitudes
      min: -1             ← Minimum value of the audio amplitudes
     mean: -5.3888e-004   ← Mean value of the audio amplitudes
   median: 0              ← Median value of the audio amplitudes
    range: 1.8906         ← Range of the audio amplitudes
      std: 0.2040         ← Standard deviation of audio amplitudes
```

As another option we just view the statistics like the figure 2.2(d) by the following actions: recall that the figure 2.2(b) is graphed for the **bird.wav** in a MATLAB window, that window is called the figure window. In the figure window, you find the menu **Tools** and **Data Statistics** in the pulldown menu. Clicking the **Data Statistics** prompts with the figure 2.2(d). The X and Y of the figure 2.2(d) correspond to t and $v[n]$ respectively.

2.7 How to get a digital audio file?

Digital audio study always requires a softcopy audio file to commence with. In various chapters we will be implementing audio fundamentals employing several digital files. The softcopy file size of these audio is not so large so it can be transferred through email attachment. Preliminary audio processing M-files and/or function files sometimes need to be written by the reader in order to gain hand-on experience about the subject. The reader can promptly get the softcopy of any digital audio or author-written M-file presented in the text by contacting at **nzaman@kfupm.edu.sa** or **nzaman@ymail.com**. However we are going to mention some options in the following so that the reader may have his own way of obtaining a digital audio.

✦ ✦ Option 1

We assume that you are familiar with the Microsoft windows operating system. This option tells you the easiest way of getting a digital test audio even if you do not have a digital camera or recorder. In Microsoft Windows, you find the **Start** menu. From there you can easily reach to the **Search** option. In the prompt window of the **Search**, check for **All files and folders** option and find a slot by title **All or part of the file name** in another prompt window. In that slot, enter the word ***.wav** from the keyboard and click the **Search**. You find dozens of digital audio files on the right half part of the prompt window which come with your PC system. Choose a smaller size file and copy and paste it through the windows explorer to your working path of MATLAB for example to C:\Documents and Settings\ Administrator\My Documents\MATLAB.

✦✦ Option 2

If you have a headset or microphone and your sound system works fine, you can record your speech by window supplied sound recorder. To reach to the Windows Sound recorder in Windows XP system, the clicking sequence is Start → All Programs → Accessories → Entertainment → Sound Recorder.

✦✦ Option 3

Following example audio is chosen all over the text for implementing fundamental digital audio processing theory. We wish to cite some descriptions of the exercised digital audio:

bird.wav	The sound of a crow which contains 19456 samples at a sampling frequency 22050 samples/sec.
pitch.wav	The sound of a single frequency sine signal which contains 6616 samples at a sampling frequency 22050 samples/sec.
test.wav	The sound of a noisy music which contains 161002 samples at a sampling frequency 22050 samples/sec.
nosound.wav	Very short duration sound of a room without any music or speech which contains 2800 samples at a sampling frequency 22050 samples/sec.
b.wav	Speech signal of the english letter b which contains 2800 samples at a sampling frequency 22050 samples/sec.
hello.wav	Speech signal of the english word hello which contains 50388 samples at a sampling frequency 22050 samples/sec.

However we close the chapter with the links of the exercised softcopies in the text.

Mohammad Nuruzzaman

-28-

Mohammad Nuruzzaman

Chapter 3

Elementary Operations on A Digital Audio

At the heart of digital audio processing, elementary mathematical operations are very much related. The intrinsic computing is mainly discrete time oriented for this reason we often illustrate the operation employing modular discrete signal which is immediately followed by a digital softcopy audio. Many of this chapter's basics the reader can easily apply to multimedia applications. The operations we are going to focus are the following:

- ✦ ✦ Direct scaling operation on a digital softcopy audio
- ✦ ✦ Linear audio range mapping through built-in functions
- ✦ ✦ Programming techniques to handle audio on frame basis
- ✦ ✦ Multimedia operations such as merging, splitting, mixing, etc
- ✦ ✦ Signal processing mathematical operations e.g. convolution

3.1 Scaling operation on a digital audio

Suppose $v[n]$ represents a digital audio (section 2.2). We wish to scale the digital audio by a factor c that is we have to find $cv[n]$. If the c is greater than 1, that is sound volume increasing operation and if the c is less than 1, that is sound volume reducing operation. Since the audio player operates within $[-1,1]$, audio amplitude values less than -1 must be sealed to

−1 and the ones more than 1 to 1. Figure 3.1(a) illustrates the clipping operation. In order to achieve such clipping operation we can apply the **find** function of appendix B.9. In the case of $c < 1$, the clipping is not required. If $c > 1$, clipping is mandatory.

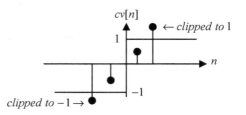

Figure 3.1(a) Scaling operation on a digital audio

Let us consider the digital audio **bird.wav** of section 2.7. We wish to implement a scaling operation by a factor 2 on the digital audio.

We know that the digital audio signal $v[n]$ is stored in the workspace **v** upon reading the wave as follows:

```
>>[v,fs]=wavread('bird.wav'); ↵    ← From section 2.4
>>y=2*v; ↵    ← y holds the cv[n] values where the y is a user-chosen variable
```

We apply the **find** on $cv[n]$ for $cv[n] > 1$ as follows:

```
>>r=find(y>=1); ↵ ← finding the integer position index as a row matrix r where
                     cv[n] ≥ 1 and the r is a user-chosen variable
>>y(r)=1; ↵    ← setting only cv[n] ≥ 1 amplitude values to 1 in y
>>r=find(y<=-1); ↵ ← finding the integer position index as a row matrix r where
                      cv[n] ≤ −1 and the r is a user-chosen variable
>>y(r)=-1; ↵    ← setting only cv[n] ≤ −1 amplitude values to 1 in y
```

The last **y** keeps scaled audio signal, so playing the audio signal by the player of section 2.3 takes place as:

```
>>wavplay(y,fs) ↵
```

Certainly you hear the sound intensity increased from what you heard by executing **wavplay(v,fs)** in section 2.3. If the reader needs the audio only to be played, then exercising the command **wavplay(2*v,fs)** is enough because the **wavplay** clips the audio data to [−1,1] by default if it fits out of the range. If the audio data needs to be stored or processed, then scaling by this sort of programming is required.

3.2 Changing the range of a digital audio

We know that if the digital audio amplitude is in between −1 and 1, we play the audio by the function **wavplay** after knowing the sampling frequency (sections 2.3 and 2.5). Sometimes sound or physical devices provide data which is not in [−1,1] range. Moreover signal processing of audio data results amplitude falling out of [−1,1]. Not only that, digital audio storage for example unsigned 8-bit integer needs the digital audio data to be in [0,255] or [0,511]. The reader might say why do not we perform the clipping operation as stated in section 3.1? Some audio information might be lost if we clip audio data indiscriminately.

Under this sort of situation we apply linear mapping. For instance x_1 is the value of a digital audio data in an existing range and which should be y_1 in the wanted range. Again considering another audio sample value, say the x_2 and the y_2 are the existing and the wanted range values respectively. All data in the audio is then linearly mapped to the new range by the equation $y = \frac{y_2 - y_1}{x_2 - x_1}x + \frac{x_2 y_1 - x_1 y_2}{x_2 - x_1}$ where the x and the y are the existing sample value and wanted sample value respectively. Usually we map the minimum of the given discrete signal to that of the new range, so do the maximum. Following examples are presented in this regard.

✦✦ Example 1

Let us say we have the discrete signal $v[n] = [-4\ \ 5\ \ 6\ \ 9\ \ 0\ \ 3]$ and wish to map the signal data in $[-1,1]$ range as required by the digital audio player of MATLAB. It is given that the mapped data in the new range should be $[-1\ \ \ 0.3846\ \ \ 0.5385\ \ \ 1\ \ -0.3846\ \ \ 0.0769]$ which we intend to obtain.

First we enter the given data of $v[n]$ to some user-chosen variable **v** as follows:
```
>>v=[-4 5 6 9 0 3];  ⏎          ← v⇔ x
```
The minimum/maximum in the given data is obtained by the appendix B.5 mentioned **min/max** respectively. The minimum of the $v[n]$ and the minimum of the new range should be the x_1 and the y_1 respectively and enter as follows:
```
>>x1=min(v); y1=-1;  ⏎    ← x1⇔ x₁ , x1 or y1 is user-chosen variable, y1⇔ y₁
```
Similarly we then enter the maximum values in the range by:
```
>>x2=max(v); y2=1;  ⏎    ← x2⇔ x₂ , x2 or y2 is user-chosen variable, y2⇔ y₂
```
Next we vector code (appendix A) the expression $\frac{y_2 - y_1}{x_2 - x_1}x + \frac{x_2 y_1 - x_1 y_2}{x_2 - x_1}$ and assign that to **y** by the following:
```
>>y=(y2-y1)/(x2-x1)*v+(x2*y1-x1*y2)/(x2-x1)  ⏎   ← y⇔ y , y is user-chosen
                                                      variable
```
```
y =
    -1.0000   0.3846   0.5385   1.0000   -0.3846   0.0769
```
Therefore the last variable **y** retains the $v[n]$ discrete data as a row matrix in $[-1,1]$ range which you can verify by inspection. The minimum -4 of $v[n]$ is mapped to -1 and the maximum 9 of $v[n]$ is mapped to 1.

✦✦ Example 2

Existing built-in function sometimes eases computing hassle. There is a function by the name **mat2gray** which maps any matrix data between 0 and 1 corresponding to minimum and maximum of the matrix data respectively. The syntax we apply is **mat2gray**(matrix name). As in example

1 the wanted range is $[-1,1]$ in contrast **mat2gray**'s is $[0,1]$. If we apply $2X-1$ on the return from the **mat2gray**, we get the range to be in $[-1,1]$ where X is any return from **mat2gray**. Let us verify that by the example 1 mentioned discrete signal as follows:

>>v=[-4 5 6 9 0 3]; ⏎   ← v⇔ v[n]
>>y=2*mat2gray(v)-1 ⏎

y =
    -1.0000   0.3846   0.5385   1.0000 -0.3846   0.0769

The result is identical to the return of the example 1, which we obtained by direct application of the mathematical expression.

✦✦ **Example 3**

While storing the data of a digital audio, frequently we keep the data in fixed level numbers rather than in decimal form for instance unsigned 8-bit integer which has the level number ranging from 0 to 255. Let us say the example 1 exercised $v[n]$ discrete signal becomes [0   177   196   255   78   137] in $[0,255]$ range which we wish to obtain.

This example basically needs the repetition of the example 1 mentioned commands with little exception and do so as follows:

>>x1=min(v); y1=0; x2=max(v); y2=255; ⏎
>>y=(y2-y1)/(x2-x1)*v+(x2*y1-x1*y2)/(x2-x1) ⏎

y =
     0   176.5385   196.1538   255.0000   78.4615   137.3077

From the last execution we see that the $v[n]$ data is mapped in $[0,255]$ range but fractional part appears due to the calculation which is not acceptable in unsigned 8-bit integer storage. Strictly the audio level should be integer and do so by the appendix B.8 cited **round** as follows:

>>y=round(y) ⏎   ← rounded y data is again assigned to y

y =
     0   177   196   255   78   137

Or if you intend to use the example 2 mentioned **mat2gray** function, the command **round(255*mat2gray(v))** can be exercised.

Now you may store the above shown data as 8-bit binary numbers for instance 255 by 11111111.

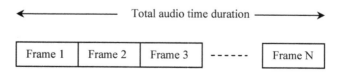

Figure 3.1(b) Splitting an audio into N frames

-32-

3.3 Frame by frame analysis of a digital audio

A given digital audio of few seconds duration contains millions of samples. Analysis or synthesis of such practical audio in a Laptop or Desktop suffers from time constraints. Let us say a softcopy digital audio has 10000 samples. If we analyze the audio by employing matrix algebra, frequently we have to handle a matrix of size in the order 10000×10000. Calculation on such matrix takes huge time. Besides every software provides some finite memory for workspace computation. Several computations like this may lead the computer workspace running out of memory. For this reason most audio algorithms work on frame by frame basis. By frame by frame we mean working on short interval rather than on the whole timeline of the digital audio.

Figure 3.1(b) depicts the frame by frame strategy of a digital audio. An audio timeline is divided into N frames. Each frame is handled separately while studying the audio. The problem is how we pick up and put in place each frame in the whole timeline.

❖ ❖ Example on a discrete signal or prototype audio

Let us consider the prototype digital audio or discrete signal $f[n]=[5\ 6\ 7\ 0\ -5\ -6\ 9\ -2\ 3\ 4\ 6]$. Assume that we are going to choose 3 samples a frame from $f[n]$. So first we should be getting [5 6 7], then [0 −5 −6], and after that [9 −2 3].

We first assign the $f[n]$ to some workspace variable f as a column matrix (because a softcopy audio is usually a column matrix) as follows:

```
>>f=[5 6 7 0 -5 -6 9 -2 3 4 6]'; ↵
>>N=numel(f); ↵          ← N holds the number of samples in f, which is 11
```

The reader also decides how the end elements of $f[n]$ should be selected in case the number of elements in $f[n]$ is not multiple of 3. Since there are 11 samples in $f[n]$, we have to decide what we will do with the last 2 samples. One way to solve the problem is pad the rest element(s) by zero(es).

Before frame selection we check the multiplicity of 3. The integer remainder followed by the integer division of N by 3 (which is 2) for the example $f[n]$ is found by the command rem (appendix B.8) as follows:

```
>>R=rem(N,3); ↵ ← R holds the integer remainder which is 2, R is user-chosen
```

If the R is 0, the number of samples in $f[n]$ is multiple of 3. If the R is not 0, we include 3-R zeroes as a column matrix to f to pad the f. Zero generation we perform by the command zeros (appendix B.10). The command zeros(3-R,1) generates 3-R zeroes as a column matrix. The checking takes place by the control statement if-end (appendix B.2). The inequality

(appendix B.1) to 0 becomes the logical statement of the **if-end**. Let us perform the last two actions as follows:
>>if R~=0 f1=[f;zeros(3-R,1)]; end ⏎

In above execution the **f1** is some user-chosen variable for holding the given **f** and **3-R** zeroes by vertical data accumulation technique (appendix B.3).

In order to gain control on the elements in **f1**, we use a for-loop (appendix B.4) in which the loop counter (later exercised as **k**) as integer provides the control therefore execute the following:
>>for k=1:3:numel(f1), g=f1(k:k+2), end ⏎

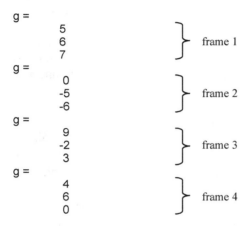

g =
```
    5
    6      }  frame 1
    7
```
g =
```
    0
   -5      }  frame 2
   -6
```
g =
```
    9
   -2      }  frame 3
    3
```
g =
```
    4
    6      }  frame 4
    0
```

In the last execution the **g** is another user-chosen workspace variable which holds the three elements as a column matrix at a time and indicates one frame of the figure 3.1(b). The last frame is padded by a single 0. The **numel(f1)** finds the number of elements in **f1** which is **12** for the example. The **k** assumes 1, 4, 7, and 10 at a time while the loop is running. The increment **3** of the for-loop is due to the frame sample number and the **2** in **k+2** is from sample number minus 1. The **f1(k:k+2)** indicates elements in **f1** from the k-th to k+2-th.

```
f=[5 6 7 0 -5 -6 9 -2 3 4 6]';
L=3;
N=numel(f);
R=rem(N,L);
if R~=0
   f1=[f;zeros(L-R,1)];
end
for k=1:L:numel(f1)
   g=f1(k:k+L-1);
end
```

Figure 3.1(c) M-file for the frame by frame selection of a discrete signal

For demonstration reason we displayed each frame contents. If we do not wish to display each frame contents, we use one semicolon as follows:
>>for k=1:3:numel(f1), g=f1(k:k+2); end ⏎
You may perform any computation on each **g** column matrix. Here we chose the frame length as 3 and if it were **L** (assign **3** to **L** by writing **L=3;** first), the command would be **for k=1:L:numel(f1), g=f1(k:k+L-1); end.**

✦ ✦ Devising an M-file for frames

So long we explained step by step programming tactic for frame based study of a discrete signal. You can open a new M-file (section 1.3), type the codes of the figure 3.1(c) in the opened M-file, save the file by some name **test**, and run the file by calling **test** at the command prompt. Just one point, each frame will be available in the variable **g** sequentially.

```
f=[5 6 7 0 -5 -6 9 -2 3 4 6]';
L=3;
N=numel(f);
R=rem(N,L);
if R~=0
    f1=[f;zeros(L-R,1)];
end
fn=[ ];
for k=1:L:numel(f1)
    g=f1(k:k+L-1);
    fn=[fn;g];
end
```

Figure 3.1(d) M-file for retrieval of the signal from successive frames

✦ ✦ How to put the frames in place?

In figure 3.1(c) we made each frame available in **g**. What if we need to retrieve the whole audio from the successive **g**s? All we need is the vertical data accumulation technique of appendix B.3 in conjunction with an empty matrix in the for-loop of figure 3.1(c).

Let us call the retrieved signal as **fn** (some user-chosen variable) so the modified commands are as shown in figure 3.1(d). Between the two line commands of the for-loop you may perform any computation on **g**.

✦ ✦ Practicalities in a softcopy digital audio

Usually audio frames are chosen of duration $10\text{-}30\,m\sec$. In window operated computer system frequently a sampling frequency $f_s = 22050\,Hz$ is applied. Let us say we choose an audio frame of $15\,m\sec$ duration. With the window system sampling frequency, the sampling period is $T_s = 1/f_s = 4.5351\times10^{-5}$ sec. In $15\,m\sec$ frame, there will be $\dfrac{15m\sec}{T_s} = 330.75$ samples. The sample number is always integer so 330.75 should be rounded to 331. The rounding is performed by the command **round**. The **L** of earlier discussion should be now 331 for this particular example.

✦ ✦ Example on a softcopy digital audio

Now we wish to address the frame by frame issue related to a softcopy audio for instance the **bird.wav** of section 2.7. From section 2.4 we know that the audio duration is $0.8823\,\sec$ meaning $882.3\,m\sec$. Sampling frequency of the audio is $f_s = 22050\,Hz$. If we consider $15\,m\sec$ frame duration, the **L** of the figure 3.1(c) or 3.1(d) described commands should be 331. Let the machine read the audio as follows (section 2.3):

```
>>[v,fs]=wavread('bird.wav'); ↵
```

In figure 3.1(d), delete the first two lines command and type **f=v; L=331**; in lieu of that. The reader might say how to verify that it works for the whole audio. You may play first **wavplay(v,fs)** and later **wavplay(fn,fs)** at the command prompt after executing the commands of the figure 3.1(d) with the last modification. Both commands play the same audio.

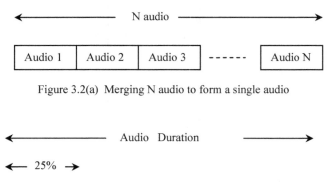

Figure 3.2(a) Merging N audio to form a single audio

Figure 3.2(b) Splitting an audio into user-defined time portion

3.4 Merging of digital audio

In multimedia applications we may need to merge several digital audio in order to form a single audio. Figure 3.2(a) shows the strategy we plan to implement. First requirement in the merging is all related digital audio should have the identical sampling frequency f_s.

In section 2.7 we mentioned three audio by the names **bird.wav**, **b.wav**, and **test.wav** having identical sampling frequency so the merging exercise one can conduct on these three audio. Let the machine read (section 2.3) the three audio one at a time as follows:

```
>>[v1,fs]=wavread('bird.wav'); ↲
>>[v2,fs]=wavread('b.wav'); ↲
>>[v3,fs]=wavread('test.wav'); ↲
```

The **v1**, **v2**, and **v3** are user-chosen variables in the last executions and hold the samples of the three audio respectively. Each of the three variables is a column matrix. If the reader wishes to listen the three audio separately, following commands are exercised:

```
>>wavplay(v1,fs) ↲        ← For the bird.wav
>>wavplay(v2,fs) ↲        ← For the b.wav
>>wavplay(v3,fs) ↲        ← For the test.wav
```

Column matrix accumulation technique is explained in appendix B.3 which we exercise here to merge the three audio on the three column matrices **v1**, **v2**, and **v3** as follows:

```
>>v=[v1;v2;v3]; ↲          ← v holds the merged three audio as a single one
```

where the v is a user-chosen variable

We easily verify whether the merging happened by:

>>wavplay(v,fs) ⏎

If there were another audio held by the variable **v4** with the same sampling frequency, the merging command would have been **v=[v1;v2;v3;v4];**.

3.5 Splitting a digital audio

In section 3.3 the frame by frame strategy is based on uniform frame sample number selection. Splitting a digital audio means selecting an audio segment by user-defined time portion from a given audio duration which we specify by percentage as shown in figure 3.2(b) for example first 25% of the audio, middle 25% of the audio, last 25% of the audio, etc.

Total sample number in a digital audio we obtain by the function **numel** let us call the total sample number as **N**. Since array indexing starts from 1 in MATLAB, we may select any number from **1** to **N** for the splitting. Let us see the following splitting:

The first 25% duration means sample index from 1 to **0.25*N**,

The last 25% duration means sample index from **0.75*N** to N,

The middle 25% duration means sample index from **0.375*N** to **0.625*N**,

From 30% to 70% duration means sample index from **0.3*N** to **0.7*N**, and so on.

One problem with the multiplication of **N** by a fractional number is the result may also be fractional for example when **N=11**, **0.25*N** becomes 2.75. Sample number is always integer so we round the multiplication result to its nearest integer by the function **round** of appendix B.8. If the variable **v** is a column array, its elements indexed from integer **N1** to **N2** are selected by the command **v(N1:N2)**.

As an example considering the **bird.wav** of section 2.7, we intend to exercise the following:

(a) play the whole audio,
(b) play first 25% of the audio duration,
(c) play last 25% of the audio duration, and
(d) play from the 30% to the 70% of the audio duration.

Its straightforward implementation is the following:

```
>>[v,fs]=wavread('bird.wav'); ⏎       ← section 2.3 for the audio reading
>>wavplay(v,fs) ⏎                     ← playing the whole bird.wav
>>N=numel(v); ⏎                       ← finding number of samples in bird.wav
>>v1=v(1:round(0.25*N)); ⏎            ← picking up 25% of the audio
>>wavplay(v1,fs) ⏎                    ← playing the 25% of the audio
>>v2=v(round(0.75*N):N); ⏎           ← picking up last 25% of the audio
>>wavplay(v2,fs) ⏎                    ← playing the last 25% of the audio
>>v3=v(round(0.3*N):round(0.7*N)); ⏎ ← picking up 30 to 70% of the audio
>>wavplay(v3,fs) ⏎                    ← playing the 30 to 70% of the audio
```

In just exercised commands, the **v**, **v1**, **v2**, and **v3** are user-chosen variables which hold the wanted audio or audio segment samples as a column matrix.

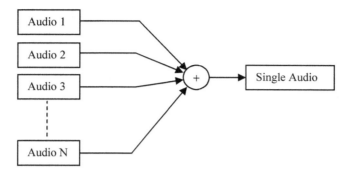

Figure 3.2(c) Mixing N audio to form a single audio

3.6 Mixing several digital audio

In multimedia applications it is a common practice that we mix audio signals of various types. Mixing is simply the superposition or mathematical summation of several digital audio. A compulsory condition in the mixing is identical sampling frequency for all given digital audio. Let us say we have N identical sampling frequency audio which are symbolically expressed as $v_1[n]$, $v_2[n]$, $v_3[n]$, $\cdots v_N[n]$. Figure 3.2(c) depicts audio mixing.

The user has to decide the dominant mode of the component audio. We express the dominance by a scalar ranging from 0 to 1. The 0 means no dominance or absence, 1 means full dominance or presence, some in-between value 0.5 means 50% dominance, and so on obviously for each audio. Mathematically we write the audio mixing operation by $v[n] = c_1 v_1[n] + c_2 v_2[n] + c_3 v_3[n] + \cdots + c_N v_N[n]$ where the c's are the user-defined dominance factors ranging from 0 to 1 and the $v[n]$ is the resultant audio we are interested in.

3.6.1 Example on two modular discrete signals

Let us say we have two discrete signals mimicking two prototype audio as $v_1[n] = [-0.1 \quad 0.4 \quad 0.5 \quad 0.9]$ and $v_2[n] = [0.8 \quad -0.7 \quad 0.6 \quad 0.9 \quad 1]$ with identical sampling frequency and each in range $[-1,1]$. The dominance factors for the two signals are $c_1 = 0.3$ and $c_2 = 0.7$ respectively. Given that the resultant signal is $v[n] = [0.4173 \quad -1 \quad 0.4803 \quad 1 \quad 0.6850]$ in range $[-1,1]$ which we wish to implement.

Addition of two or more discrete signals needs identical size of each matrix because $v_1[n]$ or $v_2[n]$ is usually a row or a column matrix. First consideration is we have to check that each matrix contains same number of samples which is not the case with the given $v_1[n]$. Under this circumstance we pad the deficit sample or samples by zeroes. Let us enter the given discrete signals as follows:

>>v1=[-0.1 0.4 0.5 0.9 0]; ↵   ← v1 holds $v_1[n]$ with padding, v1 is user-chosen

>>v2=[0.8 -0.7 0.6 0.9 1]; ↵   ← v2 holds $v_2[n]$, v2 is user-chosen

>>c1=0.3; c2=0.7; ↵   ← Assigning the dominance factor to like name variable for example c1 for c_1, where c1 is user-chosen variable

>>v=c1*v1+c2*v2; ↵   ← Performing $c_1 v_1[n] + c_2 v_2[n]$ operation and the result is assigned to v, where the v is user-chosen

Section 3.2 mentioned **mat2gray** is applied to map the range in $[-1,1]$ as follows:

>>v=2*mat2gray(v)-1 ↵ ← Performing range mapping and the result is again assigned to v, where the left v is user-chosen

v =

    0.4173 -1.0000   0.4803   1.0000   0.6850

3.6.2 Example of mixing on softcopy audio

The addition technique we applied in subsection 3.6.1 is equally applicable for softcopy audio besides sample number checking and zero padding need extra function to handle huge number of data which will be demonstrated by taking three softcopy audio of section 2.7 namely **bird.wav**, **test.wav**, and **hello.wav**, all of which has identical sampling frequency. Let us choose the dominance factors as 0.1, 1, and 0.05 respectively. We assume that you have the three audio in your working path of MATLAB.

As a first step, let the machine read and play (section 2.3) each audio as follows:

>>[v1,fs]=wavread('bird.wav'); ↵   ← audio reading of the bird.wav
>>wavplay(v1,fs) ↵   ← playing the bird.wav
>>[v2,fs]=wavread('test.wav'); ↵   ← audio reading of the test.wav
>>wavplay(v2,fs) ↵   ← playing the test.wav
>>[v3,fs]=wavread('hello.wav'); ↵   ← audio reading of the hello.wav
>>wavplay(v3,fs) ↵   ← playing the hello.wav

The given three audio signals $v_1[n]$, $v_2[n]$, and $v_3[n]$ are available in **v1**, **v2**, and **v3** respectively, each as a column matrix which you verify by looking into the workspace browser (section 1.4). Complete programming approach is somewhat cumbersome that is why we introduce slight interactive approach in mixing the audio signals.

Let us find the total sample number in each audio by the **numel** function as follows:

>>N1=numel(v1), N2=numel(v2), N3=numel(v3) ↵

N1 =
        19456   ← for the bird.wav

Mohammad Nuruzzaman

N2 =
    161002   ← for the test.wav
N3 =
    50388   ← for the hello.wav

In the last execution the **N1, N2,** and **N3** are user-chosen variables and hold the sample numbers of the three audio respectively. By inspection the **N2** is the largest so padding takes place for the other two audio up to **N2** which we perform by the **padarray** of appendix B.12 as follows:

>>y1=padarray(v1,[N2-N1,0],'post'); ↵ ← y1 holds padded v1, y1 is user-chosen
>>y3=padarray(v3,[N2-N3,0],'post'); ↵ ← y3 holds padded v3, y3 is user-chosen

How you pad the given audio signal determines the relative location of the component in the resultant signal. We assumed that the **bird.wav** is added right from the beginning in the resultant signal so post padding of **v1** should be conducted, so should be for the **v3** or **hello.wav**. Again if we assume the **bird.wav** is to be mixed in the ending timeline, prepadding must be conducted. However let us enter the given dominance factors to like name variables as follows:

>>c1=0.1; c2=1; c3=0.05; ↵

Now we need to perform $c_1 v_1[n] + c_2 v_2[n] + c_3 v_3[n]$ operation for the three audio. The $v_2[n]$ is not padded because it's having maximum sample number therefore the addition is carried out as follows:

>>y=c1*y1+c2*v2+c3*y3; ↵   ← the addition result is assigned to y, where the y is user-chosen

The last **y** holds the resultant signal with different audio amplitude range. From section 2.5 discussion we know that each audio is in range [−1,1]. If we assume 100% dominance of each audio, the range becomes [−2,2], [−3,3], [−4,4], etc for two, three, four, etc audio mixing respectively which necessitates the range mapping of the mixed signal to [−1,1] and do so by the section 3.2 mentioned **mat2gray** as follows:

>>yn=2*mat2gray(y)-1; ↵ ← Performing range mapping and the result is assigned to yn, where the yn is user-chosen

Finally the variable **yn** contains our wanted digital signal in accordance with the figure 3.2(c). We are one click away from hearing the mixed signal and do so by:

>>wavplay(yn,fs) ↵

Certainly you hear all three audio **bird.wav, test.wav,** and **hello.wav** by the last aural action and the single signal is available in the workspace variable **yn** with the same sampling frequency as each of the three audio has.

For demonstration we mixed three audio but you can add arbitrary number of audio in a similar fashion.

-40-

3.7 Down and up samplings of discrete and audio signals

The down and the up samplings of a signal mean contracting and dilating the time domain of the signal respectively. The operations are very much associated with the discrete signals. Both samplings happen according to user-defined factor and conventionally it is an integer.

$f[n]$ → [3↓] → $g[n]$ $f[n]$ → [2↑] → $g[n]$

Figure 3.3(a) Down sampling of $f[n]$ Figure 3.3(b) Up sampling of $f[n]$

The effect of sampling appears in the functional expression of the signal if it has. If the signal is downsampled like the figure 3.3(a), downsampling by a factor 3 (as an example) means $g[n] = f[3n]$. For example $f[n] = \sin n$ turns the $g[n]$ as $\sin 3n$. In the case of upsampling by a factor of 2 (as an example) like the figure 3.3(b), the $g[n]$ becomes $f\left[\dfrac{n}{2}\right]$ or $\sin\dfrac{n}{2}$ when $f[n] = \sin n$.

$f[n] = [4 \quad -4 \quad 7 \quad 0 \quad 4 \quad 8 \quad 1 \quad 3]$ → [2↓] → $g[n] = f[2n] = [4 \quad 7 \quad 4 \quad 1]$

Figure 3.3(c) Downsampling of the discrete $f[n]$

$f[n] = [4 \quad -4 \quad 7 \quad 0 \quad 4]$ → [2↑] → $g[n] = f\left[\dfrac{n}{2}\right] = [4 \quad 0 \quad -4 \quad 0 \quad 7 \quad 0 \quad 0 \quad 0 \quad 4 \quad 0]$

Figure 3.3(d) Upsampling of the discrete $f[n]$

We know that a discrete signal $f[n]$ takes the shape of a row matrix for different n. Downsampling the $f[n]$ by a factor of 2 (for example) means turning the $f[n]$ to $f[2n]$. The number of elements in $f[n]$ must be multiple of downsampling factor. For downsampling by a factor 2, the number of elements in $f[n]$ can be 2, 4, 6, etc. Again for downsampling by a factor 3, the number of elements in $f[n]$ can be 3, 6, 9, etc and so on.

Figure 3.3(c) presents an example of downsampling. The discrete signal $f[n] = [4 \quad -4 \quad 7 \quad 0 \quad 4 \quad 8 \quad 1 \quad 3]$ turns to $g[n] = f[2n] = [4 \quad 7 \quad 4 \quad 1]$ when downsampled by a factor of 2. We wish to implement this downsampling.

MATLAB built-in function **downsample** implements this sort of operation with the syntax **downsample**(signal as a row matrix, downsampling factor). First we enter the $f[n]$ as a row matrix as follows:

>>f=[4 -4 7 0 4 8 1 3]; ⏎   ← f holds the signal $f[n]$, where f is user-chosen
>>g=downsample(f,2) ⏎   ← Workspace g holds the downsampled discrete
                                signal $f[2n]$, g⟺ $f[2n]$,g is user-chosen variable

g =
        4   7   4   1

Skipping every alternate sample of $f[n]$ provides the $f[2n]$. If the factor were 3, the skipping would happen for every two alternate samples.

Upsampling is the reverse process of the downsampling. Figure 3.3(d) presents the upsampling of a discrete signal $f[n]=[4 \quad -4 \quad 7 \quad 0 \quad 4]$ by a factor of 2 which provides $g[n]=f\left[\dfrac{n}{2}\right]=[4 \quad 0 \quad -4 \quad 0 \quad 7 \quad 0 \quad 0 \quad 0 \quad 4$

0]. We intend to implement this upsampling operation.

When upsampling is conducted by a factor of 2, it is the insertion of a single zero between the samples thereby forming $f\left[\dfrac{n}{2}\right]$. If the factor were 3, the insertion would occur with two zeroes. MATLAB built-in function **upsample** conducts upsampling with the syntax **upsample**(signal as a row matrix, upsampling factor) as follows:

>>f=[4 -4 7 0 4]; ⏎   ← Workspace f holds the signal $f[n]$, f⟺ $f[n]$
>>g=upsample(f,2) ⏎   ← Workspace g holds the upsampled discrete signal
                     $f\left[\dfrac{n}{2}\right]$, g⟺ $f\left[\dfrac{n}{2}\right]$, g is user-chosen variable

g =
        4   0   -4   0   7   0   0   0   4   0

Whether down or up sampling, the n information is lost during the sampling so it is the user's accountability to keep a mark on that. Let us see another example on the sampling.

Figure 3.3(e) shown discrete signal $f[n]=2^{-n}$ when downsampled over $0 \le n \le 11$, the $g[n]$ becomes [1   0.125   0.0156   0.002] which we intend to get.

2^{-n} ⟶ [3↓] ⟶ g[n]

Figure 3.3(e) The discrete signal 2^{-n} is downsampled by a factor 3

In this sort of problem we first generate the integer n values as a row matrix (section 1.3) and then write the scalar code (appendix A) to obtain the discrete 2^{-n} values. Let us go through the following in this regard:

>>n=0:11; ⏎   ← n is a row matrix holding integers from 0 to 11
>>f=2.^(-n); ⏎   ← f is a row matrix holding discrete 2^{-n} signal values
>>g=downsample(f,3) ⏎   ← Workspace g holds the downsampled discrete
                           signal, g⟺ $g[n]$, g is user-chosen variable

g =

1.0000   0.1250   0.0156   0.0020

The down and up sampling factors are called decimation and interpolation factors respectively.

◆ ◆ Example on a softcopy digital audio

Now we wish to implement how the up or down sampling changes the aural characteristic of a digital audio. In order to do so, we assume same sampling frequency before and after the down or up sampling and consider the **bird.wav** of section 2.7. Let the machine read and play the audio as follows:

>>[v,fs]=wavread('bird.wav'); ↵   ← Reading the audio, section 2.3,workspace
                                             v holds earlier mentioned $f[n]$

>>wavplay(v,fs) ↵   ← Playing the audio

Let us choose a down and an up samplings by a factor 2 and execute the samplings as follows:

>>g=downsample(v,2); ↵   ← g holds the downsampled v, g is user-chosen
>>wavplay(g,fs) ↵   ← playing the downsampled audio
>>h=upsample(v,2); ↵   ← h holds the upsampled v, h is user-chosen
>>wavplay(h,fs) ↵   ← playing the upsampled audio

As you hear from the last executions, the downsampled **bird.wav** sounds faster than the original one whereas upsampled **bird.wav** sounds slower than the original one.

◆ ◆ Resampling on audio applications

Just discussed up and down samplings are for basic understanding where we chose the sampling factor as simple integers like 2 or 3. In audio applications the sampling factor becomes in the *KHz* range. For example in CD applications an audio sampled with 48 *KHz* needs to be resampled with 44.1 *KHz* . In this regard the built-in function **resample** is helpful which requires the syntax **resample**(discrete signal vector, new sampling rate, original sampling rate).

As an example just addressed **bird.wav** has the sampling frequency **fs** as 22050 *Hz* . We wish to resample the **bird.wav** with 15 *KHz* and play both the original and the resampled audio.

>>[v,fs]=wavread('bird.wav'); ↵   ← Reading the audio, workspace v holds the
                                      discrete audio signal, fs does the
                                      sampling frequency

>>v1=resample(v,15000,fs); ↵   ← Resampling the discrete signal stored in v
                                      and put the resampled signal to v1 where
                                      v1 is user-chosen variable

>>wavplay(v,fs) ↵   ← Playing the original audio
>>wavplay(v1,fs) ↵   ← Playing the resampled audio with the
                                      same sampling frequency

If you compare the two sounds aurally, the sharpness of the latter audio is increased due to the reduction of the sampling frequency.

3.8 Quantization of a discrete signal and a digital audio

Quantization is a process which rounds or truncates a signal amplitude whether continuous or discrete to a user-defined finite number of levels. The quantization is solely associated with the signal amplitudes or values not with the time or sample index. If the signal value levels are equally spaced, the quantization is called uniform quantization. MATLAB function **quant** performs the uniform quantization. The function accepts two input arguments, the first and second of which are the signal as a row or column matrix and the quantization step size (called resolution Δ of the quantization process in signal processing term) respectively. Let us go through the following examples on the subject of quantization.

✦✦ Example 1

Given a single signal value $V = 1.3$, quantize it by using $\Delta = 0.2$. Computationally we divide the value of V by Δ and round the division result $\dfrac{V}{\Delta}$ towards the nearest integer M. The division result is checked according to the following: $\begin{cases} \text{if the fractional part} \ge 0.5, \text{take that as } 1 \\ \text{if the fractional part} < 0.5, \text{take that as } 0 \end{cases}$. Once the integer M is found, the quantized signal value is given by $M\Delta$. For the numerical value, we have $\dfrac{V}{\Delta} = 6.5$ or $M = 7$ consequently the quantized single signal value should be $M\Delta = 1.4$. Following is the implementation:

```
>>Q=quant(1.3,0.2) ↵   ← Workspace Q holds the quantized single signal value,
                          Q is user-chosen variable
Q =
      1.4000
```

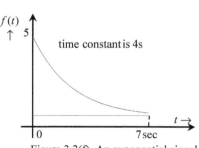

Figure 3.3(f) An exponential signal

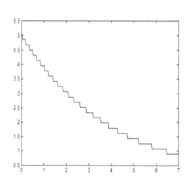

Figure 3.3(g) Quantization of the continuous signal of figure 3.3(f)

✦✦ Example 2

Example 1 mentioned $\Delta = 0.2$ is applied to the discrete signal $f[n] = $ [−1.3 1.3 0 1.8] so that we finish up with the quantized signal $q[n] = $ [−1.4 1.4 0 1.8] whose implementation is shown in the following:

>>f=[-1.3 1.3 0 1.8]; ↵ ← Workspace f holds the given discrete signal $f[n]$ as a
row matrix

>>Q=quant(f,0.2) ↵ ← Workspace Q holds the quantized discrete signal $q[n]$

Q =
    -1.4000   1.4000   0   1.8000

✦ ✦ Example 3

Figure 3.3(f) presented continuous signal has the expression $f(t) = 5e^{-\frac{t}{4}}$ over $0 \le t \le 7\,\text{sec}$. Our objective is to quantize the signal at a Δ =0.18 and to plot the quantized signal in continuous sense.

Scalar code (appendix A) generates the samples of a continuous signal so one needs to choose some t step size (must be small enough to generate $f(t)$ samples which are much greater than Δ) to have the signal samples despite the signal is a continuous one and let it be 0.02 therefore we conduct the following:

>>t=0:0.02:7; ↵ ← t holds the t sample points as a row matrix on chosen step
size over $0 \le t \le 7\,\text{sec}$ (section 1.3)

>>f=5*exp(-t/4); ↵ ← f holds the $f(t)$ samples as a row matrix at the
t points stored in t

>>Q=quant(f,0.18); ↵ ← Q holds quantized $f(t)$ sample values as a row matrix
at the t points stored in t

>>plot(t,Q) ↵ ← Figure 3.3(g) shows the response, appendix D for the plot

✦ ✦ Example 4

In analog to digital conversion, the number of levels L rather than resolution Δ is invariably used. The L is frequently a power of 2 because quantized signals are mapped to the binary form. The relationship among quantization step, number of levels, and signal data is $\Delta = \dfrac{f(t)|_{\max} - f(t)|_{\min}}{L}$. Suppose example 3 mentioned exponential signal is to be quantized by 3-bit quantizer. From the exponential function properties, we have $f(t)|_{\max} = 5$, $f(t)|_{\min} = 0$, and $L = 2^3 = 8$ whence Δ =0.625 and the relevant command would have been Q=quant(f,0.625);.

✦ ✦ Example 5

Quantization error and signal to noise power ration (SNR) are also of interest in the audio signal literature.

In example 3 (re-execute the commands of the example at the command prompt), the workspace variables f and Q hold the samples of the original and quantized signals respectively both as the identical size row matrix. The error signal (which is $e[n] = f[n] - q[n]$ in discrete sense and is also the same size row matrix) is just the f-Q. In discrete sense, the power in the original and error signals are given by $P_s = \dfrac{1}{N}\sum_{n=0}^{N-1} f^2[n]$ and $P_q = \dfrac{1}{N}\sum_{n=0}^{N-1} e^2[n]$ respectively (ignoring the sampling period effect) where the N is the number of samples in $f[n]$ or the row matrix element number in f or Q. On account of the quantization, the signal to noise ratio in dB is given by $SNR = 10\log_{10}\dfrac{P_s}{P_q}$ whose calculation is the following:

```
>>e=f-Q;            ← Workspace e holds the error signal e[n] as a row matrix
>>Ps=sum(f.^2)/length(f);   ← Workspace Ps holds the computed Ps,
                                        appendices A and B.11
>>Pe=sum(e.^2)/length(e);   ← Workspace Pe holds the computed Pq
>>R=10*log10(Ps/Pe)         ← R holds computed SNR due to quantization

R =
          34.1734           ← i.e. the SNR is 34.1734 dB due to quantization
```

In above executions the **e, Ps, Pe**, and **R** are all user-chosen variables.

♦ ♦ Quantization on a digital audio

So far whatever we exercised can be implemented on a digital audio. We should remember one point that a given softcopy digital audio is already quantized usually in 256, 512, etc levels.

As an example let us consider the softcopy **bird.wav** of the section 2.7. Just from softcopy we can not take any decision about the number of quantization levels through which it originates from. All we know is the data range of the audio, which can be between -1 and 1 upon the machine reading (section 2.3) of the audio. We wish to quantize the audio samples in 8 levels and play the audio for comparison with the original one.

```
>>[v,fs]=wavread('bird.wav');   ← Reading the audio, workspace v holds the
                                   discrete audio signal, fs does the
                                   sampling frequency
```

From the minimum and the maximum of the audio data, we obtain the Δ as in example 4 by the appendix B.5 cited functions as follows:

```
>>d=(max(v)-min(v))/8;   ← d holds Δ, where d is user-chosen variable
>>v1=quant(v,d);         ← v1 holds 8 level quantized audio data, where v1
                            is user-chosen variable
>>wavplay(v,fs)          ← Playing the original audio
>>wavplay(v1,fs)         ← Playing the quantized audio
```

Certainly the reader will hear aural distortion of the bird sound from above executions. If we assume the original audio data as held in **v** is the actual one, signal to noise ration (SNR) due to the 8 level quantization of the **bird.wav** can be found like the example 5 as follows:

```
>>e=v-v1; Ps=sum(v.^2)/length(v); Pe=sum(e.^2)/length(e);
>>R=10*log10(Ps/Pe)

R =
       11.2080
```

Above return says that the SNR due to the quantization is 11.208 dB where the symbols have example 5 mentioned meanings. The **v** and the **v1** are synonyms of the **f** and the **Q** respectively.

3.9 Interpolation of a discrete signal and digital audio

In section 3.7 we addressed that if upsampling or downsampling of a discrete signal takes place in integer numbers, skipping samples or inserting zeroes is plainly conducted. But zero insertions do not reveal actual signal variation.

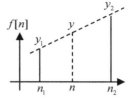

Figure 3.4(a) Linear interpolation of a discrete signal from two sample values

Figure 3.4(a) shows two sample values of a linear discrete signal coordinated by (n_1, y_1) and (n_2, y_2). Over the interval $n_1 \le n \le n_2$ if the discrete signal value is needed, the correct value is y. If we insert 0 instead of y, the signal will be corrupted that is why interpolation is necessary.

Between the two known samples of the figure 3.4(a), any sample over $n_1 \le n \le n_2$ is given by $y = \dfrac{y_2 - y_1}{n_2 - n_1}(n - n_1) + y_1$. Since the $\dfrac{n - n_1}{n_2 - n_1}$ is a ration, horizontal axis unit does not matter whether it is in integer index or absolute time domain. The bottomline is from the (n_1, y_1) and (n_2, y_2) we look for any (n, y) over $n_1 \le n \le n_2$ – this is called interpolation and specifically linear interpolation because the two given samples are connected by a straight line.

MATLAB built-in function **interp1** performs this sort of interpolation with the syntax **interp1**(given n values as a row matrix, given y values as a row matrix, n as a row matrix for the value wanted at). As soon as the syntax is known, one may proceed with the following.

♦ ♦ **Interpolation at a single point**

As an example we have (n_1, y_1)=(4,3.7) and (n_2, y_2)=(5, 6.3) and between the two samples some n=4.3 has the linearly interpolated $y = \dfrac{6.3 - 3.7}{5 - 4}(4.3 - 4) + 3.7 = 4.48$. We wish to implement this linear interpolation.

Execute the following for the interpolation:

>>x=[4 5]; ↵   ← Given n_1 and n_2 are assigned to x as a row matrix, where x is
                                          user-chosen

>>y=[3.7 6.3]; ↵ ← Given y_1 and y_2 are assigned to y as a row matrix, where y
                                       is user-chosen

>>interp1(x,y,4.3) ↵   ← Calling the function

ans =
            4.4800   ← Value of the y we are looking for

♦ ♦ **Interpolation at multiple points**

The function also keeps provision for multiple returns. Suppose we have four samples of a discrete signal given as (4,3.7), (5,6.3), (6,–2), and

(7,−7). The linearly interpolated values of the signal at 4.3, 5.2, and 6.9 are 4.48, 4.64, and −6.5 respectively which we implement as follows:

>>x=[4 5 6 7]; ↵   ← Given horizontal points are assigned to x as a row matrix, where x is user-chosen

>>y=[3.7 6.3 -2 -7]; ↵ ← Given vertical points are assigned to y as a row matrix, where y is user-chosen

>>interp1(x,y,[4.3 5.2 6.9]) ↵   ← Points of interest are as a row matrix

ans =

    4.4800   4.6400 -6.5000   ← The return is also as a row matrix

You could have executed I=interp1(x,y,[4.3 5.2 6.9]); to hold the values to I where the I is a user-chosen variable. Should the reader need each interpolated value, the command I(1), I(2), or I(3) is exercised respectively.

✦ ✦ Interpolation over a range of points

We may seek the interpolated values over a range of horizontal axis point values. For example starting from (4,3.7) and (5,6.3) over the interval $4.1 \leq n \leq 4.6$ and with a step size 0.1, the linearly interpolated values are 3.96, 4.22, 4.48, 4.74, 5, and 5.26 for 4.1, 4.2, 4.3, 4.4, 4.5, and 4.6 respectively which we wish to implement.

This implementation is similar to the earlier ones with the exception that the third input argument of the **interp1** is a range data which we feed by writing **4.1:0.1:4.6** and execute as follows:

>>x=[4 5]; ↵   ← Given n_1 and n_2 are assigned to x as a row matrix, where x is user-chosen

>>y=[3.7 6.3]; ↵ ← Given y_1 and y_2 are assigned to y as a row matrix, where y is user-chosen

>>yn=interp1(x,y,4.1:0.1:4.6) ↵ ← Calling the function with the 3rd input as a range data

yn =

    3.9600   4.2200   4.4800   4.7400   5.0000   5.2600

In above execution the **yn** is a user-chosen variable which holds the return interpolated values as a row matrix respectively.

✦ ✦ Other type of interpolation

So far the exercised interpolation examples used linear interpolation which is the default type in **interp1**. There are other types of interpolation whose theoretical discussions are beyond the scope of the text. Some other interpolation methods that are available in MATLAB are nearest neighborhood, piecewise cubic spline, and shape preserving piecewise cubic whose MATLAB indicatory reserve words are **nearest**, **spline**, and **cubic** respectively. The reserve word is put as the fourth input argument to the **interp1** but under a quote.

As an example starting from three given points (4,3.7), (5,6.3), and (6,−1.2), the piecewise cubic spline method provides interpolated sample values 3.7, 5.7589, 5.5405, 6.5805, 5.3719, and 3.8125 at the horizontal coordinate points 4, 4.35, 4.3, 4.7, 5.25, and 5.5 respectively which we wish to implement. Following is the execution:

>>x=[4 5 6]; ⏎   ← Given horizontal points are assigned to x as a row matrix,
                          where x is user-chosen variable

>>y=[3.7 6.3 -1.2]; ⏎ ← Given sample values are assigned to y as a row matrix,
                          where y is user-chosen variable

>>xn=[4 4.35 4.3 4.7 5.25 5.5]; ⏎ ← Given horizontal coordinate points are
                          assigned to xn where xn is user-chosen variable

>>yn=interp1(x,y,xn,'spline') ⏎ ← Calling the function where the 4th input
                          argument is the reserve word for piecewise cubic spline

yn =
        3.7000   5.7589   5.5405   6.5805   5.3719   3.8125

In above execution the **yn** is a user-chosen variable which holds the interpolated values as a row matrix respectively.

✦ ✦ Application on a digital audio

We have been discussing the implementation of various interpolations in MATLAB. Now we explain how the interpolation methods can be applied to a softcopy digital audio.

An interpolation may produce digital audio effects which we wish to demonstrate considering the audio **b.wav** of section 2.7. Obtain the audio in your working path of MATLAB and let the machine read (section 2.3) the audio as follows:

>>[v,fs]=wavread('b.wav'); ⏎

Therefore the workspace variables **v** and **fs** hold the audio samples as a column matrix and sampling frequency respectively. The **v** just contains the samples of the audio, there is no information about horizontal or time coordinate points unless we apply the sampling frequency information. We do need horizontal coordinates for any interpolation whether in absolute time or integer sample index domain. In linear interpolation discussion we quoted that interpolated values are independent of the absolute time unit so we may work on sample index values. Looking into the workspace browser (section 1.4) we find that there are 2800 samples in **v** or we can execute the command **length** on the **v** as follows:

>>length(v) ⏎

ans =
        2800

One way to index the samples is to assign integers staring from 1 to 2800 i.e. we are considering the integer index domain as $1 \le n \le 2800$ and whose generation takes place as follows:

>>x=[1:2800]'; ⏎   ← x holds the 2800 integers as a column matrix,
                              where the x is user-chosen variable

As an example we wish to find the interpolated audio sample values at every 0.75 unit away i.e. at 1, 1.75, 2.5, until 2800 whose generation occurs by:

>>xn=[1:0.75:2800]'; ⏎   ← xn holds the wanted horizontal coordinate points as
                              a column matrix, where the xn is user-chosen variable

At this point we have to decide the type of interpolation say linear so we execute the following:

>>yn=interp1(x,v,xn); ↵ ← Calling the function with earlier mentioned syntax
where the yn holds interpolated audio sample values
and the yn is user-chosen variable

Making the original and the interpolated audio samples available in the workspace, we easily play them by:

>>wavplay(v,fs) ↵   ← Playing the original audio
>>wavplay(yn,fs) ↵   ← Playing the interpolated audio

You must be hearing the distorted sound of the audio **b.wav** from above executions. Thus moviemakers generate a special audio effect as frequently heard in fictional Hollywood movies.

3.10 Convolution of discrete signals and digital audio

Formal definition of discrete convolution of two discrete signals $x[n]$ and $y[n]$ is given by $c[n] = x[n] * y[n] = \sum_{k=-\infty}^{k=\infty} x[k]y[n-k]$ where $c[n]$ is the discretely convolved signal. If lengths (mean numbers of samples) of the signals $x[n]$ and $y[n]$ are M and N respectively, the length of $c[n]$ is $M+N-1$. The summation $\sum_{k=-\infty}^{k=\infty} x[k]y[n-k]$ can be viewed as the polynomial multiplication of $x[n]$ and $y[n]$. The $x[n]$ or $y[n]$ is just a row or column matrix for different n and the n is always integer.

MATLAB built-in function **conv** (abbreviation for convolution) computes the convolution of two discrete signals $x[n]$ and $y[n]$ with the syntax **conv**($x[n]$ as a row matrix, $y[n]$ as a row matrix).

The function does not provide any information about the index. Suppose $x[n]$ and $y[n]$ exist over $N_1 \le n \le N_2$ and $N_3 \le n \le N_4$ respectively, what should be the $c[n]$ interval? First the user has to decide where to start the index of $c[n]$ say p then the interval of $c[n]$ should be $p \le n \le N_2 - N_1 + N_4 - N_3 + p$. For example the $x[n]$ and the $y[n]$ intervals are $-2 \le n \le 3$ and $0 \le n \le 3$ respectively, then the $c[n]$ interval is $0 \le n \le 8$ assuming the starting p is 0. Let us see the following examples on the convolution.

❖❖ **Example 1**

Given that $c[n] = [-1 \quad -2 \quad -4 \quad 16 \quad 37 \quad 12 \quad -43 \quad -6 \quad 40]$ is the discrete convolution of the following tabulated discrete signal data which we intend to verify.

n	−2	−1	0	1	2	3
$x[n]$	1	2	7	−5	−6	8
$y[n]$			−1	0	3	5

First we assign the discrete signals $x[n]$ and $y[n]$ as a row matrix to **x** and **y** (user-chosen variables) respectively as follows:

>>x=[1 2 7 -5 -6 8]; ↵   ← x⇔ $x[n]$
>>y=[-1 0 3 5]; ↵   ← y⇔ $y[n]$

Then we call the **conv** with earlier mentioned syntax and put the return to **c** (user-chosen variable) as follows:

>>c=conv(x,y) ↵   ← c holds the convolved signal $x[n] * y[n]$, c⇔ $c[n]$

c =
   -1   -2   -4   16   37   12   -43   -6   40

Should you need the index n (assuming that starts from −2), the interval of $c[n]$ becomes $-2 \le n \le 6$ and which we generate by writing the code **-2:6** (section 1.3).

✦ ✦ Example 2

This example illustrates function based discrete convolution. When $x[n]=n$ over $0 \le n \le 3$ and $y[n]=n^2$ over $2 \le n \le 5$, the discrete convolution is $c[n] = x[n] * y[n] = [0 \quad 4 \quad 17 \quad 46 \quad 84 \quad 98 \quad 75]$ which we wish to implement.

In this sort of problems we generate the n and sample generation of each signal takes place by the scalar code (appendix A) as follows:

>>n=0:3; ↵   ← n holds the integers 0, 1, 2, and 3 as a row matrix
>>x=n; ↵   ← Code of $x[n]$ is assigned to x, x⇔ $x[n]$
>>n=2:5; ↵   ← n holds the integers 2, 3, 4, and 5 as a row matrix
>>y=n.^2; ↵   ← Code of $y[n]$ is assigned to y, y⇔ $y[n]$
>>c=conv(x,y) ↵ ← Workspace c holds the discrete convolved signal $x[n] * y[n]$,
                                   c⇔ $c[n]$

c =
    0   4   17   46   84   98   75

Regarding the interval, the n should be from 0 to 6 if we assume the starting of $c[n]$ from $n=0$ and which is obtained by **0:6**.

Figure 3.4(b)
Plots of $x[n]$,
$y[n]$, and $c[n]$
versus n (right
side figure)

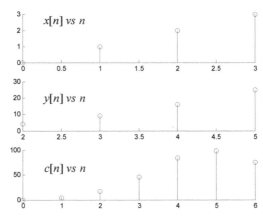

✦ ✦ Example 3

How we graph the discrete convolution is illustrated in this example. Let us consider the example 2 mentioned discrete signals. Appendix D cited **subplot** can be an option to plot the graphs of such discrete convolution. Each signal is plotted by using the **stem** of the same appendix. From example 2 we know that $x[n]$, $y[n]$, and $c[n]$ are available in the workspace **x**, **y**, and **c** respectively. Since n information is lost, we re-execute the **n** generation each time and graph the three traces on top of the other so there are three graphs. Treating each graph as a matrix, we have 3×1 and the control on each graph is obtained by 311, 312, and 313 respectively. Knowing so, graph the three signals as follows:

>>subplot(311),n=0:3; stem(n,x) ↵ ← Graphing *x*[*n*] versus *n*
>>subplot(312),n=2:5; stem(n,y) ↵ ← Graphing *y*[*n*] versus *n*
>>subplot(313),n=0:6; stem(n,c) ↵ ← Graphing *c*[*n*] versus *n*

Figure 3.4(b) shows the graphical response from MATLAB.

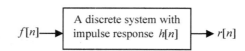

Figure 3.4(c) A discrete system has input *f*[*n*] and output *r*[*n*]

♦♦ Example 4

This example illustrates how the discrete convolution becomes useful in finding the discrete output.

Figure 3.4(c) shows a discrete system (can be a discrete time filter) which has the impulse response $h[n] = n^2$ over $2 \leq n \leq 5$. An

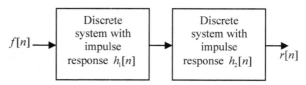

Figure 3.4(d) Two discrete systems connected in series

input $f[n]=n$ over $0 \leq n \leq 3$ is applied to the system. The output *r*[*n*] is $f[n]*h[n]$ whose implementation is the example 2 with different dependent variable name.

♦♦ Example 5

Figure 3.4(d) shows two discrete time systems with impulse responses $h_1[n]$ and $h_2[n]$ which are seriesly placed. Their equivalent system function is $h_1[n]*h_2[n]$ which you can compute like the other four examples.

♦♦ Convolution on a digital audio

Now we wish to address the convolution operation on a softcopy digital audio e.g. **bird.wav** of section 2.7. The convolution operation is schematized in the figure 3.4(e).

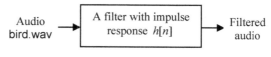

Figure 3.4(e) Filtering a digital audio

The filter needs impulse response *h*[*n*] which usually has some objective criteria. Filter details itself require another chapter (see 6). Besides most filter specifications are often in frequency domain. However following example demonstrates the convolution operation in time index domain.

Obtain the audio in your working path of MATLAB and let the machine read (section 2.3) the audio as follows:

    >>[v,fs]=wavread('bird.wav'); ↵   ← v holds $f[n]$

Therefore the workspace variables **v** and **fs** hold the audio samples as a column matrix and sampling frequency respectively. The **v** just contains the samples of the audio, there is no information about horizontal axis or time and which is also the $f[n]$.

Let us choose the $h[n]$ as $u[n]$, the unit sample sequence. The $u[n]$ we generate by the function **ones** of appendix B.10 depending on n variation. Say the n of $u[n]$ is changing over $0 \leq n \leq 25$ meaning we need to generate 26 ones either as a row or as a column matrix. Considering column matrix, we execute the following:

    >>h=ones(26,1); ↵   ← h holds $h[n]$, h is user-chosen
    >>c=conv(v,h); ↵   ← Workspace c holds the discrete convolved signal
                            $h[n] * f[n]$, c⇔$c[n]$, c is user-chosen

The data range in the **bird.wav** is $[-1,1]$ but the convolution results a new data range. We may map the new range minimum and maximum to -1 and 1 respectively for the wave play reason and do so by the section 3.2 (example 2) cited function and technique:

    >>yn=2*mat2gray(c)-1; ↵ ← yn holds newly mapped data in [-1,1], yn
                                   is user-chosen

The **yn** in above execution refers to the filtered audio of the figure 3.4(e). Making the original and the filtered audio samples available in the workspace, we easily play them by:

    >>wavplay(v,fs) ↵   ← Playing the original audio
    >>wavplay(yn,fs) ↵   ← Playing the filtered audio

In any audio sharp intensity sound manifests high frequency contents. From the last two executions, the high frequency components of the **bird.wav** are reduced.

Anyhow the convolution discussion brings an end to the chapter 3.

Mohammad Nuruzzaman

Chapter 4

 Discrete Fourier
Transform and Digital
Audio

The subject matter in this chapter is to address the issues related to the discrete Fourier transform (DFT) – both in forward and inverse domains. The DFT is the perfect tool for spectral analysis-synthesis of short duration digital audio. DFT is also needed to understand many speech enhancement and noise reduction algorithms. Dedicated MATLAB functions make the study so easy on whose account we plan to explain the following:

❖ ❖ DFT implementation for discrete time as well as digital audio signals in both domains
❖ ❖ Tools to verify the characteristic difference for single or multiple sine signals in the transform domain
❖ ❖ Ways to obtain digital audio frequency components and statistical characteristics

4.1 Discrete Fourier transform of a discrete signal

The discrete Fourier transform (DFT) is merely relevant to a discrete function or sequence $f[n]$. The independent variable n is integer only. The values of $f[n]$ are the samples of a function but the $f[n]$ does not hold any sampling information. For different values of n, the $f[n]$ is just a row or column matrix.

Forward transform:

The forward discrete Fourier transform $F[k]$ of the discrete signal $f[n]$ is defined as $F[k] = \sum_{n=1}^{N} f[n] e^{-j2\pi(k-1)\frac{(n-1)}{N}}$ where N is the number of samples in $f[n]$. The transform $F[k]$ is discrete and possesses N samples as well. The discrete frequency variable k is integer.

Inverse transform:

The recovery of $f[n]$ from $F[k]$ takes place through the inverse discrete Fourier transform (IDFT) which is defined as $f[n]=\frac{1}{N}\sum_{k=1}^{N}F[k]e^{j2\pi(k-1)\frac{(n-1)}{N}}$ where the symbols have just mentioned meanings.

The index n or k can vary from 1 to N in the presented expressions. But

$$f[n]\xrightarrow{\text{forward discrete Fourier transform}}F[k]\xrightarrow{\text{inverse discrete Fourier transform}}f[n]$$

Figure 4.1(a) The concept behind the discrete Fourier transform

it can vary from 0 to $N-1$ (most textbook follows this convention) in that case the modified expressions become $F[k]=\sum_{n=0}^{N-1}f[n]e^{-j2\pi k\frac{n}{N}}$ and $f[n]=\frac{1}{N}\sum_{k=0}^{N-1}F[k]e^{j2\pi k\frac{n}{N}}$ for DFT and IDFT respectively. Usage of either expression does not change the signal or its transform contents at all.

The $F[k]$ is in general complex. Figure 4.1(a) presents the concept behind the discrete Fourier transform. When we handle the $f[n]$ of length N being the power of 2, the transform is called the fast Fourier transform. In MATLAB the DFT and IDFT are simulated by using the built-in functions **fft** (abbreviation for the fast Fourier transform but also handles the sequence whose length is other than power of 2) and **ifft** (abbreviation for the inverse fast Fourier transform) respectively.

Example on the transform:

Let us consider the discrete signal $f[n]=[1\ \ 2\ \ 7\ \ -5\ \ -6\ \ 8]$. There are 6 samples in the $f[n]$. The choice of index n is the user's. Let us say the n varies from 1 to 6, so does k. The expression $F[k]=\sum_{n=1}^{N}f[n]e^{-j2\pi(k-1)\frac{(n-1)}{N}}$ yields the computation of $F[k]$ as [7 $\quad$ 10.5 $-j6.0622$ $\quad$ $-9.5+j16.4545$ $\quad$ -3 $-9.5-j16.4545$ $\quad$ 10.5 $+j6.0622$]. Our objective is to obtain the complex sequence $F[k]$ from $f[n]$ and we do so as follows:

>>f=[1 2 7 -5 -6 8]; ↵ ← $f[n]$ is assigned to the workspace f as a row matrix, f is user-chosen variable

>>F=fft(f) ↵ ← DFT on $f[n]$ is taken and assigned to the workspace F, F⇔$F[k]$, F is user-chosen variable

F =

7.0000 10.5000 - 6.0622i -9.5000 +16.4545i -3.0000 -9.5000 -16.4545i 10.5000 + 6.0622i

Starting from the computed values of $F[k]$, the expression $f[n]=$ $\frac{1}{N}\sum_{k=1}^{N}F[k]e^{j2\pi(k-1)\frac{(n-1)}{N}}$ brings the signal $f[n]$ back. Having the discrete forward transform values available in the workspace F, one can verify that as follows:

>>f=ifft(F) ↵ ← IDFT on $F[k]$ is taken and assigned to the workspace f, f⇔ $f[n]$, f is user-chosen variable

f =

    1.0000   2.0000   7.0000 -5.0000 -6.0000 8.0000

Component separation:

The variation of the $F[k]$ versus integer k is termed as the discrete Fourier spectrum. Since the $F[k]$ is complex, there are four discrete component spectra namely real, imaginary, magnitude, and phase whose symbolic representations are $\text{Re}\{F[k]\}$, $\text{Im}\{F[k]\}$, $|F[k]|$, and $\angle F[k]$ and whose MATLAB counterparts are **real**, **imag**, **abs**, and **angle** respectively. For instance the discrete phase angle spectrum (in radians) $\angle F[k]$ can be picked up from $F[k]$ by employing the command **angle(F)**.

More examples:

Let us see two more examples on DFT computation in the following.

⌸ Example 1

The beginning example presents the $f[n]$ from some observational data. A discrete signal may follow some functional pattern for example sine.

Consider a sine signal $f[n]=5\sin\frac{2\pi n}{5}$ where the n is integer and the n exists from 0 to 5 thus making $f[n]$ discrete. We intend to find the discrete magnitude and phase spectra of the finite duration discrete sine signal.

Applying the $F[k]=\sum_{n=0}^{N-1}f[n]e^{-j2\pi k\frac{n}{N}}$, one obtains $|F[k]|=$ [0 12.4495   5.0904   3.6327   5.0904   12.4495] and $\angle F[k]$=[180⁰ -60^0 150^0 180^0 -150^0 60^0] for which the implementation is as follows:

>>n=0:5; ↵ ← Placing the index n as a row matrix to the workspace n, n is user-chosen variable (section 1.3)

>>f=5*sin(2*pi*n/5); ↵ ← Scalar code (appendix A) of $5\sin\frac{2\pi n}{5}$ is assigned to f, f⇔discrete $f[n]$

>>F=fft(f); ↵ ← DFT on $f[n]$ is taken and workspace F holds $F[k]$

>>A=abs(F) ↵ ← A holds $|F[k]|$ as a row matrix, A is user-chosen variable

A =

    0.0000 12.4495 5.0904 3.6327 5.0904 12.4495

>>P=180/pi*angle(F) ⏎ ← P holds $\angle F[k]$ in degrees, default return of
angle is in radian from $-\pi$ to π, P is user-chosen variable

P =

    180.0000 -60.0000 150.0000 180.0000 -150.0000 60.0000

⊟ Example 2

Suppose we wish to find the same discrete spectra as in
example 1 but for the discrete function $f[n]=2^{-n}$ on the same n
duration. Only change in previous commands do we need is in the
scalar code of $f[n]$ which now should be 2.^(-n).

4.2 Graphing a discrete Fourier transform

Graphing a discrete Fourier
transform happens by the discrete
function plotter **stem** (appendix D).
Referring to the beginning example (i.e.
the $f[n]$ with six samples) of $F[k]$
computation in the last section, the
workspace variable F holds the $F[k]$ for
the data based $f[n]$.

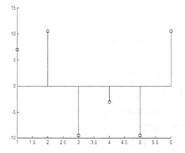

We wish to plot Re{$F[k]$} versus
k for the $f[n]$ i.e. real discrete spectrum
of $f[n]$.

Figure 4.1(b) Plot of the discrete real
spectrum Re{$F[k]$} versus k

Let us obtain the discrete real
spectrum from the complex values stored in F as follows:
>>R=real(F) ⏎

R =

    7.0000 10.5000 -9.5000 -3.0000 -9.5000 10.5000

In above execution the R holds the Re{$F[k]$} values as a row matrix where
the R is a user-chosen variable. In order to plot the real spectrum, the integer
index k needs to be known. The k changes from 1 to 6 therefore the
graphing is as follows:
>>k=1:6; ⏎   ← k holds the index integers as a row matrix
>>stem(k,R) ⏎   ← first and second input arguments are index and real spectrum
              both as a row matrix respectively

Above execution results the figure 4.1(b). The reader may execute the
commands **stem(k,imag(F))**, **stem(k,abs(F))**, and **stem(k,angle(F))** for the
discrete spectra Im{$F[k]$}, $|F[k]|$, and $\angle F[k]$ respectively.

4.3 DFT implications on a discrete sine signal

The sine function is the most addressed one in sound engineering
systems. In discrete analysis, always we are focused on the sample. Behind
the discrete terminology, a physical sound is continuous. The $f[n]$ is just the

samples taken at different index or integer n. Samples do not reveal actual sound system until we relate those with the sampling information. We know (chapter 2) that the discrete function $f[n]$ is generated after sampling the function $f(t)$ with a specific frequency called sampling frequency f_s. Or in other words the function $f(t)$ is sampled with a step size T_s where $T_s = \dfrac{1}{f_s}$.

Now we demonstrate what allied implication holds the sampling frequency regarding the discrete Fourier transform.

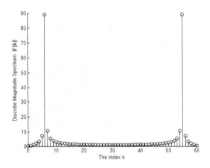

Figure 4.1(c) Discrete magnitude spectrum of a pure discrete sine wave

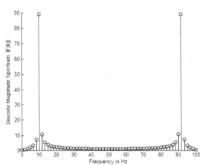

Figure 4.1(d) Discrete magnitude spectrum of the pure discrete sine wave plotted against frequency

Let us consider a single frequency sine function of $\left\{\begin{array}{l} amplitude \ \pm 3 \\ frequency \ 10Hz \end{array}\right\}$

where t is in second and which has the continuous expression $x(t) = 3\sin 2\pi 10t$ and exists over $0 \le t \le 0.6\sec$. The user has to decide the sampling frequency to discretize this continuous function.

Let us say the sampling frequency is $f_s = 100\,Hz$ with that the sampling period or the step size must be $T_s = \dfrac{1}{f_s} = 0.01\sec$ which implies that we must choose the t step as $0.01\sec$ while discretizing the $x(t)$. The moment we choose the t step size, we take the samples of the continuous function. Let us conduct the following:

>>t=0:0.01:0.6; ↵ ← t holds a row matrix whose elements are points on t at which the samples to be taken (section 1.3)
>>x=3*sin(2*pi*10*t); ↵ ← x holds the discrete sine wave as a row matrix in which t information is lost

The **x** corresponds to the DFT theory mentioned $f[n]$. In order to decide the N, the number of elements in **x** is to be found as follows:

>>numel(x) ↵ ← The function numel finds the number of elements in the matrix x

ans =

61

Last return indicates that we must take $N=61$ in the DFT computation thereby changing the index n or k from 0 to 60.

An interesting fact is revealed from the following implementation. Let us graph (section 4.2) the discrete magnitude spectrum $|F[k]|$ against the index k as follows:

```
>>k=0:60; ⌐    ← k holds the DFT mentioned indexes from 0 to 60 as a row matrix
>>A=abs(fft(x)); ⌐    ← A holds the discrete magnitude spectrum |F[k]| for
                         ongoing discrete sine function
>>stem(k,A) ⌐       ← Graphing the magnitude spectrum
```

Figure 4.1(c) presents the magnitude spectrum of the discrete sine wave. The DFT has the half index symmetry that is $|F[k]|$ is symmetric about $\frac{N-1}{2}$ or $k=30$. For $k=0$ to 30, there is only one strong peak which corresponds to the frequency of the continuous sine function $x(t)$ we started with.

Knowing the sampling frequency, it is also convenient to display the magnitude spectrum plot in terms of the discrete frequency f (not $f[n]$) rather than the sample index k. The relationship is given by $f=\frac{f_s}{N-1}k$. For ongoing example, we have $\begin{cases} f_s = 100Hz \\ N = 61 \\ k = 0 \text{ to } 60 \end{cases}$ thereby providing $f=\frac{100}{60}k$ hence the frequency f changes from 0 to 100 Hz with a step $\frac{100}{60}$ Hz. Now we plot the horizontal axis of the figure 4.1(c) in terms of the discrete frequency as follows:

```
>>f=0:100/60:100; ⌐     ← f holds the discrete frequencies as a row matrix
>>stem(f,A) ⌐           ← The first input argument of the stem is now the
                           discrete frequency row matrix
```

Above execution results the graph of the figure 4.1(d) in which you see the horizontal axis in terms of frequency in Hertz. In the plot exactly at $f=10$ Hz does the peak magnitude appear – which is the signal frequency of $x(t)$ we started with.

♦ ♦ What about two sine frequency signal?

The computing and graphing we applied just now are to be checked for a two frequency sine signal on $f_s=100$ Hz for example signal consisting of two identical phase sinusoids which have $\begin{cases} amplitude \ \pm 3 \\ frequency \ 10Hz \end{cases}$ and $\begin{cases} amplitude \ \pm 2 \\ frequency \ 20Hz \end{cases}$ and exists over $0 \le t \le 0.3 \sec$.

What do we wish to achieve? We expect two strong peaks in the magnitude spectrum at exactly 10 Hz and 20 Hz.

The signal has the continuous expression $x(t)=3\sin 2\pi 10t + 2\sin 2\pi 20t$ and we generate the samples over the interval as follows:

>>t=0:1/100:0.3; ↵ ← t holds the points of *t* at which the samples to be taken
>>x=3*sin(2*pi*10*t)+2*sin(2*pi*20*t); ↵ ← x holds the two frequency signal
samples at the points in t
>>A=abs(fft(x)); ↵   ← A holds the |*F*[*k*]| for the two frequency signal
>>N=numel(x); ↵   ← The numel finds the number of elements N in x
>>f=0:100/(N-1):100; ↵   ← f holds the discrete frequencies as a row matrix
>>stem(f,A) ↵   ← Calling plotter for graphing

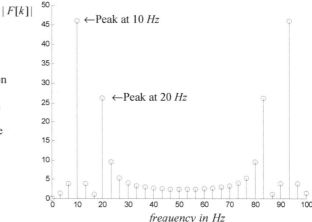

Figure 4.1(e) Illustration of two strong peaks in the magnitude spectrum for a two frequency signal – right side figure

Figure 4.1(e) is the result from the last line execution in which we have shown the two strong peaks by an arrow indication.

4.4 Half index flipping of the DFT

The discrete magnitude spectrum |*F*[*k*]| of *f*[*n*] is related with the information content in the continuous counterpart *f*(*t*) that is why it receives ample attention in the Fourier literature.

The |*F*[*k*]| has half index symmetry when the total number of samples in it is N. Not only the magnitude one but the other spectra may also have even or odd symmetry about the half index.

If a signal is composed of many frequencies, the significant frequency components of |*F*[*k*]| are located at the smaller and larger values of *k* (for example *k* =0, 1, 2,... and *k* = *N* −1, *N* − 2,...). Flipping the |*F*[*k*]|

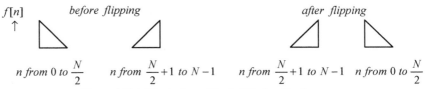

Figure 4.2(a) Illustration of the half index flipping

about the half index gathers significant frequencies in the middle of the k axis which exhibits more perceptibleness in frequency in the $|F[k]|$ versus k plot.

Considering a discrete function $f[n]$ whose integer index n varies from 0 to $N-1$ and where N is the number of samples in the $f[n]$. Flipping the $f[n]$ about its half index $\dfrac{N}{2}$ is graphically interpreted in figure 4.2(a). MATLAB built-in function **fftshift** helps us obtain the half index flipping as illustrated in the figure 4.2(a).

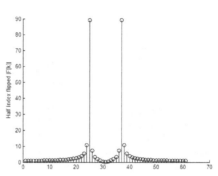

Figure 4.2(b) Half indexedly flipped magnitude spectrum of a discrete sine wave

Let us say we have the odd and even sample number discrete signals $x[n] =[10 \quad 2 \quad 1 \quad 1 \quad 2]$ and $y[n]=[10 \quad 2 \quad 1 \quad 1 \quad 2 \quad 10]$ respectively. Following the half index flipping, one should obtain the flipped signals as $[1 \quad 2 \quad 10 \quad 2 \quad 1]$ and $[1 \quad 2 \quad 10 \quad 10 \quad 2 \quad 1]$ for $x[n]$ and $y[n]$ respectively whose implementations are shown below:

Flipping the odd number sample,	Flipping the even number sample,
>>x=[10 2 1 1 2]; ↵	>>y=[10 2 1 1 2 10]; ↵
>>F=fftshift(x) ↵	>>F=fftshift(y) ↵
F =	F =
1 2 10 2 1	1 2 10 10 2 1

In either case in the last execution the **F** holds the flipped result as a row matrix where the **F** is a user-chosen variable. We know that any discrete one dimensional function takes the form of a row or column matrix. Working on a row or column matrix in a sense is working on the discrete function.

Concerning the **fftshift**, if the number of samples in a discrete function is odd, the half index flipping happens about the index $k=\dfrac{N-1}{2}$ assuming k changes from 0 to $N-1$. Elements whose indexes are from $k=\dfrac{N+1}{2}$ to $k=N-1$ are placed in front of the other half. If the N is even, elements whose indexes are from $k=\dfrac{N}{2}$ to $k=N-1$ are placed in front of the other.

Two times half index flipping brings back the discrete signal we start with provided that the number of samples in the signal is even. That is to say, the **fftshift(fftshift(y))** returns **y**. Let us see one example on the half index flipping.

⊟ Example

From the section 4.3, the workspace variable A (re-execute the commands for single frequency until you get A) holds the discrete magnitude spectrum for the single frequency sine signal whose plot is presented in the figure 4.1(c) and in which you find the two strong peaks are located close to $k=0$ and $k=60$. The half index flipping operation should bring these two peaks in the middle of the k axis for which we conduct the following:

>>F=fftshift(A); ↵ ← The return is assigned to F, where F is user-chosen
>>stem(F) ↵   ← Graphs the figure 4.2(b)

The workspace F holds the half indexedly flipped $|F[k]|$. As the figure 4.2(b) displays, the two strong peaks are brought in the middle.

Note that we used only the flipped spectrum as the input argument of the **stem**. The **fftshift** does not hold the k information. It is the user who keeps the information about the index k. By default the **stem** numbers the horizontal indexes from 1 to the number of samples present in A (which is here 61). If the information about the k is necessary, use the **fftshift** on **k** and assign the **fftshift(k)** to some workspace variable.

4.5 Discrete Fourier transform of a digital audio

In this section we compute the discrete Fourier transform of a digital audio and display its spectrum by considering the softcopy **bird.wav**.

In sections 2.3 and 2.7 we explained how one obtains the discrete audio data so re-execute the following command:

>>[v,fs]=wavread('bird.wav'); ↵

The discrete audio signal $v[n]$ is available in **v**. The $v[n]$ is basically section 4.1 cited $f[n]$. In order to take the section 4.1 mentioned forward DFT, we simply exercise:

>>F=fft(v); ↵

Figure 4.2(c)
Magnitude spectrum of
the digital audio
bird.wav
(right side figure)

$|F[k]|$

↑

$k \rightarrow$

In the last execution the **F** is some user-chosen variable which holds the data of $F[k]$, is certainly complex, and is a column matrix of the same size as that of **v**. Section 4.2 mentioned **stem** is not suitable for huge amount of data instead we apply the **plot** (appendix D) to see the $|F[k]|$ i.e.

>>plot(abs(F)) ⏎

Figure 4.2(c) appears before you as the magnitude spectrum in which the horizontal and vertical axes correspond to k and $|F[k]|$ respectively. Clearly the k changes from 1 to 19456, the total sample number in **bird.wav**. If you say the starting of k is zero which is often the case in most textbooks, the variation will be from 0 to 19455. The command **plot** assumes $k=1$.

It is also a matter of interest that we plot the $|F[k]|$ versus frequency to understand the spectral or frequency behavior of the audio. The index to frequency conversion takes place in accordance with the section 4.3:

>>N=numel(F); ⏎   ← Computing total number of elements in F and put to N
                             where N is a user-chosen variable

>>f=0:fs/(N-1):fs; ⏎   ← Generating discrete frequencies as a row matrix and
                             holding those to f where f is a user-chosen variable

>>plot(f,abs(F)) ⏎   ← Calling the plot to see the figure 4.2(d)

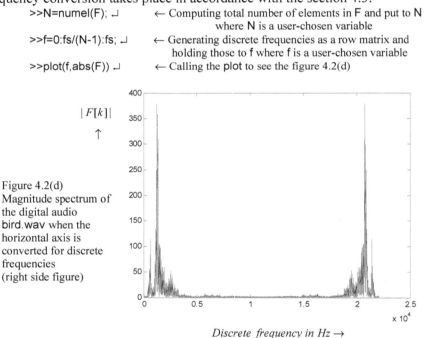

$|F[k]|$

↑

Figure 4.2(d)
Magnitude spectrum of
the digital audio
bird.wav when the
horizontal axis is
converted for discrete
frequencies
(right side figure)

Discrete frequency in Hz →

Now the horizontal axis of the figure 4.2(d) is in terms of discrete Hertz frequencies rather than integer index k.

As regard to section 4.3 a strong peak in the magnitude spectrum is the indication of sine component in the audio signal. In figure 4.2(d) we see many local peaks meaning there are many sine components in the **bird.wav**. By inspection significant sine frequencies cover approximately the range 0 to

$0.5 \times 10^4 Hz$ in the spectrum. For analysis of the digital audio we concentrate only over $0 \le f \le \dfrac{f_s}{2}$ because the spectrum over the $\dfrac{f_s}{2} \le f \le f_s$ is the repetition to that over $0 \le f \le \dfrac{f_s}{2}$.

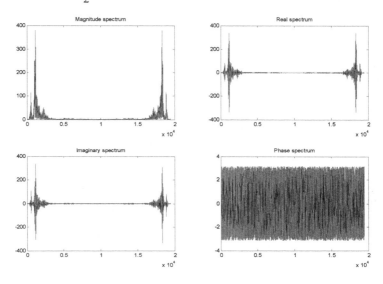

Figure 4.2(e) Four spectra of the digital audio bird.wav

If we concentrate on a full-fledged DFT, it is not just the magnitude spectrum besides three more components are associated with the transform as regard to section 4.1. If the reader is interested to see all four spectra in a single plot, appendix D mentioned **subplot** can be helpful and which we exercise as follows:

```
>>subplot(221),plot(abs(F)),title('Magnitude spectrum') ⌐     ← For | F[k]|
>>subplot(222),plot(real(F)),title('Real spectrum') ⌐         ← For Re{F[k]}
>>subplot(223),plot(imag(F)),title('Imaginary spectrum') ⌐    ← For Im{F[k]}
>>subplot(224),plot(angle(F)),title('Phase spectrum') ⌐       ← For ∠F[k]
```

Outcome of the last four lines is shown in figure 4.2(e) which depicts the four spectra of the digital audio **bird.wav** and in which you easily identify the respective spectrum by looking into the title. The built-in command **title** puts user-supplied text to a graph once the graph has been drawn. For example in the magnitude spectrum we chose **Magnitude spectrum** and the text must be under the quote. From ongoing discussion, each of the four horizontal axes is in terms of the integer frequency index k.

4.6 Hidden single frequency in a digital audio

In section 4.3 we elaborately explained how single frequency sine signal behaves in the magnitude spectrum. Extrapolating the concept one easily identifies the hidden frequency in a given digital audio. If we do not have any sampling frequency information in a given discrete signal, we start with some sampling frequency and look for the strongest peak as done in the figure 4.1(d). If it does not appear, we can try with another sampling frequency. Fortunately most digital audio comes with a predecided sampling frequency. Appendix B.5 cited **max** helps us determine the maximum frequency point.

The fundamental frequency present in a digital audio is called the pitch of the audio.

We wish to determine the hidden frequency in the digital audio **pitch.wav** of section 2.7. Obtain the audio, place it in your working path of MATLAB, and then carry out the following:

>>[v,fs]=wavread('pitch.wav'); ↵   ← Reading the audio file (section 2.3)
>>wavplay(v,fs) ↵   ← Playing the audio file

You must be hearing the tone of a single sine frequency signal. Our objective is to determine the pitch of the tone. We know that the discrete signal is stored in workspace **v** as column matrix. As a procedure we first perform the DFT on the signal as follows:

>>F=abs(fft(v)); ↵ ← F holds $|F[k]|$ as a column matrix

>>N=numel(F); ↵ ← N holds number of samples where N is user-chosen variable

We know that the sampling frequency f_s is stored in **fs** so discrete frequencies as a row matrix from 0 to f_s with increment $\dfrac{f_s}{N-1}$ we then generate by:

>>f=0:fs/(N-1):fs; ↵   ← f holds the discrete frequencies

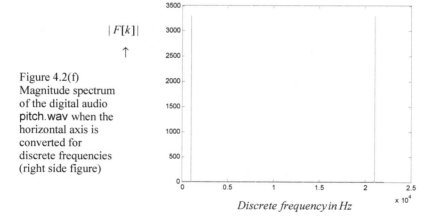

$|F[k]|$

↑

Figure 4.2(f)
Magnitude spectrum
of the digital audio
pitch.wav when the
horizontal axis is
converted for
discrete frequencies
(right side figure)

Discrete frequency in Hz

In section 4.4 we mentioned the **stem** plot for plotting, that grapher is not suitable for huge number of data. Appendix D cited **plot** is helpful in this regard. Next we see the $|F[k]|$ versus discrete frequency plot by:

>>plot(f,F) ⏎   ← Figure 4.2(f) is the plot of the magnitude spectrum

Looking into the spectrum in figure 4.2(f), one strong peak is present in the audio **pitch.wav**. After that we apply the **max** as follows:

>>[M,I]=max(F); ⏎   ← M holds the maximum and I holds its integer index
>>f(I) ⏎   ← Calling the discrete frequency corresponding to the
                                              index stored in I

ans =
         1000

Above output says that frequency hidden in the **pitch.wav** is 1000 Hz . Even if we have multiple peaks in a given magnitude spectrum, the strongest one corresponds to the pitch or fundamental frequency.

4.7 Half spectrum of a digital audio

From sections 4.4 and 4.5, it is obvious that the DFT spectrum has repetition about the half index. While working on a realistic audio, it is convenient that we work only on one half of the spectrum or may need to avoid redundant processing. In that event we concentrate only on the first half magnitude spectrum i.e. half of $|F[k]|$.

Let us say there are M samples in $|F[k]|$ which we obtain by the function **numel** of MATLAB. If the M is even, we take the integer index k range from 0 to $\frac{M}{2}-1$ and if the M is odd, we take the range from 0 to $\frac{M-1}{2}$ considering k changes from 0. While programming in MATLAB we add 1 to each k for array indexing reason. We apply the same indexing to select half angle, real, and imaginary spectra.

As an example, select the half magnitude spectrum of the **pitch.wav** which we implemented in the last section and display the spectrum.

We know that the $|F[k]|$ and M are stored in **F** and **N** respectively. Just call the **N** to see the total sample number as follows:

>>N ⏎

N =
         6616

The last return indicates that the M is even. If only the integer index related to the half spectrum is needed, we execute the following:

>>k1=1:N/2; ⏎   ← k1 holds the half spectrum related integer indexes where k1
                                    is user-chosen variable

If the discrete frequencies (which are stored in **f** from the last section) related to the spectrum are needed, we execute the following:

>>f1=f(k1); ⏎   ← f1 holds the discrete frequencies related to the half
                                    spectrum where the f1 is user-chosen variable

If the magnitude spectrum values related to the half spectrum are needed, we execute the following:

>>F1=F(k1); ↵   ← F1 holds the $|F[k]|$ values related to the half
                       spectrum where the F1 is user-chosen variable

>>plot(f1,F1) ↵   ← Just to see the plot of the half spectrum

Figure 4.3(a) is the response from the last command. By inspection figure 4.3(a) is the half spectrum for that of the figure 4.2(f).

Figure 4.3(a) Half magnitude spectrum of the digital audio pitch.wav (right side figure)

$|F[k]|$
↑

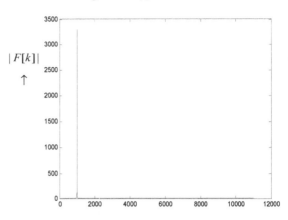

Discrete frequency in Hz →

It is not feasible that we check the total sample number visually. Built-in function **rem** (appendix B.8) helps us check the total sample number being odd or even. The **if-else** statement (appendix B.2) of MATLAB performs the checking. If the total sample number is divisible by 2, the number is even otherwise odd. For ongoing example we could have performed the **k1** deciding after obtaining the N as follows:

>>if rem(N,2)==0 k1=1:N/2; else k1=(N+1)/2; end ↵

⌷ What about full spectrum from half spectrum?

Despite half spectrum eases the computational burden or removes redundancy, recovery of original signal from the forward transform requires full spectrum. The half spectrum can be in row or column matrix form. If it is in row matrix form, we apply the **fliplr** and if it is in column matrix form, we apply the **flipud** for the other half of the full spectrum (appendix B.7).

One problem is associated with the full spectrum formation. From the half spectrum total sample number, we can not decide whether the original spectrum has odd or even symmetry. Another problem is the magnitude spectrum does not have perfect half index symmetry. Except $|F[0]|$ and $|F[M/2-1]|$ values, other values of $|F[k]|$ have perfect half index symmetry considering even sample number on $|F[k]|$ and starting k from 0. Let us call the first half spectrum of $|F[k]|$ as $|F_1[k]|$.

So to form the complete spectrum from the half spectrum $|F_1[k]|$, we organize the samples as follows: $|F_1[k]|$, $|F[M/2]|$, and flipped $|F_1[k]|$ except the first sample of $|F_1[k]|$.

Hence one has to apply the command Fn=[F1;F(N/2+1); flipud(F1(2:end))]; where Fn holds the spectrum and it is a user-chosen variable. The F1 is the half spectrum meaning $|F_1[k]|$. The command 2:end indicates second element through last element in F1.

If we had the sample number N being odd, the command would be Fn=[F1;flipud(F1(2:end))];.

The reader can verify that the graphs conducted by the commands plot(f,F) and plot(f,Fn) are identical.

4.8 Decibel value on a digital audio spectrum

It is a common practice that we display the magnitude spectrum of an audio signal in terms of the decibel (dB) scale which needs the normalization with respect to the maximum of $|F[k]|$ and we apply the **max** of appendix B.5 to determine the $|F[k]|_{max}$.

In the decibel scale, the $|F[k]|$ values are mapped between 0 and negative ∞.

Let us concentrate on the spectrum $F[k]$ of the digital audio **bird.wav** (section 4.5) which we stored in the workspace F (re-execute the commands until you get the F). Mathematically we need to perform the $20\log_{10}\dfrac{|F[k]|}{|F[k]|_{max}}$ operation on the $F[k]$. Let us assign the $|F[k]|$ to workspace A as follows:

```
>>A=abs(F); ↵
>>M=max(A); ↵              ← M holds the |F[k]|max
```

For the decibel scale conversion the necessary command is **20*log10(A/M)**.

There is some intricacy while using the decibel scale. Since $\log_{10} 0$ is minus infinity, machine returns a very large value of the dB.

In most audio spectrum the dB values are confined between −60dB and 0dB. Emphasis should be given on the following points.

Bypassing $\log_{10} 0 = -\infty$ point in the dB audio spectrum:

In the decibel plot it is very common that the $|F[k]|$ value is 0 for some k. The computer prints some error message when $\log_{10} 0 = -\infty$. Under this type of situation, we add a negligible positive quantity epsilon to the $|F[k]|$ or A values, which has the MATLAB code **eps**. For the ongoing spectrum, we should use **20*log10(eps+A/M)** for the decibel conversion.

Clipping high dB values in the dB audio spectrum:

Even though $\log_{10} 0 = -\infty$ is overcome by using the **eps**, the values of the **20*log10(eps)** become too much negative like -300 dB or less. This sort of dB manipulation needs some programming technique. Before we do that, let us assign the calculated dB values to some variable **D** (any user-chosen variable) as follows:

>>D=20*log10(eps+A/M); ↵ ← D holds the dB values as a row matrix

Let us say any dB value stored in the **D** less than or equal to -60 will be set to -60. For this purpose the function **find** (appendix B.9) becomes useful as follows:

>>r=find(D<=-60); D(r)=-60; ↵

Concerning above implementation, the **r** (any user-chosen variable) holds the integer position index of the column matrix **D** where $20\log_{10} | F[k]| \leq -60$ (conducted by the command **r=find(D<=-60);**). Only $20\log_{10} | F[k]| \leq -60$ elements in the **D** are set to -60 by writing the command **D(r)=-60;**.

Figure 4.3(b)
Clipped decibel
magnitude spectrum
of the digital audio
bird.wav (right side
figure)

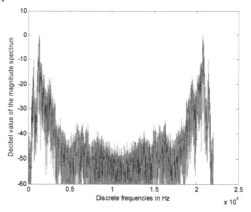

Finally exercising the command **plot(f,D)** returns the decibel as well as clipped magnitude spectrum of the **bird.wav** which is depicted in the figure 4.3(b).

4.9 Frequencies in a digital audio

In section 4.3 we elaborately explained how single frequency sine signal behaves in the magnitude spectrum. Section 4.6 demonstrates the finding of the pitch or fundamental frequency in a digital audio. The audio **pitch.wav** is a simple one. Musical audio or speech does not hold a single frequency rather a series of frequencies. Obviously the highest peak in the magnitude spectrum corresponds to the pitch or fundamental frequency. In addition other peaks in the magnitude spectrum are the frequency components present in the digital audio. The function **max** determines only the pitch. Author written **mmax** (appendix B.17) determines approximate significant frequencies present in a digital audio.

We intend to determine the pitch and significant frequencies present in the digital audio **bird.wav**.

From section 4.5 we know that the $F[k]$ and discrete frequencies of the **bird.wav** are stored in the workspace **F** and **f** as a column and as a row matrices respectively. As a procedure, we first determine the magnitude spectrum $|F[k]|$ as follows:

  >>A=abs(F); ⏎   ← A holds the $|F[k]|$ values as a column matrix
  >>[M,I]=max(A); ⏎ ← For finding the pitch, section 4.6
  >>f(I) ⏎

  ans =
     1.2637e+003

Above return says that the pitch or fundamental frequency in the **bird.wav** is 1.2637 KHz. Its corresponding $|F[k]|_{max}$ we find by:

  >>A(I) ⏎

  ans =
     379.7681   ← Meaning $|F[k]|_{max}$=379.7681

Looking into the magnitude spectrum of the figure 4.2(d) for **bird.wav**, there are many peaks in the spectrum. Then just to avoid redundancy, we select only the half spectrum from the **A** as well as its corresponding frequencies from **f** (section 4.7) as follows:

  >>F1=A(1:N/2); f1=f(1:N/2); ⏎ ← Symbols have section 4.7 cited meanings

Next we call the author-written function **mmax** based on the half spectrum data with two output arguments as follows:

  >>[fr,Fr]=mmax(f1,F1); ⏎   ← The fr and Fr are user-chosen variables

Above execution says that discrete frequencies corresponding to local maxima are returned to **fr** as a row matrix and local maxima values of $|F[k]|$ to **Fr** as a column matrix.

After that we look for the **fr** and **Fr** in the workspace browser (section 1.4) and find their dimensions as 1×3605 and 3605×1 respectively meaning the digital audio **bird.wav** contains 3605 frequencies – so many frequencies indeed. That is, the

  fr(1) and **Fr(1)**: first frequency and its corresponding $|F[k]|$ value,

  fr(2) and **Fr(2)**: second frequency and its corresponding $|F[k]|$ value, and so on respectively.

If the reader wishes to see the first five frequencies present in bird.wav, following command can be executed:

>>[fr(1:5)' F1(1:5)] ⌐

ans =

3.4002	10.4844	← i.e. $\mid F[k] \mid = 10.4844$ at $f = 3.4002\ Hz$
5.6669	2.8148	
10.2005	3.0643	$\vdots$
13.6006	4.0312	
15.8674	3.0842	← i.e. $\mid F[k] \mid = 3.0842$ at $f = 15.8674\ Hz$

In the last execution we applied 1:5 meaning the first five frequencies, transpose operator is used because the fr is a row matrix (section 1.3 and appendix B.3).

Finally make no mistake that these frequencies are not the pitch of the audio signal. The pitch, strongest peak, we found as $1.2637\ KHz$ earlier. Obviously the pitch will be within the fr. We investigated and found that the 451-th frequency in fr is the pitch. Let us verify the two values as follows:

>>fr(451) ⌐   >>Fr(451) ⌐

ans =   ans =
    1.2637e+003   379.7681

As expected.

4.10 Basic statistical properties of an audio spectrum

In section 2.6 we discussed how the datastats determines the basic statistical properties of an index or a time domain signal. The same function one can apply in the frequency domain. For example we wish to see the basic statistical properties of the magnitude spectrum on the digital audio bird.wav. The spectrum $\mid F[k] \mid$ is stored in the workspace A as in the last section so simply execute the following:

>>datastats(A) ⌐

ans =

num: 19456	← Number of $\mid F[k] \mid$ amplitudes
max: 379.7681	← Maximum value of $\mid F[k] \mid$ amplitudes
min: 0.0219	← Minimum value of $\mid F[k] \mid$ amplitudes
mean: 9.3029	← Average value of $\mid F[k] \mid$ amplitudes
median: 1.5936	← Median value of $\mid F[k] \mid$ amplitudes
range: 379.7462	← Range of the $\mid F[k] \mid$ amplitudes
std: 26.8840	← Standard deviation of $\mid F[k] \mid$ amplitudes

With this article we terminate the discrete Fourier transform exercises on a digital audio.

Chapter 5

 ## Discrete Short Time Fourier Transform and Digital Audio

In this chapter discrete short time Fourier transform (STFT) implementation fundamentals are discussed. In DFT as addressed in chapter 4 the time information is lost completely and all information is spectral i.e. frequency oriented. If signal spectral characteristic is sought against time, then the STFT is the correct tool for the study i.e. STFT is a time-frequency based analysis-synthesis approach. In order to exercise the STFT related theory matter, we introduce the following:

- ✦ ✦ Short time Fourier transform – STFT for discrete signals
- ✦ ✦ Forward and inverse STFT counterparts implementation through fundamental programming
- ✦ ✦ MATLAB tools available for STFT and its linked windows
- ✦ ✦ STFT on a digital audio and spectrographic display tools for the audio

5.1 Short time Fourier transform of a discrete signal

When a signal contains millions of samples, discrete Fourier transform (DFT) as discoursed in chapter 4 is not feasible at all. DFT can be useful for short audio segments for example few milliseconds to seconds duration. If the audio properties change with time or duration is very long, we can not apply the DFT. Short time Fourier transform (STFT) is the analysis-synthesis tool for this sort of signal. Since a digital audio is discrete in time, we address the STFT implementation only in discrete form. Like the DFT, STFT also shares the notion of forward-inverse as depicted in the figure 5.1(a).

The discrete signal $f[n]$ with the help of the window function $w[m]$ is transformed to $F[n,k]$ by the forward STFT which

$$f[n] \xrightarrow{\text{forward discrete STFT}} F[n,k] \xrightarrow{\text{inverse discrete STFT}} f[n]$$

on $w[m]$

Figure 5.1(a) The concept behind the discrete short time Fourier transform (STFT)

is one to two dimensional mapping whereas the $F[n,k]$ is transformed to $f[n]$ by the inverse STFT which is two to one dimensional mapping. The representation in the transform is time-frequency based in contrast the DFT is only frequency based. The expression we apply for the transform is the following.

Forward short time Fourier transform (FSTFT):

The forward transform is given by the relationship $F[n,k] = \sum_{m=0}^{N-1} f[m]w[n-m]e^{-j\frac{2\pi km}{N}}$ where the N is the number of samples in $f[n]$. The m, n, and k are all integers. For different n and k, the $F[n,k]$ is a square matrix of size $N \times N$. The $F[n,k]$ is in general complex. The n and k are called time and frequency variables respectively. The $f[n]$, $w[m]$, and $F[n,k]$ are all discrete functions. The $w[m]$ is user-chosen. For different values of n or m, the $f[n]$ or $w[m]$ is just a row or column matrix. Both the n and k change from 0 to $N-1$. The change of m depends on window selection and the $w[m]$ is an even function with $w[0] \neq 0$.

Inverse short time Fourier transform (ISTFT):

The recovery of $f[n]$ from $F[n,k]$ takes place through the inverse discrete STFT which is defined as $f[n] = \frac{1}{Nw[0]} \sum_{k=0}^{N-1} F[n,k]e^{j\frac{2\pi kn}{N}}$ where the symbols have forward transform mentioned meanings and assuming $w[0] \neq 0$.

MATLAB implementation:

Like the DFT, discrete STFT does not have a direct MATLAB counterpart because of variable user-defined window. We demonstrate the basic programming approach both for the forward and the inverse STFTs.

Summation (Σ) is implemented by the for-loop (appendix B.4). The $f[n]$ is entered as a row or column matrix. Window function $w[m]$ is written as a function file (appendix C).

```
function y=win(m)
if -2<=m&m<=2
    y=1;
else
    y=0;
end
```

Figure 5.1(b) Function file for the window function

Defining the window function $w[m]$ **:**

Suppose we wish to apply the discrete rectangular window $w[m] = \begin{cases} 1 & -2 \leq m \leq 2 \\ 0 & elsewhere \end{cases}$ in computing the forward STFT.

The interval $-2 \leq m \leq 2$ is written by operators as **-2<=m&m<=2** (appendix B.1). The interval checking is performed by the **if-else** statement (appendix B.2). Figure 5.1(b) shows the complete function file **win(m)** where the **win, m,** and **y** are user-chosen variables. The **win(m)** takes one **m** at a time and returns its corresponding $w[m]$ value. Open a new M-file (section 1.3), type the codes of the figure 5.1(b) in the file, and save the file by the name **win**. The $w[0]$ should be 1 which we verify just by calling:

>>win(0) ↵

ans =
                    1

With that our window function programming is finished.

Forward STFT example on a single point:

Suppose $F[2,3]=16-j\,27$ is the forward discrete STFT coefficient of the discrete signal $f[n]=[9\ \ -32\ \ \ -7\ \ -5]$ based on the rectangular window $w[m]=\begin{cases}1 & -2\leq m\leq 2\\0 & elsewhere\end{cases}$ where $n\geq 0$ and which we wish to obtain in MATLAB.

As a first step, we enter the given $f[n]$ to **f** as a row matrix as follows:

>>f=[9 -32 -7 -5]; ↵   ← f is a user-chosen variable

Since the $F[2,3]$ is required, we have $n=2$ and $k=3$. Machine finds the number of elements or samples in **f** by dint of the function **numel** so we execute the following:

>>N=numel(f); n=2; k=3; ↵ ← Assigning the values of n, k, and N to like
                                                name variables e.g. 2 to n

Implementation of the forward transform expression $\sum_{m=0}^{N-1}f[m]w[n-m]e^{-j\frac{2\pi km}{N}}$

needs us to execute the following:

>>s=0; for m=0:N-1, s=s+f(m+1)*win(n-m)*exp(-j*2*pi*k*m/N); end ↵
>>s ↵   ← calling the s to see its content

s =
             16.0000 -27.0000i

A programming tactic is applied above. The summation expression starts from $m=0$ but in MATLAB array indexing starts from 1 so theory mentioned $f[0]$ is basically **f(1)**, $f[1]$ is **f(2)**, and so on. That is how the **f(m+1)** is equivalent to $f[m]$. Due to the difference, the **win(n-m)** is tantamount to $w[n-m]$ and we assume that you must have written the **win(m)** function file in your working path of MATLAB by using earlier quoted instructions. The expression **exp(-j*2*pi*k*m/N)** is the code of $e^{-j\frac{2\pi km}{N}}$ (appendices A and B.6).

The for-loop (appendix B.4) counter m simulates $\sum_{m=0}^{N-1}$. For each m we calculate the expression $f[m]w[n-m]e^{-j\frac{2\pi km}{N}}$ by the code f(m+1)*win(n-m)* exp(-j*2*pi*k*m/N) and put the result to the last s by the statement s=s+ f(m+1)*win(n-m)*exp(-j*2*pi*k*m/N) where the s is a user-chosen variable and holds the computed $F[2,3]$ coefficient. Before entering to the for-loop we must assign 0 to s by writing the command s=0. We called the s just to see its content which is $F[2,3]=16-j\,27$.

Forward STFT example on the whole domain:

$$F[n,k] = \begin{bmatrix} -30 & 16+j32 & 34 & 16-j32 \\ -35 & 16+j27 & 39 & 16-j27 \\ -35 & 16+j27 & 39 & 16-j27 \\ -44 & 7+j27 & 30 & 7-j27 \end{bmatrix} \text{ is the forward discrete}$$

STFT of the discrete signal $f[n]=[9 \quad -32 \quad -7 \quad -5]$ based on the rectangular window $w[m] = \begin{cases} 1 & -2 \leq m \leq 2 \\ 0 & elsewhere \end{cases}$ where $n \geq 0$ and which we wish to obtain in MATLAB.

Just now we finished the computation of a single coefficient of $F[n,k]$. Since the $f[n]$ has 4 elements, both the n and k change from 0 to 3 and we control these two variable change by using two more for-loops – one for each. Exercising of multiple command statements is imminent; as such it is not feasible that we run the program statements in the command window instead use an M-file.

```
f=[9 -32 -7 -5];
N=numel(f);
for n=0:N-1
    for k=0:N-1
        s=0;
        for m=0:N-1
            s=s+f(m+1)*win(n-m)*exp(-j*2*pi*k*m/N);
        end
        F(n+1,k+1)=s;
    end
end
```

Figure 5.1(c) M-file for computation of the forward STFT $F[n,k]$ of a discrete signal $f[n]$

Figure 5.1(c) presents the complete M-file code for computing the $F[n,k]$ square matrix from $f[n]$. Referring to the code, the for-loop statements n=0:N-1 and k=0:N-1 stand for the variations $0 \leq n \leq N-1$ and $0 \leq k \leq N-1$ respectively. As discussed earlier, single element computation through s is placed in the innermost for-loop. Output of the innermost loop is assigned to F(n+1,k+1) which is equivalent to the $F[n,k]$. If you write the command F(1,1), machine itself opens a two dimensional array by the name F and considers the first position index which starts from 1 rather than 0. The F is a user-chosen variable. Obviously the F(1,1), F(2,1), F(1,2), etc refer to

$\begin{Bmatrix} n=0 \\ k=0 \end{Bmatrix}$, $\begin{Bmatrix} n=1 \\ k=0 \end{Bmatrix}$, $\begin{Bmatrix} n=0 \\ k=1 \end{Bmatrix}$, etc respectively. The **s** computed by **n** and **k** is put to **F(n+1,k+1)** which is what we require.

As a continuation, we open a new M-file (section 1.3), type the codes of the figure 5.1(c) in the M-file, and save the file by the name **stft** (any user-chosen file name). Calling the file and the STFT forward transform coefficients takes place as follows:

```
>>stft ↵
>>F ↵
```

F =

-30.0000	16.0000 +32.0000i	34.0000 + 0.0000i	16.0000 -32.0000i
-35.0000	16.0000 +27.0000i	39.0000 + 0.0000i	16.0000 -27.0000i
-35.0000	16.0000 +27.0000i	39.0000 + 0.0000i	16.0000 -27.0000i
-44.0000	7.0000 +27.0000i	30.0000 + 0.0000i	7.0000 -27.0000i

We also assume that the window function **win(m)** resides in the working path of **stft**. At the end we made the $F[n,k]$ available to **F** as displayed above.

Inverse STFT example on a single point:

Here our objective is to compute $f[n]$ starting from the matrix of $F[n,k]$. For example $f[n] = -32$ for $n = 1$ which we wish to obtain for ongoing

example of $F[n,k] = \begin{bmatrix} -30 & 16+j32 & 34 & 16-j32 \\ -35 & 16+j27 & 39 & 16-j27 \\ -35 & 16+j27 & 39 & 16-j27 \\ -44 & 7+j27 & 30 & 7-j27 \end{bmatrix}$.

It goes without saying that the complete $F[n,k]$ matrix of the size $N \times N$ must be available for the recovery of $f[n]$ and which one reaches by calling just implemented variable **F**.

Consequently we need to implement $f[n] = \dfrac{1}{Nw[0]} \sum_{k=0}^{N-1} F[n,k] e^{\frac{j2\pi kn}{N}}$ for

the inverse discrete STFT and do so as follows:

```
>>n=1; f1=0; ↵
>>for k=0:N-1, f1=f1+F(n+1,k+1)*exp(j*2*pi*k*n/N)/N/win(0); end ↵
>>f1 ↵                                    ← Calling the f1
```

f1 =

-32.0000 + 0.0000i

Like the forward counterpart the **F(n+1,k+1)** in above execution indicates the $F[n,k]$ due to array indexing reason. The **exp(j*2*pi*k*n/N)/N/win(0)** is

the code of $\dfrac{e^{\frac{j2\pi kn}{N}}}{Nw[0]}$. The programming tactic of the forward counterpart is also

applied here by using a for-loop and through the **f1** where the **f1** is a user-chosen variable. After the completion of the for-loop, the **f1** holds the $f[1]$ which we called just to see its content. Some negligible imaginary data appears due to the discrete computational reason.

Inverse STFT example on the whole domain:

Now we wish to recover the whole $f[n]$ starting from the earlier mentioned $F[n,k]$ matrix.

For the single element recovery we passed the value of n by writing n=1 before entering to the for-loop. That n is now controlled by another for-loop and the output from the single element for-loop (which occurred by the variable f1) is assigned to another variable o by writing the command o(n+1) for the same indexing reason where the o is a user-chosen variable hence we carry out the following:

```
>>for n=0:N-1 ↵
>>f1=0; ↵
>>for k=0:N-1, f1=f1+F(n+1,k+1)*exp(j*2*pi*k*n/N)/N/win(0); end ↵
>>o(n+1)=f1; end ↵
>>o ↵                          ← Calling the o
```

o =

  9.0000 + 0.0000i   -32.0000 + 0.0000i   -7.0000   -5.0000 - 0.0000i

```
for n=0:N-1
        f1=0;
        for k=0:N-1
                f1=f1+F(n+1,k+1)*exp(j*2*pi*k*n/N)/N/win(0);
        end
        o(n+1)=f1;
end
```

Figure 5.1(d) M-file for computation of the inverse STFT
$f[n]$ from $F[n,k]$

If the reader wishes to place the commands in an M-file, figure 5.1(d) shown codes should be written in a new M-file and saved the file by some name istft. After that the reader would call the istft first and the o afterwards at the command prompt.

As we have found, the samples in f and o must be identical from the STFT theory instead the o contents are having some imaginary components. Computer always performs computations in fixed number of bits which have some preselected accuracy. The imaginary components can be ignored because those will be of very small in magnitude compared to the real counterpart. Some function might be sensitive even with small imaginary numbers in that case we select only the real components of the elements in o by writing the command real(o).

What about discrete STFT for a mathematical function?

Just now we finished the discussion on discrete forward-inverse STFT by taking the example of a $f[n]$ which has four discrete values. The

$f[n]$ may follow some functional variation for example $f[n] = 2^{-n} \sin 2n$ over $0 \le n \le 5$.

In this sort of discrete function we generate first the n variation as a row or column matrix (section 1.3) and then the functional data of $f[n]$ by writing its scalar code (appendix A) as follows:

```
>>n=0:5; ↵
>>f=2.^n.*sin(2*n); ↵
```

Therefore the last f holds the discrete data of $2^{-n} \sin 2n$. Now this f can be used as the earlier mentioned f which we exercised to determine the forward STFT.

5.2 Discrete window functions for STFT

From the discussion of STFT in last section it is mandatory that we apply a discrete window function for the forward STFT computation. There are at least a dozen of window functions available, four of which are addressed in the following.

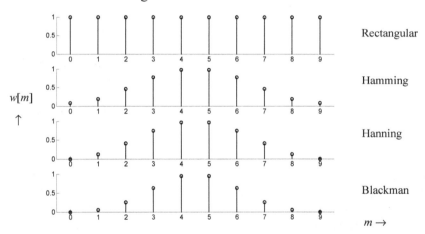

Figure 5.2(a) Plots of the four discrete window functions – rectangular, Hamming, Hanning, and Blackman over $0 \le m \le 9$

Rectangular window:

A rectangular window is defined by $w[m] = 1$ over $0 \le m \le M - 1$ where M is the number of samples required for $w[m]$. Its MATLAB counterpart is rectwin which needs the syntax rectwin(M).

Hamming window:

A Hamming window is defined by $w[m] = 0.54 - 0.46\cos\dfrac{2\pi m}{M-1}$ over $0 \le m \le M - 1$ where M is the number of samples required for $w[m]$. Its MATLAB counterpart is hamming which needs the syntax hamming(M).

-79-

Hanning window:

A Hanning window is defined by $w[m] = 0.5\left(1 - \cos\dfrac{2\pi m}{M-1}\right)$ over $0 \le m \le M-1$ where M is the number of samples required for $w[m]$. Its MATLAB counterpart is hann which needs the syntax hann(M).

Blackman window:

A Blackman window is defined by $w[m] = 0.42 - 0.5\cos\dfrac{2\pi m}{M-1} + 0.08\cos\dfrac{4\pi m}{M-1}$ over $0 \le m \le M-1$ where M is the number of samples required for $w[m]$. Its MATLAB counterpart is blackman which applies the syntax blackman(M).

General properties of these four windows:

(1) Other than the discrete interval $0 \le m \le M-1$, each of the four windows has functional value 0 that is $w[m] = 0$ elsewhere.

(2) The m and M are an integer.

(3) The window is defined for a single integer m but the MATLAB return from each of the rectwin, hamming, hann, and blackman is a column matrix.

(4) Figure 5.2(a) shows the plots of the four windows considering $0 \le m \le 9$ requiring each of which should have 10 samples i.e. $M = 10$.

(5) All of the four windows are symmetric about the half point over $0 \le m \le 9$.

Example on a window function return:

Let us consider the Hamming window over $0 \le m \le 4$. If we insert the m values in the $w[m]$ expression, we obtain $w[m] = 0.08$, 0.54, 1, 0.54, and 0.08 for the $m = 0$, 1, 2, 3, and 4 respectively. Obviously the M is 5. We obtain the window functional values as a column matrix as follows:

```
>>w=hamming(5) ↵
```

```
w =
        0.0800
        0.5400
        1.0000
        0.5400
        0.0800
```

The functional values of the $w[m]$ are stored in the workspace **w** where the **w** is a user-chosen variable.

Note that we wrote the function by the name win(m) for rectangular window in section 5.1 considering one value of m at a time. Above four MATLAB window functions can not be substituted directly to section 5.1 mentioned STFT computation owing to column matrix output. Modification in section

5.1 programming is required as follows: vectorizing the for-loop, removing the if-else statement by vector checking, and considering certain number of samples out of the whole signal as in the overlap and add technique [4], each of which is beyond the scope of the text.

How to use these window functions in the discrete STFT?

If the built-in window functions are to be used, first execute **w= hamming(5)**; for example over the interval $0 \le m \le 4$ for a Hamming window and then apply **w(n-m)** instead of **win(n-m)** in the program statements of figure 5.1(c). Again there is another obstacle – if the **n-m** is negative and the negative index is not defined to machine, then a checking by using **if-else** must be conducted.

```
function y=win(m)
if 0<=m&m<=4
   y=0.54-0.46*cos(2*pi*m/4);
else
   y=0;
end
```

Figure 5.2(b) Function file for the Hamming window

As another option, write the definition of the Hamming window like the figure 5.2(b) and use it for the forward STFT computation.

5.3 Readymade option of MATLAB for discrete STFT

Section 5.1 illustrated programming technique is applicable for basic STFT understanding. For a practical digital audio, that technique is not suitable at all for the following reasons:

(1) for-loop programming approach is time consuming specially when we have millions of samples which is often the case with a digital audio for example an audio $f[n]$ having 100000 samples needs the $F[n,k]$ to be of the size 100000×100000 – a huge square matrix indeed moreover involvement of complex numbers turns the situation worse

(2) for a particular value of n, the STFT turns to DFT as explained in section 4.1, the redundant DFT computation is not considered in section 5.1 that is computations between 0 to $\frac{f_s}{2}$ and $\frac{f_s}{2}$ to f_s have redundancy where f_s is the sampling frequency

(3) our window function as implemented in section 5.1 works for one sample at a time which is also not convenient for large sample number audio signals

(4) checking statement like **if-else** is also time-ineffective while handling huge amount of discrete data

MATLAB computes discrete STFT slight differently but that needs to bring some other quantities onboard. Since we are focused on implementation, hidden algorithm is beyond the scope of the text.

MATLAB built-in function **specgram** computes the discrete STFT of a discrete signal but only the forward counterpart that is $F[n,k]$ from $f[n]$. The function conceives different arguments and has many options, our concentration will be only on the STFT data obtaining.

One of the syntaxes we apply is [user-supplied variable for $F[n,k]$, user-supplied variable for frequency f in Hz, user-supplied variable for time t in sec]=**specgram**(given discrete or audio signal $f[n]$ as a row or column matrix, user-supplied point number N for DFT computation, user-supplied sampling frequency f_s, user-supplied window length M, user-supplied number of samples P for overlapping the discrete sample segments in $f[n]$). The N, M, and P are integers and must be power of 2 for example 32, 64, 128, etc. Selection of N, M, and P must comply $N > M > P$. For example we can use $N=8$, $M=4$, and $P=2$ for applying the **specgram**. The **specgram** employs a Hanning window as explained in section 5.2 for forward STFT computation.

Frequency variation in the computation is from 0 to $\frac{f_s}{2}$ in order to avoid redundancy. The time variation is in accordance with the section 2.4 cited concept but not for the whole time interval due to window selection and segment overlapping reason. Actual time interval should be from 0 to (number of samples in $f[n]-1$)/f_s but the last t computation point might be less than the last bound.

As an example we have the discrete signal $f[n]$ as [0   9   78   −8   −3   9   3   0   4   −5]. With $N=8$, $M=4$, and $P=2$ and a sampling frequency $f_s=500\,Hz$, the discrete forward STFT matrix $F[n,k]$ is given as follows:

$$
\begin{bmatrix}
75.9283 & 20.1082 & 9.8176 & 2.9271 \\
7.7107-j\,74.3535 & 19.6330+j\,5.6315 & 4.7198-j\,8.4698 & 2.2580-j\,2.3965 \\
-70.5517-j\,10.9045 & 29.6619+j\,10.3455 & -3.7500-j\,8.1406 & -2.5816-j\,1.7275 \\
-7.7107+j\,66.7498 & 34.2637+j\,0.2045 & -6.7927-j\,3.0427 & -0.1850+j\,4.8395 \\
65.1750 & 28.3615 & -6.4635 & 6.3820
\end{bmatrix}
$$

We wish to obtain the $F[n,k]$ matrix by using the **specgram**.

We first enter the given signal $f[n]$ as a row matrix to workspace variable v (user-chosen variable) as follows:
>>v=[0 9 78 -8 -3 9 3 0 4 -5]; ↵   ← v holds the $f[n]$
Then we call the **specgram** with earlier mentioned syntax as follows:
>>[F f t]=specgram(v,8,500,4,2); ↵

We call the **F**, **f**, and **t** to see each of the variables contents:

```
>>F↵
```

```
F =

        75.9283              20.1082              9.8176            2.9271
         7.7107 -74.3535i  19.6330 + 5.6315i   4.7198 - 8.4698i  2.2580 - 2.3965i
       -70.5517 -10.9045i  29.6619 +10.3455i  -3.7500 - 8.1406i -2.5816 - 1.7275i
        -7.7107 +66.7498i  34.2637 + 0.2045i  -6.7927 - 3.0427i -0.1850 + 4.8395i
        65.1750              28.3615             -6.4635            6.3820
```

```
>>f ↵                          >>t ↵
```

```
f =                            t =
        0                              0
   62.5000                        0.0040
  125.0000                        0.0080
  187.5000                        0.0120
  250.0000
```

We have chosen the sampling frequency as 500 *Hz* . Half of the frequency is 250 *Hz* that is why the contents of **f** are from 0 to 250 i.e. 0, 62.5, 125, 187.5, and 250. Again the total sample number in $f[n]$ is 10 therefore the time duration should be from 0 to (10-1)/500 or 0.018sec but the last bound in the **t** is **0.012** which is less than the 0.018sec as explained earlier. The time points considered for the forward STFT are 0sec, 0.004sec, 0.008sec, and 0.012sec. For processing reason one may need the last time point which can be obtained by the command **t(end)**.

There is consistency among the **f** points, **t** points, and **F** points. The orders of (referring to the workspace browser, section 1.4) **f**, **t**, and **F** are 5×1, 4×1, and 5×4 respectively. Each row of **F** and each column of **F** correspond to frequencies in **f** and times in **t** respectively. Also matrix or array indexing starts from 1 in row or column direction. For example left uppermost element in **F** or $F[n,k]$ is 75.9283 which is called by **F(1,1)** and corresponds to $t = 0$ or **t(1)** and $f = 0$ or **f(1)**. Again the fourth row and third column intersecting element in $F[n,k]$ is $-6.7927 - j\,3.0427$ which is called by **F(4,3)** and corresponds to $t = 0.008$sec or **t(3)** and $f = 187.5\ Hz$ or **f(4)** and so on.

5.4 Forward STFT of a digital audio

Section 5.1 exercised programming technique can not be applied to a digital audio since a digital audio contains millions of discrete data instead we apply the section 5.3 mentioned MATLAB function **specgram** to determine the forward STFT. We obtain only the forward transform not the inverse counterpart by the **specgram**.

Let us consider the digital audio **test.wav** of section 2.7. Our objective is to find the discrete forward STFT of the audio signal.

As a first step we make sure that the file is in our working path of MATLAB. Then reading (section 2.3) of the audio file takes place as follows:

>>[v,fs]=wavread('bird.wav'); ↵

Therefore the workspace variables **v** and **fs** hold the $f[n]$ discrete signal as a column matrix and sampling frequency f_s respectively.

In the last section we mentioned that in order to make **specgram** operational, user-supplied point number N for DFT computation, user-supplied window length M, and user-supplied number of samples P for overlapping the discrete sample segments in $f[n]$ should be provided and let us choose those values as $N=512$, $M=256$, and $P=64$ respectively.

We call the **specgram** with section 5.3 mentioned syntax as follows:

>>[F f t]=specgram(v,512,fs,256,64); ↵

Therefore the $F[n,k]$, discrete frequencies corresponding to the rows of $F[n,k]$, and discrete times corresponding to the columns of $F[n,k]$ for the digital audio **bird.wav** are available in the workspace **F**, **f**, and **t** respectively. If we wish to see the discrete frequency variations in $F[n,k]$ for the digital audio, we may execute the following:

>>[f(1) f(end)] ↵   ← Forming a two element row matrix

ans =

        0   11025   ← Meaning frequencies vary from 0 to 11025 Hz

Similarly the time variations in the $F[n,k]$ for the digital audio are seen by:

>>[t(1) t(end)] ↵   ← Forming a two element row matrix

ans =

    0   0.8707   ← Meaning times vary from 0 to 0.8707sec

5.5 Spectrogram of a digital audio

In order to understand the spectrogram of a digital audio signal $x[n]$ (which is the synonym of previous sections mentioned $v[n]$ or $f[n]$), first we need to understand the short time Fourier transform $X[n,k]$ whose discussion is seen in section 5.1 ($X[n,k]$ is the synonym of $F[n,k]$). The spectrogram of a digital audio is the plot of $|X[n,k]|^2$ on the n-k plane or discrete t-f plane where the n and the k are integer time and frequency indices and the t and the f are time and frequency respectively.

The quantity $|X[n,k]|^2$ is viewed as a two dimensional power spectral density. A spectrogram pictures the audio signal's relative energy concentration in frequency as a function of time. The graph is basically a three dimensional one but drawn on two dimensional convenience. Typically a spectrogram is displayed in gray or color scale. If it is a gray scale, darker display in the spectrum indicates more energy in the audio. If it is a color scale, the same property as the gray scale shows is displayed by the color code.

Let us say we wish to see the spectrogram of the digital audio **pitch.wav** (section 2.7).

From the section 4.6 we know that the **pitch.wav** is an audio in which the fundamental or pitch frequency is $f = 1\ KHz$. We do not know the frequency variation of the **pitch.wav** with time but if we assume the frequency f is constant over time t, what should we expect as the f versus t characteristic? Certainly figure 5.2(c) is the answer.

Figure 5.2(c) Frequency versus time plot for a frequency which is constant over time

When we are interested only in the spectrogram, we apply the command **specgram** of section 5.3 with syntax **specgram**(audio signal $v[n]$ or $f[n]$ as a row or column matrix, user-supplied number of points for discrete Fourier transform computation, sampling frequency). For the **pitch.wav**, first let the machine read the audio as follows:

>>[v,fs]=wavread('pitch.wav'); ↵ ← Reading the audio file, v holds $v[n]$, fs holds the sampling frequency

For the DFT point number selection (should be power of 2) let us choose 256 on account of that calling of the syntax takes place as:

>>specgram(v,256,fs) ↵

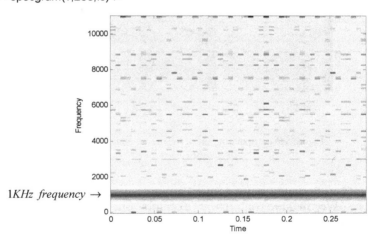

Figure 5.2(d) Spectrogram of the digital audio pitch.wav

Last execution shows the figure 5.2(d) as the output spectrogram of the audio **pitch.wav**. In the figure a dark thick line parallel to the time axis corresponds to the 1 KHz frequency in which most audio energy is concentrated (shown by an arrow indication in the figure). The frequency axis covers up to half of the sampling frequency i.e. **fs/2** (section 5.3). The time axis covers less than or equal to (total sample number in the audio-1)/**fs**. If you are interested to find the frequency and the time variations, exercise the section 5.4 explained syntax:

```
>>[F f t]=specgram(v,256,fs); ⌐
>>t(end) ⌐                  ← To see the last time point

ans =
        0.2844              ← Meaning time variation from 0 to 0.2844sec
>>f(end) ⌐                  ← To see the last frequency point

ans =
        11025               ← Meaning frequency variation from 0 to 11025 Hz
```

The horizontal – time and vertical – frequency axes of the figure 5.2(d) cover just found bounds respectively.

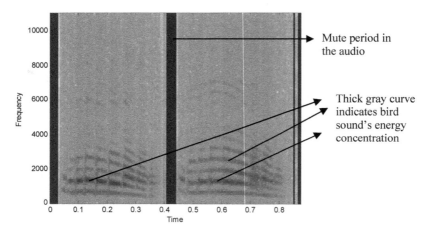

Figure 5.2(e) Spectrogram of the digital audio
bird.wav

Spectrogram example on a bird's sound:

We include one more example on spectrogram for the audio **bird.wav** of section 2.7.

Applying earlier mentioned function and parameter we obtain the spectrogram of the **bird.wav** as follows:

```
>>[v,fs]=wavread('bird.wav'); ⌐ ← Reading the audio file, v holds v[n], fs
                                   holds the sampling frequency
>>specgram(v,256,fs) ⌐        ← Calling the grapher
```

Figure 5.2(e) depicts the spectrogram of the **bird.wav**. In contrast to the **pitch.wav**, the **bird.wav** has many frequencies, some of which are shown by arrows in figure 5.2(e). The thick gray curve in the spectrogram indicates bird sound's energy concentration. If you listen the audio by using **wavplay** of section 2.3, you hear some pause in the audio that mute period is also shown by an arrow in the figure 5.2(e). By inspection we can say that frequencies in the audio change over time. At a particular instant of time the audio signal energy is concentrated in several frequencies. In the text you find the spectrogram in gray scale but in MATLAB that appears in color.

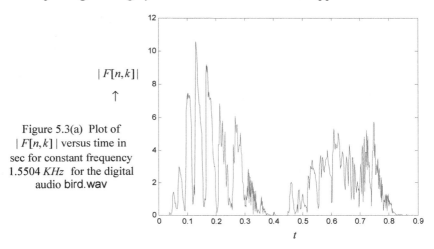

$|F[n,k]|$

↑

Figure 5.3(a) Plot of $|F[n,k]|$ versus time in sec for constant frequency 1.5504 KHz for the digital audio bird.wav

5.6 Constant time or frequency variation in STFT

While working on forward STFT sometimes freezing one variable – time or frequency, variation of the other is sought for example variation of $F[n,k]$ for $n=3$ symbolically $F[3,k]$. Concerning the section 5.3, the workspace variables F, f, and t hold the information for $F[n,k]$, k, and n respectively. Each row and each column in F correspond to constant frequency and constant time respectively. For graphing of the variation, appendix D cited plot is applied.

We wish to consider the bird.wav (section 2.7) for the two variations demonstration.

Let the machine read the audio file as follows:
```
>>[v,fs]=wavread('bird.wav'); ↵      ← Reading the audio file
```
Section 5.3 cited function is applied so we choose $N=128$, $M=64$, and $P=32$ to obtain the discrete forward STFT:
```
>>[F f t]=specgram(v,128,fs,64,32); ↵
```
That is we obtained the $F[n,k]$, k, and n through the workspace variables F, f, and t respectively. The rows and columns of F refer to discrete frequencies and discrete times respectively.

As an example let us get $|F[n,k]|$ versus t for the 10th row of the F. The 10th row from the F is selected by the command F(10,:) and assign that to some user-chosen variable F1 as follows:
```
>>F1=F(10,:); ↵
```

The 10th row of the **F** corresponds to 1.5504 *KHz* frequency which we verify by calling:

>>f(10) ⏎

ans =
                1.5504e+003

The integer frequency k of $F[n,k]$ is 9 for the 10th row assuming that the k changes from 0 and the **F1** holds the values of $F[n,9]$. Note that rows in **F** become columns in $F[n,k]$. The discrete time information is available in the **t** so we exercise the **plot** as follows:

>>plot(t,abs(F1)) ⏎

Figure 5.3(a) depicts the $|F[n,k]|$ versus time variation for 1.5504 *KHz*. The last bound of the **t** is obtained as:

>>t(end) ⏎

ans =
                0.8795   ← Meaning the interval in figure 5.3(a) is $0 \le t \le 0.8795$ sec

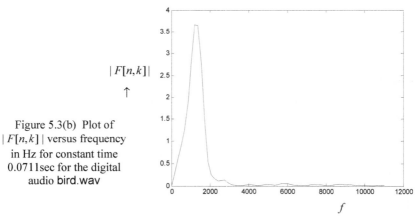

Figure 5.3(b) Plot of
$|F[n,k]|$ versus frequency
in Hz for constant time
0.0711sec for the digital
audio bird.wav

Just now we finished the graphing of $|F[n,k]|$ versus time for constant frequency. What if we need $F[n,k]$ versus frequency for constant time? As an example let us consider the 50th column of **F**. The 50th column from **F** is selected by the command **F(:,50)** and assign that to some user-chosen variable **F2** as follows:

>>F2=F(:,50); ⏎

As we have exercised for the constant frequency, execution of the command **plot(f,abs(F2))** at the prompt presents the $|F[n,k]|$ versus frequency variation before you like the figure 5.3(b). Executions of the **t(50)** and the **f(end)** separately at the prompt return 0.0711 and 11025 respectively meaning $|F[n,k]|$ versus frequency over $0 \le f \le 11025Hz$ for $t = 0.0711$sec.

With this graphing we bring an end to the STFT chapter.

Chapter 6

Z Transform and Digital Audio

Digital audio devices, filtering, representation, or analysis-synthesis tools employ Z transform based terminology or definition on whose account this chapter is planned. The transform is the discrete counterpart of Laplace transform which frequently handles continuous time signals and systems. Digital audio systems which need prediction, monitoring, or stability checking apply the Z transform as an indispensable mathematical tool. Whether analytical audio expression or stacked audio data, one can easily study the following Z transform topics in friendly platform of MATLAB:

- ❖ ❖ Analytical forward-inverse Z transform of discrete signals along with the numeric counterpart
- ❖ ❖ Symbolic analysis of difference equation based systems
- ❖ ❖ Audio filter frequency response obtaining functions and tools
- ❖ ❖ Filter design example of varying types with built-in functions
- ❖ ❖ Audio filter applications on digital audio both in time and frequency domains

6.1 Forward Z transform

Z transform is applicable for the discrete functions. Like the Fourier and Laplace, the Z transform also shares the strategy of the forward-inverse as shown in the figure 6.1(a). The function $f[n]$ is discrete but the transform $F(z)$ is continuous. Depending on the integer index n variation, the Z transform can be bilateral or unilateral. In the former the n changes from minus infinity to plus infinity whereas the n changes from 0 to plus infinity in the latter. In MATLAB we implement the unilateral one as addressed in the following.

Table 6.A   Unilateral forward Z transforms of some discrete functions

Mathematical form	MATLAB commands

$Z\{\sin[na]\}=\dfrac{z\sin a}{z^2-2z\cos a+1}$	>>syms a n, F=ztrans(sin(n*a)); ↵
	>>pretty(F) ↵
	
	     sin(a) z
	  -------------------------- 	
	           2
	  - 2 z cos(a) + z + 1

$Z\{[a^n]\}=\dfrac{z}{z-a}$	>>syms a n, F=ztrans(a^n); ↵
	>>pretty(simple(F)) ↵
	      z
	  - ----------
	    -z + a

$Z\{\delta[n]\}=1$	>>d=sym('charfcn[0](n)'); ↵ >>F=ztrans(d); ↵ ← d⇔ $\delta[n]$ >>pretty(F) ↵ 1 The $\delta[n]$ is represented by sym('charfcn[0](n)') not by dirac(n)

$Z\{[n^2 a^n]\}=\dfrac{za(z+a)}{(z-a)^3}$	>>syms a n, F=ztrans(n^2*a^n); ↵
	>>pretty(F) ↵
	     z a (z + a)
	  - -------------
	         3
	    (-z + a)

$Z\{\cos[na]\}=\dfrac{z(z-\cos a)}{z^2-2z\cos a+1}$	>>syms a n, F=ztrans(cos(n*a)); ↵
	>>pretty(F) ↵
	   z (-cos(a) + z)
	  -------------------------- 	
	           2
	  - 2 z cos(a) + z + 1

$Z\{u[n]\}=\dfrac{z}{z-1}$	>>syms n, F=ztrans(heaviside(n)); ↵
	>>pretty(F) ↵
	   z
	  ------- 	
	  z - 1

$Z\{[r^n \sin na]\}=$
$\dfrac{rz\sin a}{z^2-2zr\cos a+r^2}$
	>>syms a n r, F=ztrans(r^n*sin(n*a)); ↵
	>>pretty(simple(F)) ↵
	     r sin(a) z
	  ------------------------------- 	
	             2 2
	  -2 z r cos(a) + z + r

$Z\{a^{n-3}\,u[n-3]\}=\dfrac{1}{z^2(z-a)}$	>>syms a n, F=ztrans(a^(n-3)*heaviside(n-3)); ↵
	>>pretty(simple(F)) ↵
	     1
	  ------------ 	
	        2
	  (z - a) z

$$\textit{discrete } f[n]\xrightarrow{\textit{forward Z transform}} F(z)\xrightarrow{\textit{inverse Z transform}} f[n]$$

Figure 6.1(a) The concept behind the Z transform

❖ ❖ Forward Z transform of discrete functions

Unilateral forward Z transform $F(z)$ of the discrete function $f[n]$ is defined as $Z \{f[n]\} = F(z) = \sum_{n=0}^{n=\infty} f[n]z^{-n}$ where $n \geq 0$ and n is integer, and Z is the forward transform operator. MATLAB function **ztrans** (abbreviation for the $\underline{Z}$ transform) provides many symbolically known forward transforms. To have the unilateral Z transform of the discrete function $f[n]$ (assuming that the envelope of $f[n]$ follows some functional variation), we employ the command **ztrans**(vector code of $f[n]$ – appendix A, independent variable of $f[n]$, wanted transform variable). Since the computation is merely symbolic, declaration of the associated variables before by using the **syms** is mandatory.

Let us implement the transform with $f[n] = e^{na}$. Its unilateral Z transform is $F(z) = Z \{[e^{an}]\} = \sum_{n=0}^{n=\infty} e^{an}z^{-n} = \dfrac{z}{z - e^{a}}$. Starting from e^{na}, we intend to obtain $\dfrac{z}{z - e^{a}}$ – that is the problem statement for which we conduct the following:

```
>>syms n a z ↵ ← Declaring related and wanted transform variables as symbolic,
                        a⇔a , n⇔n , z⇔z
>>F=ztrans(exp(n*a),n,z); ↵ ← F holds the Z transform on the vector code of
                        eⁿᵃ , F⇔F(z) , F is user-chosen variable
>>pretty(simple(F)) ↵       ← Displaying the readable form of the string for
                        F(z) stored in F, simple for simplification
                        appendix B.13 for pretty
         z
    --------------
    z - exp(a)
```

The command **ztrans(exp(n*a))** also brings about the same result because of the default return. If the independent and transform variables were **p** and **w** respectively, the command would be **ztrans(exp(p*a),p,w)**. Discrete function like $f[n] = e^{-n^2}$ does not have the forward Z transform, we see then **ztrans(exp(-n^2),n,z)** as the return which is the code for the definition $Z \{[e^{-n^2}]\}$. Applying the same symbols and functions, shown table 6.A presents the mathematical and MATLAB correspondence of some unilateral Z transforms.

❖ ❖ Forward Z transform of finite sequences

Most examples of $f[n]$ set in the table 6.A conceive n in the range $0 \leq n \leq \infty$. Discrete functions may exist only for few values of n then the $f[n]$ is called a finite sequence. We present two examples on the finite sequence transform in the following.

Let us consider the first finite sequence $f[n]=\begin{cases}2^{-\frac{n}{4}} & for\ 0\le n\le 7\\ 0 & elsewhere\end{cases}$.

By dint of the unit step sequence $u[n]$, the $f[n]$ is expressed as $f[n]=$

$[2^{-\frac{n}{4}}]\{u[n]-u[n-8]\}$ on that $F(z)=Z\{[2^{-\frac{n}{4}}]u[n]\}-Z\{[2^{-\frac{n}{4}}]u[n-8]\}=\dfrac{2^{\frac{1}{4}}z}{2^{\frac{1}{4}}z-1}-$

$\dfrac{1}{4}\dfrac{2^{\frac{1}{4}}}{(2^{\frac{1}{4}}z-1)z^7}=\dfrac{1}{4z^7}\left[4z^7+2\times2^{\frac{3}{4}}z^6+2\times2^{\frac{1}{2}}z^5+2\times2^{\frac{1}{4}}z^4+2z^3+2^{\frac{3}{4}}z^2+2^{\frac{1}{2}}z+2^{\frac{1}{4}}\right]$ and

we execute that as follows:

>>syms n ↵
>>f=2^(-n/4)*(heaviside(n)-heaviside(n-8)); ↵ ← Vector code of $f[n]$ is
assigned to f, f⇔$f[n]$
>>F=ztrans(f); ↵ ← F holds the transform on $f[n]$, F⇔$F(z)$
>>pretty(simple(F)) ↵ ← Displaying readable form of $F(z)$ following
simplification on the string stored in F

```
              7    3/4  6      1/2  5      1/4  4      3    3/4  2    1/2      1/4
   1/4 (4 z  + 2 2    z  + 2 2    z  + 2 2    z  + 2 z  + 2    z  + 2    z + 2
                                                                       / 7
                                                                      ) / z
                                                                       /
```

The next illustrative finite sequence is $\begin{cases}f[n]\ \rightarrow\ 6\ -4\ 3\ 0\ 2\ 6\ 8\ 2\\ n\ \rightarrow\ 3\ 4\ 5\ 6\ 7\ 8\ 9\ 10\end{cases}$

which tells us that the $f[n]$ does not follow any specific function and that $f[n]$ exists for $3\le n\le10$. At $n=3$, $f[n]=6$ can be represented as $f[3]=6\delta[n-3]$. As a functional form, we write $f[n]=6\delta[n-3]-4\delta[n-4]+3\delta[n-5]+0\delta[n-6]+$

$2\delta[n-7]+6\delta[n-8]+8\delta[n-9]+2\delta[n-10]$. Applying $Z\{\delta[n-n_0]\}=\dfrac{1}{z^{n_0}}$ makes us

get $F(z)=\dfrac{6z^7-4z^6+3z^5+2z^3+6z^2+8z+2}{z^{10}}$ which we wish to implement.

This sort of problem needs slight programming tactics. As a first step we consider the $f[n]$ values as symbolic elements by the following:

>>y=sym([6 -4 3 0 2 6 8 2]); ↵ ← Declare $f[n]$ values as symbolic and put to a
row matrix y

The common delta function in the $f[n]$ is $\delta[n-m]$ which we define as symbolic then by:

>>d=sym('charfcn[m](n)'); ↵ ← $\delta[n-m]$ as symbolic, d⇔$\delta[n-m]$

The function **charfcn[m](n)** is equivalent to discrete $\delta[n-m]$ in which the **m** and **n** are under the third and first brackets respectively. In the $\delta[n-m]$, the m is also symbolic and we define that as follows:

>>syms m ↵

After that we apply the data accumulation technique of appendix B.3 to obtain the $\delta[n-m]$ as a row matrix f noticing that the m varies from 3 to 10:

```
>>f=[ ]; ↵          ← Assigning empty matrix to f before
>>for k=3:10 f=[f subs(d,m,k)]; end ↵ ← Create all delta functions from δ[n − 3]
```
to $\delta[n-10]$ and assign those to f, where f is a row matrix. For-loop counter k gives the control on the location of the delta functions

In the for-loop (appendix B.4), the command **subs(d,m,k)** indicates substitution of the value of **m** by **k** to **d**. In the last **f** we have the delta functions without amplitude. For the delta function with amplitude we have to use the scalar code (appendix A) as follows:

```
>>f=y.*f; ↵ ← The f contains the delta functions with amplitude for different n as a
```
row matrix element

In order to form the complete $f[n]$, we need to sum (appendix B.11) all elements in **f** which occurs by:

```
>>f=sum(f); ↵
```

The last **f** holds the $f[n]$ so the forward transform we get by:

```
>>F=ztrans(f); ↵                    ← F⇔ F(z)
>>pretty(simplify(F)) ↵             ← Displaying readable form of F(z) following
```
simplification on the string stored in **F**

$$
\frac{6z^7 - 4z^6 + 3z^5 + 2z^3 + 6z^2 + 8z + 2}{z^{10}}
$$

6.2 Inverse Z transform

Given a Z transform function $F(z)$, its inverse Z transform (by operator Z^{-1}) is obtained in terms of the contour integral as $f[n]=$ $\frac{1}{2\pi j}\oint_C F(z)z^{n-1}\,dz$ where the C is a counterclockwise closed contour in the region of the convergence of $F(z)$ and the contour encircles the origin of the z-plane.

MATLAB counterpart for the inverse Z transform is **iztrans** (abbreviation for the inverse z transform). To perform the inverse Z transform on $F(z)$, we apply the command **iztrans**(vector string of the transform $F(z)$ – appendix A, transform variable, inverse transform variable). We declare the related variables of $F(z)$ by using the **syms** before applying the **iztrans**. The command **pretty** (appendix B.13) shows the readable form on the inverse return. It is understood that the return function is discrete. From the table 6.A, we have $Z\{[a^n]\}=F(z)=\frac{z}{z-a}$ so the inverse Z transform is $Z^{-1}[\frac{z}{z-a}]=[a^n]=f[n]$ which happens through the following:

-93-

```
>>syms z a n ⏎   ← Defining variables of F(z) and return variable as symbolic,
                    z⟺ z , a⟺ a , and n⟺ n
>>F=z/(z-a); ⏎   ← Assigning vector code of F(z) to F, F⟺ F(z) , F is user-
                    chosen variable
>>f=iztrans(F,z,n); ⏎ ← Assigning inverse transform on F to f, f⟺ f[n] , f is user-
                    chosen variable
>>pretty(f) ⏎        ← Displaying the readable form of the string stored in f
        n
      a
```

The command **iztrans(F)** also returns above result because of the default definition. If the **F** were in **w** and the return were required in terms of **x**, the command would be **iztrans(F,w,x)**. Following the same functions and symbology, few more examples are illustrated in the following.

⊟ Example A

$$X(z)=\frac{z^2}{(z-a)(z-c)}$$ has the inverse transform $$x[n]=\frac{a^{n+1}-c^{n+1}}{a-c}$$ and is implemented as follows:

```
>>syms z a c n ⏎ ← Defining variables of X(z) and return variable as symbolic,
                    z⟺ z , a⟺ a , n⟺ n , c⟺ c
>>X=z^2/(z-a)/(z-c); ⏎    ← Assigning vector code of X(z) to X, X⟺ X(z)
>>x=iztrans(X,z,n); ⏎     ← Assigning inverse transform on X to x, x⟺ x[n]
>>pretty(x) ⏎             ← Displaying the readable form of the string stored in x
        n     n
      a a  - c c
      ------------
       - c + a
```

⊟ Example B

Given that the inverse transform of $$F(z)=\frac{z^{-1}+z^{-2}}{\left(1-\frac{1}{2}z^{-1}\right)\left(1+\frac{1}{3}z^{-1}\right)}$$ is $f[n]=$

$-6\delta[n]+\frac{18}{5}\left(\frac{1}{2}\right)^n+\frac{12}{5}\left(-\frac{1}{3}\right)^n$. It is permissible that we assign the numerator and denominator of $F(z)$ separately and combine afterwards so that less mistake happens in typing or coding. Let us carry out the following:

```
>>syms z ⏎                ←Defining variables of F(z) as symbolic, z⟺ z
>>N=1/z+1/z^2; ⏎ ← Assigning only the numerator code of F(z) to workspace N
>>D=(1-1/2/z)*(1+1/3/z); ⏎ ← Assigning only the denominator code of F(z) to
                              workspace D
>>f=iztrans(N/D); ⏎ ← F(z) is formed by N/D, f holds inverse transform, f⟺ f[n]
>>pretty(f) ⏎       ← Displaying the readable form of the string stored in f
                            n            n
   -6 charfcn[0](n) + 18/5 (1/2)  + 12/5 (-1/3)        ← charfcn[0](n)⟺ δ[n]
```

⊟ **Example C**

Transform look-up table of MATLAB does not contain the inverse of the function like $F(z) = \ln(1-4z)$ therefore the response would be iztrans(log(1-4*z),z,n) while applying the function.

⊟ **Example D**

The transform function $F(z) = \dfrac{3}{(2-\frac{2}{3}z^{-1})^2(2-3z^{-1})(1-4z^{-1})}$ has the inverse

counterpart $f[n] = \dfrac{195}{5929}3^{-n} + \dfrac{3}{308}n3^{-n} - \dfrac{729}{1960}\left(\dfrac{3}{2}\right)^n + \dfrac{432\times4^n}{605}$. It is better if we

assign the numerator and denominator strings separately and combine those afterwards as follows:

>>syms z ↵   ← Defining variables of $F(z)$ as symbolic, z⇔ z
>>N=3; ↵   ← Assigning only the numerator string of $F(z)$ to workspace N
>>D=(2-2/3/z)^2*(2-3/z)*(1-4/z); ↵   ← Assigning only the denominator string of
                                          $F(z)$ to workspace D
>>f=iztrans(N/D); ↵ ← Z^{-1} on $F(z)$ formed by N/D, f holds the string of $f[n]$
>>pretty(f) ↵   ← Displaying the readable form of the string stored in f
```
    195        n                n       729       n    432   n
    ------- (1/3)  + 3/308 (1/3)  n -   -------  (3/2)  + ------ 4
    5929                                1960              605
```

6.3 Z transform on difference equations

The difference equations always involve dependent variable terms like y_k, y_{k+1}, y_{k+2}, ... etc or $y[k]$, $y[k+1]$, $y[k+2]$, ... etc (another representation) where $y[k]$ is a discrete function and function of integer index k. The equivalent MATLAB symbolic writing is sym('y(k)'), sym('y(k+1)'), sym('y(k+2)'), ... etc respectively. From the theory of unilateral Z transform we know that $F(z) = Z\{y[k+3]\} = z^3Y(z) - y[0]z^3 - y[1]z^2 - y[2]z$ where $Y(z)$ is the Z transform of $y[k]$ and implement that as follows:

>>y3=sym('y(k+3)'); ↵ ← Workspace variable y3 holds the code of $y[k+3]$,
                          y3⇔ $y[k+3]$, y3 is user-chosen
>>F=ztrans(y3) ↵   ← Workspace variable F holds the Z transform on
                          $y[k+3]$, F⇔ $F(z)$, F is user-chosen

F =

z^3*ztrans(y(k),k,z)-y(0)*z^3-y(1)*z^2-y(2)*z   ← Vector code (appendix A)
                                                   of the string for $F(z)$

In above string the equivalence is ztrans(y(k),k,z)⇔ $Y(z)$, y(0)⇔ $y[0]$, y(1) ⇔ $y[1]$, and y(2)⇔ $y[2]$ (the last three obviously the initial conditions).

The **ztrans** equally applies for the negative indices for example $F(z) = Z\{y[k-3]\} = \frac{1}{z^3}Y(z)$ is implemented by using the commands **y3=sym('y(k-3)');** and **F=ztrans(y3).**

The difference equation $y_{k+2} + 4y_{k+1} = -4y_k$ along with the initial conditions $y_0 = 1$ and $y_1 = 2$ has the $Y(z) = \frac{z^2 + 6z}{(z+2)^2}$ and $y[k] = (-2)^k - 2(-2)^k k$ — we intend to obtain it from MATLAB.

As a procedure we first assign the given equation to workspace **E** but turning the right hand side of the equation to 0 as follows:

>>E=sym('y(k+2)+4*y(k+1)+4*y(k)'); ↵   ← E is a user-chosen variable
>>T=ztrans(E); ↵ ← Z transform on E and assign that to workspace T, T is any
                                                          user-chosen variable

From earlier discussion we know that the **T** holds the string **ztrans(y(k),k,z)** which had better be replaced (by using the command **subs**) by some user-chosen variable **Y** as follows:

>>syms Y ↵   ← Defining Y as symbolic
>>T=subs(T,'ztrans(y(k),k,z)',Y); ↵ ← In T now Y⇔ $Y(z)$ ⇔ztrans(y(k),k,z),
                                                          the result is again assigned to T

The initial condition has to be entered which is also conducted by the **subs** as follows:

>>T=subs(T,{'y(0)','y(1)'},{1,2}); ↵ ← The result is again assigned to T

In above execution the second bracket inside the **subs** holds both the initial condition and its value separated by a comma. Appendix E cited **solve** provides the solution of $Y(z)$ by turning the string to an equation:

>>Y=solve(T,Y); ↵ ← Finding $Y(z)$ by forming an equation T=0 and the output
                                           is assigned to Y again
>>pretty(Y) ↵   ← Displaying the readable form of the string for the $Y(z)$
                                           stored in Y

```
     z (z + 6)
    ---------------
         2
     z + 4 z + 4
```

Finally we apply the **iztrans** of section 6.2 to determine the $y[k]$ from the string of $Y(z)$ which is stored in **Y**:

>>syms k ↵   ← We wish to obtain the output in terms of k
>>y=iztrans(Y,k); ↵   ← y holds inverse Z transform on string stored in Y
>>pretty(y) ↵   ← Displaying the readable form of the string for the
                                           $y[k]$ stored in y

```
         k              k
    -2 (-2)   k + (-2)
```

Thus we can obtain the Z transform system function and its inverse from any difference equation which describes the digital audio related analysis.

6.4 Input-output on Z transform system

Discrete time filter and discrete time device are often modeled by the Z transform system function $H(z)$. Discrete output from such system subject to some discrete input is frequently required as depicted in figure 6.1(b). We explain in this section how one obtains $y[n]$ starting from the $f[n]$ and $H(z)$.

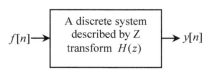

Figure 6.1(b) A single input - single output Z transform system

Both the input $f[n]$ and output $y[n]$ are function of integer index n. Theoretically speaking, we take the forward Z transform of $f[n]$ by section 6.1 mentioned **ztrans**, multiply the transform with the $H(z)$, and take the inverse Z transform of the multiplied function by section 6.2 mentioned **iztrans**. We may encounter two types of problems – expression oriented and data oriented.

⊟ **Expression based input-output**

Suppose we apply a discrete signal $f[n] = 2^{-n}$ to the Z transform system of the figure 6.1(b) with $H(z) = \dfrac{2z}{z-2}$. We wish to see the output in transform z as well as in index n domain. Maintaining ongoing symbology and function we execute the following at the command prompt:

```
>>syms n z ↵        ← Declaring related variable by the syms
>>f=2^(-n); ↵       ← f holds the code of f[n]
>>F=ztrans(f); ↵    ← F holds the Z transform of f[n] or F(z)
>>H=2*z/(z-2); ↵    ← H holds the code of H(z)
>>Y=F*H; ↵          ← Y holds the code of Y(z) from F(z)×H(z)
>>pretty(simple(Y)) ↵  ← Appendix B.13 cited pretty after simplification by
                          the command simple
```

```
             2
            z
    4 --------------------
       (2 z - 1) (z - 2)
```

As the return says we have the $Y(z)$ as $\dfrac{4z^2}{(2z-1)(z-2)}$ from which the $y[n]$ is obtained by the following:

```
>>y=iztrans(Y,z,n); ↵   ← Section 6.2 for the inverse Z transform where y
                           holds the code of y[n]

>>pretty(y) ↵
          n            n
   8/3 2  -  2/3 (1/2)
```

By expression we write the above as $y[n] = \dfrac{8 \times 2^n}{3} - \dfrac{2}{3}\left(\dfrac{1}{2}\right)^n$.

⊟ **Sample or data based input-output**

Just cited example presents expression based input-output on the Z transform system. This kind of expression manipulation is all right for academic purpose. Realistic audio needs data based manipulation. The input to and output from the system of figure 6.1(b) are simply samples or data.

For this type of computation we apply the built-in function **filter** with the syntax **filter(B,A,f)** where the input arguments of the **filter** implement any Z transform system function which has the form $H(z) = \dfrac{\sum\limits_{k=0}^{M} b_k z^{-k}}{1 - \sum\limits_{k=1}^{N} a_k z^{-k}}$ and that

evolves from the general difference equation $y[n] - \sum\limits_{k=1}^{N} a_k y[n-k] = \sum\limits_{k=0}^{M} b_k x[n-k]$.

In the **filter** input argument, we feed the coefficients of the numerator and denominator of $H(z)$ as a row matrix but in ascending power of z^{-1} to the **B** and **A** respectively. Any missing coefficient is set to 0 and normalization is necessary if $H(z)$ is not given in proper form. The **f** is the input discrete signal $f[n]$ of the figure 6.1(b) as a row matrix. The return from the **filter** is the discrete signal $y[n]$ as a row matrix.

For example $H(z) = \dfrac{3 + 2z^{-1}}{1 - 2z^{-1} + z^{-2}}$ has **B=[3 2]** and **A=[1 −2 1]**.

For example $H(z) = \dfrac{3 + 2z^{-1}}{3 - 2z^{-1} + z^{-2}}$ has **B=[3 2]/3** and **A=[3 −2 1]/3** after normalization by 3 both the denominator and numerator.

For example $H(z) = \dfrac{3z + 2z^2}{2 + 3z^4}$ needs normalization by z^4 (highest power in numerator and denominator) first to have $\dfrac{3z^{-3} + 2z^{-2}}{2z^{-4} + 3}$ or

$\dfrac{2z^{-2} + 3z^{-3}}{3 + 2z^{-4}}$ and by 3 afterwards to have **B=[0 0 2 3]/3** and **A=[3 0 0 0 2]/3**.

With these definitions of the discrete input-output we wish to present one example.

Example:

In the figure 6.1(b) suppose the discrete signal $f[n]$ is [1 1 1 1 1] and the Z transform system is $H(z) = \dfrac{3 + 2z^{-1}}{1 - 2z^{-1} + z^{-2}}$. It is given that the output $y[n]$ is [3 11 24 42 65] – which we wish to obtain.

Just mentioned explanation prompts us to proceed with the following:

```
>>f=[1 1 1 1 1]; ↵        ← Entering the f[n] to f as a row matrix
>>B=[3 2]; ↵       ← Entering the numerator coefficients of H(z) to B
>>A=[1 -2 1]; ↵    ← Entering the denominator coefficients of H(z) to A
>>y=filter(B,A,f) ↵        ← Calling the filter for solution

y =
        3   11   24   42   65
```

6.5 Numerical Z transform

Section 6.1 mentioned forward Z transform technique is suitable for theoretical understanding. When we have practical digital audio which literally contains thousands of discrete data, the same analysis-synthesis functions or tools are not helpful at all. A realistic audio always requires the numerical approach which we highlight in this section.

⊟ Numerical forward Z transform

Recall that the $F(z) = \dfrac{6z^7 - 4z^6 + 3z^5 + 2z^3 + 6z^2 + 8z + 2}{z^{10}}$ is the forward Z

transform of the finite sequence $\left\{ \begin{matrix} f[n] & \rightarrow & 6 & -4 & 3 & 0 & 2 & 6 & 8 & 2 \\ n & \rightarrow & 3 & 4 & 5 & 6 & 7 & 8 & 9 & 10 \end{matrix} \right\}$ (section

6.1). It is preferable that we express the $F(z)$ in the form $\dfrac{\sum_{k=0}^{M} b_k z^{-k}}{1 - \sum_{k=1}^{N} a_k z^{-k}}$ while

working with numerical problems where the symbols have the section 6.4 cited meanings. That is we need to conduct normalization by z^{10} hence the $F(z)$ becomes $6z^{-3} - 4z^{-4} + 3z^{-5} + 2z^{-7} + 6z^{-8} + 8z^{-9} + 2z^{-10}$. On comparison, the

denominator coefficient of the $\dfrac{\sum_{k=0}^{M} b_k z^{-k}}{1 - \sum_{k=1}^{N} a_k z^{-k}}$ is just 1 and the numerator

coefficients are the given $f[n]$. The power of z^{-1} continues from 3 to 10 which is the given variation of the n that is $3 \le n \le 10$. The coefficients before z^{-3} are simply zero. Most filter functions are written for handling unilateral situations so $n \ge 0$ is maintained.

As another example, the discrete function $f[n] = [6 \quad 0 \quad 5 \quad 2]$ over $2 \le n \le 5$ have the $F(z)$ representation as follows:

Numerator coefficient as a row matrix: $[0 \quad 0 \quad 6 \quad 0 \quad 5 \quad 2]$
Denominator coefficient: 1

⊟ Numerical inverse Z transform

If the $F(z)$ is given in just described numerator-denominator format, the inverse Z transform numerically we obtain by the built-in function **impz**. The function keeps different provisions, numerical transform of which is addressed in the following:

Table 6.A says that $Z\{[2^n \sin 3n]\} =$ $\dfrac{2z \sin 3}{z^2 - 4z \cos 3 + 4} = F(z)$. Let us compute the $[2^n \sin 3n]$ over

Table 6.B Discrete $[2^n \sin 3n]$ vs n over $0 \le n \le 4$	
n	$[2^n \sin 3n]$
0	0
1	0.2822
2	−1.1177
3	3.2969
4	−8.5852

$0 \le n \le 4$. Table 6.B presents the computed values of $[2^n \sin 3n]$ over $0 \le n \le 4$.

If we are provided with $F(z) = \dfrac{2z \sin 3}{z^2 - 4z \cos 3 + 4}$, our problem is to determine the

table 6.B shown values.

The **impz** operates with the syntax [user-supplied variable for $f[n]$, user-supplied variable for n]=**impz**(numerator coefficients of $F(z)$ as a row matrix, denominator coefficients of $F(z)$ as a row matrix, number of samples required). Following the normalization by the z^2, the $F(z)$ becomes

$\dfrac{2z^{-1}\sin 3}{1-4z^{-1}\cos 3+4z^{-2}}$ or $\dfrac{[0\quad 2\sin 3]}{[1\quad -4\cos 3\quad 4]}$ in coefficient form. The implementation

is shown as follows:

>>N=[0 2*sin(3)]; ↵　　← The numerator coefficients of $F(z)$ are assigned to N
　　　　　　　　　　　　where N is user-chosen variable
>>D=[1 -4*cos(3) 4]; ↵ ← The denominator coefficients of $F(z)$ are assigned to
　　　　　　　　　　　　D where D is user-chosen variable
>>[f,n]=impz(N,D,5); ↵ ← Calling the function where f⇔ $f[n]$ and n⇔ n and
　　　　　　　　　　　　both the f and n are user-chosen variables
>>[n f] ↵　　　　　　　← Calling the n and f side by side to see the contents
　　　　　　　　　　　　　　(appendix B.3)

ans =

```
        0        0
   1.0000   0.2822
   2.0000  -1.1177
   3.0000   3.2969
   4.0000  -8.5852
     ↑        ↑
     n       f[n]
```

Since over the interval $0\le n\le 4$ five values of n are associated, we used 5 as the third input argument to the **impz**.

Thus the reader finds the forward and the inverse Z transforms numerically.

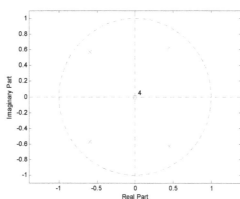

Figure 6.1(c) Pole-zero map of a discrete system – right side figure

6.6 Poles and zeroes of a discrete system

The Z plane pole-zero description of a discrete system represents the system behavior indicating stability. The reader may design any filter or digital device. Whether the filter or device will be stable is validated by the pole-zero analysis. Regardless of signal or system, mostly the stability testing commences with a Z transform system function. If all poles of a system function are within the unit circle, the system is stable.

The built-in function **zplane** graphs the pole-zero map of a system $F(z)$ with the syntax **zplane**(numerator polynomial coefficients of $F(z)$ as a row matrix, denominator polynomial coefficients of $F(z)$ as a row matrix). Note that the polynomial coefficients are as a function of z^{-1} and in ascending order for example the polynomial $1 + 0.3z^{-1} + 0.2z^{-2}$ is written as [1 0.3 0.2]. Any missing coefficient is set to zero for example $1 + 0.3z^{-3}$ as [1 0 0 0.3].

🖱 **Example**

Draw the pole-zero map of a system which is characterized by $F(z) =$
$$\frac{1}{1 + 0.3z^{-1} + 0.2z^{-2} + 0.1z^{-3} + 0.4z^{-4}}.$$

By definition the roots of the numerator and the denominator polynomials of $F(z)$ are termed as the zeroes and the poles respectively. But the polynomial must be in terms of z. Rearranging the $F(z)$, one writes the system function in terms of z as $\dfrac{z^4}{z^4 + 0.3z^3 + 0.2z^2 + 0.1z + 0.4}$. The polynomial order is 4 both in numerator and denominator meaning four roots in each or in other words four zeroes – four poles. However we graph the map as follows:

>>N=1; ↵   ← Entering numerator coefficients of $F(z)$ to N, N is user-chosen
>>D=[1 0.3 0.2 0.1 0.4]; ↵   ← Entering denominator coefficients of $F(z)$ to D,
                                      D is user-chosen
>>zplane(N,D) ↵   ← Calling the function for the pole-zero map

Figure 6.1(c) shows the return from the last line execution. The poles and the zeroes are indicated by × and o with a dot inside respectively. The horizontal and vertical axes of the figure 6.1(c) refer to real of z and imaginary of z respectively. The dotted circle in figure 6.1(c) indicates $|z|=1$ which is the boundary between stable and unstable systems. The 4 at the center of the map indicates the multiplicity of the zeroes (which is 4). Since all four poles of the system are within the unit circle, the system is stable.

The function merely graphs the map. If you need the poles and the zeroes of the system function, you have to apply the **roots** of appendix B.15.

6.7 Frequency response of a Z transform system

Starting from a system function $F(z)$, we may need various frequency related quantities which are the subject matter of this section. The built-in function **freqz** keeps a lot of computing options on the $F(z)$. Polynomial entering of the last section is also applicable for the **freqz**. In this section our example system function is $F(z) = \dfrac{3z^{-1} + 4z^{-2}}{2 + 0.3z^{-1} - 0.2z^{-2} - 0.1z^{-3}}$. Let us enter the Z transform system as follows:

>>N=[0 3 4]; ↵ ← Entering numerator coefficients of $F(z)$ to N, N is user-chosen

>>D=[2 0.3 -0.2 -0.1]; ⌐   ← Entering denominator coefficients of $F(z)$ to D,
D is user-chosen

Frequency response means we substitute $z = e^{j\omega}$ to $F(z)$ and get $F(e^{j\omega}) =$ $\dfrac{3e^{-j\omega} + 4e^{-j2\omega}}{2 + 0.3e^{-j\omega} - 0.2e^{-j2\omega} - 0.1e^{-j3\omega}}$. All computations on the frequency response are based on the complex quantity $F(e^{j\omega})$.

⊟ **Frequency spectrum** $F(e^{j\omega})$ **based on angular frequency vector**

It is given that the $F(e^{j\omega})$ we mentioned just now has the complex values 3.5, $3.4287 - j\,0.6124$, $3.2241 - j\,1.1812$, $2.9115 - j\,1.6723$, and 0.625 at $\omega = 0, 0.1, 0.2, 0.3$, and π radians/sec respectively which we intend to get.

The ω frequencies in radian are entered as a row matrix as follows:
>>w=[0 0.1 0.2 0.3 pi]; ⌐ ← w is user-chosen variable which holds frequencies

Then we apply the syntax **freqz**(numerator coefficients as a row matrix, denominator coefficients as a row matrix, the ω values as a row matrix for $F(e^{j\omega})$ wanted at) and the return from the **freqz** is the $F(e^{j\omega})$ complex values as a row matrix. Let us execute the following:
>>F=freqz(N,D,w) ⌐   ← F is user-chosen variable which holds $F(e^{j\omega})$ values

F =
    3.5000   3.4287 - 0.6124i   3.2241 - 1.1812i   2.9115 - 1.6723i   0.6250 + 0.0000i

Note that the w must be a vector with at least two elements. The $F(e^{j\omega})$ is 2π periodic and has symmetricity about $\omega = \pi$ that is why the ω values had better be over $0 \le \omega \le \pi$. Consecutive values of the $F(e^{j\omega})$ you access by the commands F(1), F(2), F(3), F(4), and F(5) respectively. The last element in the F (which is 0.6250+0.0000i) bears some negligible imaginary part that is due to the digital nature of the computation. The reader can ignore the imaginary part within 4 or 5 digit accuracy.

The ω values might be over a range for example $0 \le \omega \le \pi$ with an increment 0.1 radians. In that case we generate the w by writing w=0:0.1:pi; (section 1.3).

⊟ **Frequency spectrum** $F(e^{j\omega})$ **based on point number**

The **freqz** keeps another option for the number of frequency points but the frequency points cover $0 \le \omega < \pi$ (the π is not inclusive). If we wish to compute the $F(e^{j\omega})$ for N points, the considered frequencies are going to be from 0 to $\dfrac{(N-1)\pi}{N}$ with $\dfrac{\pi}{N}$ increment. As an example $N = 5$ should provide the earlier $F(e^{j\omega})$ values as 3.5, $1.5284 - j\,2.5925$, $-0.7653 - j\,2.3027$, $-1.9312 - j\,0.7009$, and $-0.8917 + j\,1.0806$ at $\omega = 0$, $\dfrac{\pi}{5}$, $\dfrac{2\pi}{5}$, $\dfrac{3\pi}{5}$, and $\dfrac{4\pi}{5}$ respectively. We have these values computed as follows:
>>F=freqz(N,D,5) ⌐   ← Assuming N or D exists from previous executions

F =

3.5000
1.5284 - 2.5925i
-0.7653 - 2.3027i
-1.9312 - 0.7009i
-0.8917 + 1.0806i

The return takes place as a column matrix. If we use two output arguments (section 1.3), the ω values are also returned so execute the following:

>>[F,w]=freqz(N,D,5); ↵   ← F or w is user-chosen variable

The first and the second output arguments (i.e. **F** and **w**) of the **freqz** in above command line refer to $F(e^{j\omega})$ and ω respectively. The five ω values in decimal form become 0, 0.6283, 1.2566, 1.8850, and 2.5133 (each in radian) respectively we may verify by calling the **w** as follows:

>>w ↵

w =

0
0.6283
1.2566
1.8850
2.5133

⊟ Frequency spectrum $F(e^{j\omega})$ based on discrete frequencies

Just addressed ω is also linked to the sampling frequency f_s. The $\omega=0$, $\omega=\pi$, and $\omega=2\pi$ correspond to $f=0$, $f=\frac{f_s}{2}$, and $f=f_s$ respectively where f is the Hertz frequency. Any ω translates to f by the relationship $\frac{\omega f_s}{2\pi}$. Again f to ω transformation needs $\frac{2\pi f}{f_s}$.

In earlier $F(e^{j\omega})$ some example frequencies are $f=0\ Hz$, $f=1\ KHz$, $f=2\ KHz$, and $f=4\ KHz$ where sampling frequency $f_s=16\ KHz$. Given that the $F(e^{j\omega})$ values are 3.5, $2.5541-j\,2.0408$, $0.8599-j\,2.7035$, and $-1.52-j\,1.64$ respectively which we intend to obtain.

The **freqz** also keeps the option for Hertz frequency input with the syntax **freqz(N,D,** f as a row matrix, f_s) where the **N** and the **D** have earlier meanings and values and the return from the **freqz** is the $F(e^{j\omega})$ values as a row matrix so we carry out the following:

>>f=[0 1e3 2e3 4e3]; ↵ ← Hertz frequencies assigned to f as a row matrix, f is
                                         user-chosen variable, appendix A for scale factor

>>F=freqz(N,D,f,16e3) ↵   ← F holds $F(e^{j\omega})$ values

F =

3.5000   2.5541 - 2.0408i   0.8599 - 2.7035i -1.5200 - 1.6400i

⊟ **Separation of** $F(e^{j\omega})$ **parts**

If various parts of the $F(e^{j\omega})$ are required i.e. $\text{Re}\{F(e^{j\omega})\}$, $\text{Im}\{F(e^{j\omega})\}$, $|F(e^{j\omega})|$, and $\angle F(e^{j\omega})$, appendix B.6 mentioned commands **real, imag, abs,** and **angle** are exercised over the **F** respectively.

Figure 6.1(d) Magnitude and phase spectra of the system function $F(z)$ – right side figure

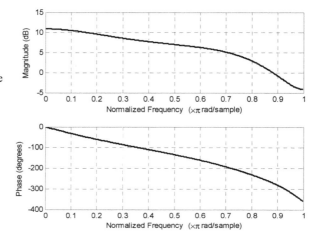

⊟ $F(e^{j\omega})$ **spectrum plotting**

Graphical viewing of magnitude and phase spectra for the system function $F(z)$ is conducted by employing the same **freqz** but without any output argument. The syntax we need is **freqz(N,D)** where **N** and **D** have earlier mentioned meanings.

As an example ongoing $F(z)$ spectra we wish to plot so just execute the following at the command prompt:

>>freqz(N,D) ↵   ← Assuming N or D exists from earlier executions

Figure 6.1(d) presents the spectra from last line execution. The top and the bottom spectra of figure 6.1(d) are basically the $20\log_{10}|F(e^{j\omega})|$ versus ω and the $\angle F(e^{j\omega})$ versus ω respectively. Unlike conventional spectrum the horizontal ω is normalized between 0 and 1 which proportionally corresponds to $0 \le \omega \le \pi$ for radian frequency or $0 \le f \le \frac{f_s}{2}$ for Hertz frequency.

As another alternative frequency response for the same system over $0 \le f \le 2KHz$ with $f_s = 4\,KHz$ is sought. In that case the syntax is **freqz(N,D,** frequencies as a row matrix, sampling frequency) with earlier **N** and **D** meanings. The user has to decide some frequency step say 0.01 KHz then the Hertz frequency vector we generate by **[0:0.01:2]*1e3** so following command is needed for the frequency spectra in which the horizontal frequency axis is labeled in Hertz:

>>f=[0:0.01:2]*1e3; ↵   ← f holds the Hertz frequencies as a row matrix where
                             f is user-chosen variable
>>freqz(N,D,f,4e3) ↵   ← Calling the function for graphing

The graph is not shown for the space reason.

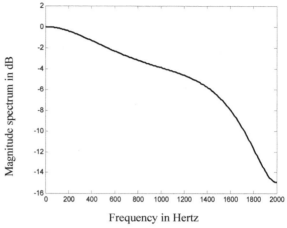

Figure 6.1(e)
Normalized magnitude
spectrum of the system
function $F(z)$ – right
side figure

Magnitude spectrum in dB

Frequency in Hertz

⊟ $F(e^{j\omega})$ **spectrum normalization**

One problem is associated with the spectrum so graphed by the
freqz. The magnitude scale is not normalized consequently starting dB is not
0. In section **4.8** we explained the technique for normalization of the
magnitude spectrum which is also applicable here. Before we apply the
technique we have to have the $F(e^{j\omega})$ values. With a single output argument,
the **freqz** returns the complex $F(e^{j\omega})$ values.

Considering ongoing $F(e^{j\omega})$ over $0 \le f \le 2KHz$ with $f_s = 4\,KHz$ and
0.01 KHz step, we exercise the following:

>>F=freqz(N,D,[0:0.01:2]*1e3,4e3); ↵   ← F holds the complex $F(e^{j\omega})$ values
>>A=abs(F); ↵   ← A holds the $|F(e^{j\omega})|$ values
>>FdB=20*log10(eps+A/max(A)); ↵   ← FdB holds the normalized $|F(e^{j\omega})|$
                                          dB values

In above executions **F**, **A**, and **FdB** are user-chosen variables. Appendix D
cited **plot** can now be exercised over **f** (Hertz frequency stored in) and **FdB**
(normalized $|F(e^{j\omega})|$ values stored in) to see the variation of normalized
$|F(e^{j\omega})|$ versus Hertz frequency over $0 \le f \le 2KHz$:

>>plot(f,FdB) ↵

Figure 6.1(e) depicts the normalized magnitude spectrum in dB in which the
horizontal and the vertical axes correspond to Hertz frequency and
$20\log_{10}\dfrac{|F(e^{j\omega})|}{|F(e^{j\omega})|_{max}}$ respectively.

6.8 Linear time invariant digital filters

Section 6.4 explained Z transform system is used to define the linear time invariant digital filters which are frequently used in processing of digital audio. The filters have different types, most standard of which are addressed in the following.

Linear time invariant digital filters as seen in figure 6.2(a) have the difference equation representation $y[n] = \sum_{k=1}^{N} a_k y[n-k] + \sum_{k=0}^{M} b_k x[n-k]$ connecting the input to (which is $x[n]$) and output from (which is $y[n]$) the digital filter.

The Z transform system function of the filter is $H(z) = \dfrac{\sum_{k=0}^{M} b_k z^{-k}}{1 - \sum_{k=1}^{N} a_k z^{-k}}$.

The system function $H(z)$ is the product of $H_1(z)$ and $H_2(z)$ where $H_1(z) = \sum_{k=0}^{M} b_k z^{-k}$ and $H_2(z) = \dfrac{1}{1 - \sum_{k=1}^{N} a_k z^{-k}}$ and called all-zero and all-pole filters respectively. In order to define the $H(z)$ we apply the techniques of the section 6.4 in conjunction with the function **filter**.

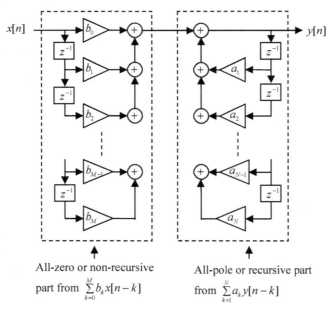

$x[n]$ $y[n]$

All-zero or non-recursive
part from $\sum_{k=0}^{M} b_k x[n-k]$

All-pole or recursive part
from $\sum_{k=1}^{N} a_k y[n-k]$

Figure 6.2(a) Filter structure of a direct form linear time invariant digital filter

Figure 6.2(a) depicted filter structure has two distinct components as indicated by the dotted lines. The all-zero and all-pole filters are also called non-recursive and recursive components respectively. Filter in figure 6.2(a)

has another synonym which is infinite duration impulse response filter (IIR) so to say a filter represented by $H(z)$ function is an IIR filter.

If a filter is represented by only the $H_1(z)$, the structure is called finite duration impulse response (FIR) filter. Figure 6.2(b) shown filter structure is called direct form FIR filter.

We first have to extract z^{-1} polynomial coefficients from given $H(z)$, $H_1(z)$, or structure and then feed the coefficients to the built-in function **filter** for the application of the filter.

Let us see the following examples on the filter definition. We assume that the input signal $x[n]$ to the filter is available in the workspace variable **f**. In addition output signal $y[n]$ is sought from the filtering operation.

🗗 **Example 1**

Define the IIR filter in MATLAB which is characterized by the system function $H(z) = \dfrac{2 + 2z^{-2}}{3 - 2z^{-2} + z^{-5}}$.

This system function is similar to the filter structure in figure 6.2(a). In section 6.4 we normalized the numerator and the denominator but the function **filter** itself keeps the provision for normalization so we just have to pick up the coefficients i.e. the numerator and the denominator polynomial coefficients of the filter are [2 0 2] and [3 0 −2 0 0 1] respectively so we conduct the section 6.4 cited technique for entering the numerator and the denominator coefficients of $H(z)$ to the user-chosen workspace variables B and A respectively as follows:

```
>>B=[2 0 2]; ↵
>>A=[3 0 -2 0 0 1]; ↵
>>y=filter(B,A,f); ↵
```

In the last line we called the **filter** for solution and the **y** would hold the $y[n]$.

🗗 **Example 2**

Define the FIR filter in MATLAB which is characterized by the system function $H(z) = 1 - 2z^{-2} + z^{-5}$.

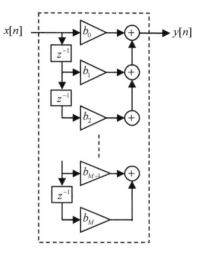

Figure 6.2(b) Direct form FIR digital filter structure

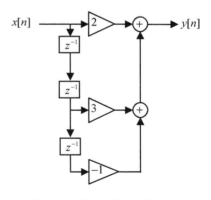

Figure 6.2(c) A digital filter

In FIR filter there is no denominator in $H(z)$ so we just assume 1 instead of the denominator. The rest is like the example 1 as follows:

```
>>B=[1 0 -2 0 0 1]; ↵
>>A=1; ↵
>>y=filter(B,A,f); ↵    ← Calling the filter for solution, y would hold the y[n]
```

⊟ **Example 3**

Often a digital filter is supplied with block diagram like the figure 6.2(c) which is to be defined in MATLAB.

We can compare the filter of figure 6.2(c) with that of figure 6.2(b) and by inspection we have $b_0 = 2$, $b_1 = 0$, $b_2 = 3$, and $b_4 = -1$. Knowing so, the filter is defined in the following where the symbols bear earlier mentioned meanings:

```
>>B=[2 0 3 -1]; ↵
>>A=1; ↵
>>y=filter(B,A,f); ↵    ← Calling the filter for solution, y would hold the y[n]
```

⊟ **Example 4**

Figure 6.2(d) shown IIR filter is to be defined. One compares figure 6.2(d) to the figure 6.2(a) and easily discovers that the z^{-1} polynomial coefficients are $b_0 = 2$, $b_1 = -1$, $a_1 = 3$, and $a_2 = 5$ so like the other examples we define the filter as follows:

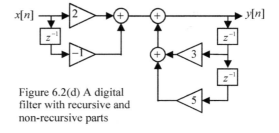

Figure 6.2(d) A digital filter with recursive and non-recursive parts

```
>>B=[2 -1]; A=[1 3 5]; ↵
>>y=filter(B,A,f); ↵    ← Calling the filter for solution, y would hold the y[n]
```

These four examples are the basic standard form of the digital filters.

6.9 Special digital filters

Last section mentioned digital filters are not the only used ones in audio processing. Other digital filters are also seen in audio literature for specific purpose, some of which are addressed in the following. We assume that you must have gone through previous sections.

Salient features about the digital filter design are the following:

(a) In accordance with the sampling theory frequency present in any given signal must not be more than the half of the sampling frequency for successful recovery.

(b) In digital filters Fourier transform of the filter function is 2π periodic and has symmetricity between 0 to π and π to 2π pertaining to the angular frequency ω. For this reason in most filters only 0 to π interval of the angular frequency is considered i.e. $0 \le \omega \le \pi$.

(c) Frequently normalization is used in the ω interval. The 0 to π is proportionately stated as 0 to 1.

(d) In order to relate the normalized angular frequency (i.e. 0 to 1 scale) with the sampling frequency f_s we apply linear mapping between 0 to 1 and 0 to $f_s/2$ i.e. 1 corresponds to $f_s/2$.

In all filtering it is about the finding of $H(z)$ which serves an objective criterion. The reader might ask why do we need $H(z)$ expression for the filter implementation? DSP chip or computer hardware is linked to the

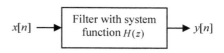

Figure 6.3(a) A digital filter

$H(z)$, not to the $h[n]$ – the time domain impulse response. In the following filters, figure 6.3(a) depicted system function $H(z)$ is sought.

⊞ Periodic notch filters

A periodic notch filter has the system function $H(z) = \dfrac{1-z^{-N}}{1-\left(\dfrac{z}{R}\right)^{-N}}$

where the N and the R are user-defined. The N is called the filter order which is an integer and the R must be less than 1 for stability concern. Closer R to unity provide sharper notches.

As an example we wish to form the $H(z)$ of notch filter as in figure 6.3(a) with $N=5$ and $R=0.8$.

By comparison of the above $H(z)$ to the one in last section for standard direct form, one gets $b_0=1$, $b_1=0$, $b_2=0$, $b_3=0$, $b_4=0$, and $b_5=-1$ for the numerator and $a_1=0$, $a_2=0$, $a_3=0$, $a_4=0$, and $a_5=0.8^5$ for the denominator. The notch filter implementation then takes place as follows:

```
>>B=[1 0 0 0 0 -1]; A=[1 0 0 0 0 -0.8^5]; y=filter(B,A,f); ↵
```

where the symbols have last section mentioned meanings.

The next query might be how we relate the frequency issue to the ideal definition of the notch filter. The periodic notch filters out the desired frequency and its harmonics. For example we wish to remove the $f=50\,Hz$ and its harmonics. The selected sampling frequency is $f_s=600\,Hz$. The N is simply then $\dfrac{f_s}{f}$ or 12.

⊞ Allpass filters

Allpass filters are a well-known class of IIR filters (last section) having a system function of the form:

$$H(z) = \frac{a_0 + a_1 z^{-1} + a_2 z^{-2} + \ldots\ldots + a_{N-1} z^{-N+1}}{a_{N-1} + a_{N-2} z^{-1} + a_{N-3} z^{-2} \ldots + a_1 z^{-N+2} + a_0 z^{-N+1}}.$$

Interestingly the numerator and the denominator polynomial coefficients are reversed in relation to each other. An example of allpass filter can be

$$H(z) = \frac{1 - 0.4 z^{-1} + 0.8 z^{-2} + 0.9 z^{-3}}{0.9 + 0.8 z^{-1} - 0.4 z^{-2} + z^{-3}}.$$

The numerator polynomial coefficients are entered as:
>>B=[1 -0.4 0.8 0.9]; ↵
Coefficient reversal of the numerator may be exercised by the command **fliplr** of appendix B.7:
>>A=fliplr(B); ↵
Figure 6.3(a) filtering operation is then conducted by:
>>y=filter(B,A,f); ↵
where the symbols have ongoing meanings.

Parametric filters

This sort of filter has the specifications in terms of the gain G in dB, center frequency f_c in Hz, sampling frequency f_s in Hz, and quality factor Q and is based on the second order system function $H(z) = \frac{b_0 + b_1 z^{-1} + b_2 z^{-2}}{a_0 + a_1 z^{-1} + a_2 z^{-2}}$

where related parameters are given by;

$$\omega_c = \frac{2\pi f_c}{f_s}, \quad \beta = \frac{2\omega_c}{Q} + \omega_c^2 + 4, \text{ and } g = 10^{\frac{G}{20}} \text{ wherefrom}$$

the filter numerator coefficients:

$$b_0 = \frac{\frac{2g\omega_c}{Q} + 4 + \omega_c^2}{\beta}, \quad b_1 = \frac{2\omega_c^2 - 8}{\beta}, \text{ and } b_2 = \frac{4 - \frac{2g\omega_c}{Q} + \omega_c^2}{\beta} \text{ and}$$

the filter denominator coefficients:

$$a_0 = 1, \quad a_1 = b_1, \text{ and } a_2 = \frac{4 - \frac{2\omega_c}{Q} + \omega_c^2}{\beta}.$$

As an example with $f_c = 200\,Hz$, $f_s = 16\,KHz$, $G = 5\,dB$, and $Q = 0.7$, the filter coefficients should be $b_0 = 1.0413$, $b_1 = -1.8881$, $b_2 = 0.8526$, $a_0 = 1$, $a_1 = -1.8881$, and $a_2 = 0.8939$ which need the following executions:
>>G=5; fc=200; fs=16e3; Q=0.7; ↵ ← Assigning given parameters G⇔G, fc⇔ f_c, fs⇔f_s, Q⇔Q, G, fc, etc are user-chosen variables
>>g=10^(G/20); ↵   ← Computing g, g⇔g, appendix A for coding
>>wc=2*pi*fc/fs; ↵   ← Computing ω_c, wc⇔ω_c
>>bt=2*wc/Q+wc^2+4; ↵   ← Computing β, bt⇔β
>>b0=(2*g*wc/Q+4+wc^2)/bt; ↵   ← Computing b_0, b0⇔b_0
>>b1=(2*wc^2-8)/bt; ↵   ← Computing b_1, b1⇔b_1

```
>>b2=(4-2*g*wc/Q+wc^2)/bt; ↵        ← Computing b₂ , b2⇔ b₂
>>a0=1; a1=b1; ↵                     ← Assigning a₀ and a₁ , a0⇔ a₀ , a1⇔ a₁
>>a2=(4-2*wc/Q+wc^2)/bt; ↵           ← Computing a₂ , a2⇔ a₂
>>B=[b0 b1 b2]; ↵       ← Forming a three element row matrix from b₀ , b₁ , and b₂
                          and assigning to B
>>A=[a0 a1 a2]; ↵       ← Forming a three element row matrix from a₀ , a₁ , and a₂
                          and assigning to A
```

The **B** and the **A** have earlier mentioned meanings. All other variables (**wc, bt, b0,** etc) are user-chosen. Now you may apply **y=filter(B,A,f);** for the filtering of figure 6.3(a). If you wish to verify the filter coefficients for example the numerator ones, just call the **B**:

```
>>B ↵
```

B =

         1.0413   -1.8881   0.8526
           ↑   ↑   ↑
          b_0 b_1 b_2

⧉ Low frequency shelving filters

   Low frequency shelving filter also employs the second order system $H(z) = \dfrac{b_0 + b_1 z^{-1} + b_2 z^{-2}}{a_0 + a_1 z^{-1} + a_2 z^{-2}}$. The filter may have two types of gain – boosting and cutting accordingly the filter coefficients are defined. The filter has the specifications in terms of gain G in dB, crossover frequency f_c in Hz, and sampling frequency f_s in Hz. Following mathematical relationships (appendix A for coding) are applied in the filter.

   Intermediate variables:

$$g = 10^{\frac{G}{20}}, \ \Omega_c = 2\pi f_c, \ T_s = \frac{1}{f_s}, \ \text{and} \ P = \tan\left(\frac{\Omega_c T_s}{2}\right)$$

   When boosting is needed:

$$b_0 = \frac{1 + \sqrt{2g}\,P + gP^2}{1 + \sqrt{2}\,P + P^2}, \quad b_1 = \frac{2(gP^2 - 1)}{1 + \sqrt{2}\,P + P^2}, \quad b_2 = \frac{1 - \sqrt{2g}\,P + gP^2}{1 + \sqrt{2}\,P + P^2}, \quad a_0 = 1,$$

$$a_1 = \frac{2(P^2 - 1)}{1 + \sqrt{2}\,P + P^2}, \ \text{and} \ a_2 = \frac{1 - \sqrt{2}\,P + P^2}{1 + \sqrt{2}\,P + P^2}$$

   When cutting is needed:

$$b_0 = \frac{1 + \sqrt{2}\,P + P^2}{1 + \sqrt{2g}\,P + gP^2}, \quad b_1 = \frac{2(P^2 - 1)}{1 + \sqrt{2g}\,P + gP^2}, \quad b_2 = \frac{1 - \sqrt{2}\,P + P^2}{1 + \sqrt{2g}\,P + gP^2}, \quad a_0 = 1,$$

$$a_1 = \frac{2(gP^2 - 1)}{1 + \sqrt{2g}\,P + gP^2}, \ \text{and} \ a_2 = \frac{1 - \sqrt{2g}\,P + gP^2}{1 + \sqrt{2g}\,P + gP^2}$$

As an example of the boosting case and with $f_c = 200\ Hz$, $f_s = 16\ KHz$, and $G = 5\ dB$, the filter coefficients should be $b_0 = 1.0187$, $b_1 = -1.8868$, $b_2 = 0.8785$, $a_0 = 1$, $a_1 = -1.889$, and $a_2 = 0.8949$ which need the following executions:

```
>>G=5; fc=200; wc=2*pi*fc; fs=16e3; ts=1/fs; ↵ ← Assigning given parameters
                                   G⇔G , fc⇔ f_c , fs⇔ f_s , wc⇔ Ω_c , ts⇔ T_s
>>g=10^(G/20); P=tan(wc*ts/2); ↵    ← Computing g and P , g⇔g , P⇔P
>>b0=(1+sqrt(2*g)*P+g*P^2)/(1+sqrt(2)*P+P^2); ↵ ← Computing b_0 , b0⇔ b_0
>>b1=2*(g*P^2-1)/(1+sqrt(2)*P+P^2); ↵          ← Computing b_1 , b1⇔ b_1
>>b2=(1-sqrt(2*g)*P+g*P^2)/(1+sqrt(2)*P+P^2); ↵ ← Computing b_2 , b2⇔ b_2
>>a0=1; ↵                           ← a0⇔ a_0
>>a1=2*(P^2-1)/(1+sqrt(2)*P+P^2); ↵     ← Computing a_1 , a1⇔ a_1
>>a2=(1-sqrt(2)*P+P^2)/(1+sqrt(2)*P+P^2); ↵    ← Computing a_2 , a2⇔ a_2
>>B=[b0 b1 b2] ↵   ← Forming a three element row matrix from b_0 , b_1 , and b_2
```

B =

  1.0187 -1.8868   0.8785
    ↑   ↑   ↑
   b_0 b_1 b_2

```
>>A=[a0 a1 a2] ↵   ← Forming a three element row matrix from a_0 , a_1 , and a_2
```

A =

  1.0000 -1.8890   0.8949
    ↑   ↑   ↑
   a_0 a_1 a_2

Now you may apply the command y=filter(B,A,f); for the operation of the figure 6.3(a). We addressed the boosting gain case similarly you may carry out the cutting gain operation by the filter.

⊟ High frequency shelving filters

High frequency shelving filter makes use of the second order system $H(z)$ of the low frequency counterpart with identical intermediate variables. The filter coefficients are the following:

When boosting is needed:

$$b_0 = \frac{g + \sqrt{2g}\,P + P^2}{1 + \sqrt{2}\,P + P^2}, \quad b_1 = \frac{2(P^2 - g)}{1 + \sqrt{2}\,P + P^2}, \quad b_2 = \frac{g - \sqrt{2g}\,P + P^2}{1 + \sqrt{2}\,P + P^2}, \quad a_0 = 1,$$

$$a_1 = \frac{2(P^2 - 1)}{1 + \sqrt{2}\,P + P^2}, \text{ and } a_2 = \frac{1 - \sqrt{2}\,P + P^2}{1 + \sqrt{2}\,P + P^2}$$

When cutting is needed:

$$b_0 = \frac{1+\sqrt{2}P+P^2}{g+\sqrt{2g}P+P^2}, \qquad b_1 = \frac{2(P^2-1)}{g+\sqrt{2g}P+P^2}, \qquad b_2 = \frac{1-\sqrt{2}P+P^2}{g+\sqrt{2g}P+P^2}, \qquad a_0 = 1,$$

$$a_1 = \frac{2\left(\dfrac{P^2}{g}-1\right)}{1+\sqrt{\dfrac{2}{g}}P+\dfrac{P^2}{g}}, \text{ and } a_2 = \frac{1-\sqrt{\dfrac{2}{g}}P+\dfrac{P^2}{g}}{1+\sqrt{\dfrac{2}{g}}P+\dfrac{P^2}{g}}$$

Its example and implementation are very similar to the one conducted for the low frequency shelving filter.

Butterworth filters

Butterworth is one kind of IIR filter. Primarily it has high and low pass types. The two types are again applied to form stop and pass band filters. These filters are widely used in audio systems as bass management filters. Bass management ensures loudspeaker audio reproduction with minimal distortion. The magnitude frequency response of the Butterworth filter is given by:

Low pass: $|H(e^{j\omega})| = \dfrac{1}{\sqrt{1+\left(\dfrac{\omega}{\omega_c}\right)^{2N}}}$

High pass: $|H(e^{j\omega})| = 1 - \dfrac{1}{\sqrt{1+\left(\dfrac{\omega}{\omega_c}\right)^{2N}}}$

where N is the filter order and ω_c is called the crossover frequency.

The built-in function **butter** helps us determine the digital system function $H(z)$ with the syntax [variable for the numerator coefficients of $H(z)$, variable for the denominator coefficients of $H(z)$]=**butter**(filter order, normalized frequency between 0 and 1,filter type under quote).

The $H(z)$ polynomial coefficients are in terms of the z^{-1} and in the descending order of z. For the $H(z)$, the $\dfrac{\omega}{\omega_c}$ had better be in 0-1. When the Hertz frequency f and the sampling frequency f_s in Hertz are given, the $\dfrac{\omega}{\omega_c}$ is tantamount to $\dfrac{2f}{f_s}$.

There are four probable filters namely lowpass, highpass, bandpass, and bandstop accordingly the function reserve words are **low**, **high**, **bandpass**, and **stop** respectively. While using the bandpass and the bandstop options, the lower and the upper normalized crossover frequencies must be provided as a two element row matrix in the second input argument of the **butter**. Let us see the following examples about the filter design.

Lowpass example:

In consumer electronics $N=2$ and $f_c=80\,Hz$ are often used for lowpass Butterworth filter. We need to select the sampling frequency say $8\,KHz$ therefore the normalized frequency $\dfrac{\omega}{\omega_c}$ becomes $\dfrac{2\times80}{8\times10^3}$ (which has the code **2*80/8e3**). Knowing so, we execute the following:

```
>>[B,A]=butter(2,2*80/8e3,'low') ↵    ← B,A is user-chosen variable

B =
        0.0009    0.0019    0.0009    ← Numerator coefficients
A =
        1.0000   -1.9112   0.9150     ← Denominator coefficients
```

As soon as we have the numerator and the denominator polynomial coefficients of $H(z)$ available in **B** and **A** respectively, the $H(z)$ one writes as $\dfrac{0.0009 + 0.0019z^{-1} + 0.0009z^{-2}}{1 - 1.9112z^{-1} + 0.915z^{-2}}$.

Bandpass example:

We intend to find the $H(z)$ of a bandpass Butterworth filter with order 2 when lower and upper crossover frequencies are $100\,Hz$ and $300\,Hz$ respectively at a sampling frequency $4\,KHz$.

Given information says that the two crossover frequencies as a two element row matrix are $[100\ 300]/2\,KHz$ or **[100 300]/2e3** in code form of coarse in normalized scale. Lucid specifications make us execute the following:

```
>>[B,A]=butter(2,[100 300]/2e3,'bandpass') ↵    ← B, A is user-
                                                   chosen variable
B =
        0.0201    0    -0.0402    0    0.0201
A =
        1.0000   -3.4289   4.5303   -2.7383   0.6414
```

Like the lowpass counterpart we easily discover the system function of the required bandpass Butterworth filter as $H(z)=$ $\dfrac{0.0201 - 0.0402z^{-2} + 0.0201z^{-4}}{1 - 3.4289z^{-1} + 4.5303z^{-2} - 2.7383z^{-3} + 0.6414z^{-4}}$.

Having found the $H(z)$, the command **y=filter(B,A,f)**; might be executed for the filtering operation of the figure 6.3(a).

⏚ **Chebyshev type I filter**

This IIR filter exhibits equiripple error in the passband and monotonically decreasing/increasing frequency response in the stopband. A lowpass N -th order Chebyshev filter has the squared magnitude response:

$$|H(e^{j\omega})|^2 = \frac{1}{1 + R^2 T_N^2\left(\dfrac{\omega}{\omega_c}\right)}$$

where $T_N(x)$ is the Chebyshev polynomial of order N and the polynomial is defined by the expression $T_N(x) = \begin{cases} \cos(N\cos^{-1}x) & |x|\le 1 \\ \cosh(N\cosh^{-1}x) & |x|\ge 1 \end{cases}$. The magnitude response oscillates between 1 and $\dfrac{1}{1+R^2}$ in the passband thereby showing ripple.

MATLAB built-in function **cheby1** implements the filter with the syntax [variable for the numerator coefficients of $H(z)$, variable for the denominator coefficients of $H(z)$]=**cheby1**(filter order, peak to peak ripple in dB in the passband, normalized frequency between 0 and 1, filter type under quote). Filter type of the Butterworth is equally applicable here i.e. lowpass, highpass, bandpass, and bandstop. See the following two examples regarding the Chebyshev type I filter:

Lowpass example:
Design a lowpass Chebyshev type I filter which has the following characteristics: order $N=2$, 3 dB peak to peak ripple in the passband, crossover frequency $f_c=400\,Hz$, and sampling frequency $f_s=2\,KHz$.

The normalized frequency $\dfrac{\omega}{\omega_c}$ becomes $\dfrac{400}{1\times 10^3}=0.4$. Its straightforward execution like the Butterworth counterpart is the following:

>>[B,A]=cheby1(2,3,0.4,'low') ↵   ← B,A is user-chosen variable

B =
          0.1436   0.2872   0.1436   ← Numerator coefficients
A =
          1.0000 -0.6799   0.4913   ← Denominator coefficients

Knowing the numerator and the denominator polynomial coefficients, one writes the $H(z)$ as $\dfrac{0.1436 + 0.2872z^{-1} + 0.1436z^{-2}}{1 - 0.6799z^{-1} + 0.4913z^{-2}}$.

Bandpass example:
Find the $H(z)$ of a bandpass Chebyshev type I filter with order 2 and 5 dB peak to peak ripple in the passband. The lower and the upper crossover frequencies should be $100\,Hz$ and $300\,Hz$ respectively at a sampling frequency $4\,KHz$.

The two crossover frequencies as a two element row matrix become [100 300]/2 KHz or **[100 300]/2e3** in code form. Having all specifications available, we execute the following:

>>[B,A]=cheby1(2,5,[100 300]/2e3,'bandpass') ↵   ← B, A is user-
chosen variable

B =

   0.0078   0 -0.0157   0   0.0078   ← Numerator

A =

   1.0000 -3.6696 5.2209 -3.4130 0.8667   ← Denominator

System function of the required bandpass Chebyshev type I filter we write then as $H(z) = \dfrac{0.0078 - 0.0157z^{-2} + 0.0078z^{-4}}{1 - 3.6696z^{-1} + 5.2209z^{-2} - 3.413z^{-3} + 0.8667z^{-4}}$.

Having found the $H(z)$, the command y=filter(B,A,f); might be executed for the filtering operation of the figure 6.3(a).

⊟ **Chebyshev type II filter**

    Contrasting to the type I, this type of filters exhibits equirriple error in the stopband and the frequency response increases or decays monotonically in the passband. The squared magnitude frequency response of a lowpass filter system is given by:

$$|H(e^{j\omega})|^2 = \dfrac{1}{1 + \dfrac{1}{R^2 T_N^2\left(\dfrac{\omega_c}{\omega}\right)}}$$

Built-in function **cheby2** implements the Chebyshev type II filters. Whatever syntax we applied for the type I counterpart by employing **cheby1** can be applied by **cheby2** for the type II.

⊟ **Filters in factored form**

    Sometimes filters in factored form are given and single filter expression might be sought. For example $H(z) = \dfrac{5.43(z^{-1} - 3)}{(z^{-1} - 1)(z^{-1} + 4)}$ turns to $H(z)$

$= \dfrac{-16.29 + 5.43z^{-1}}{-4 + 3z^{-1} + z^{-2}}$ which we intend to obtain.

    Since convolution is basically polynomial multiplication, section 3.10 mentioned **conv** is used to get the multiplication through. Be concerned about the polynomial arrangement – all representations must be in descending power of z or ascending power of z^{-1}. See the following how the multiplication is done:

>>B=[-3 1]*5.43 ↵   ← Numerator of the given $H(z)$

B =

   -16.2900 5.4300   ← Numerator of the latter $H(z)$

>>A=conv([-1 1],[4 1]) ↵   ← Denominator of the given $H(z)$

A =

   -4   3   1   ← Denominator of the latter $H(z)$

As another example, turning of $H(z) = \dfrac{5.43(3+z^{-2})(z^{-1}+7)}{(z^{-1}-1)(z^{-2}+4)}$ to $H(z) =$

$\dfrac{114.03+16.29z^{-1}+38.01z^{-2}+5.43z^{-3}}{-4+4z^{-1}-z^{-2}+z^{-3}}$ is exercised by the following:

```
>>B=5.43*conv([3 0 1],[7 1]) ↵
```
← Numerator of the given $H(z)$

```
B =
    114.0300  16.2900  38.0100  5.4300
```
← Numerator of the latter $H(z)$

```
>>A=conv([-1 1],[4 0 1]) ↵
```
← Denominator of the given $H(z)$

```
A =
    -4   4   -1   1
```
← Denominator of the latter $H(z)$

When three factor terms are given for instance $5.43(3+z^{-2})(z^{-1}+7)(1-z^{-3})$, you may exercise **5.43*conv(conv([3 0 1],[7 1]),[1 0 0 -1])**.

⊟ **Filters as an object in MATLAB**

If composite filter structure is required starting from prototype filter, it is a standard practice that we form the prototype filter as an object with the help of the built-in command **filt**. The function forms an object with the syntax **filt**(numerator polynomial coefficients of $H(z)$ as a row matrix, denominator polynomial coefficients of $H(z)$ as a row matrix) obviously in descending power of z. For instance $H(z) = \dfrac{-16.29+5.43z^{-1}}{-4+3z^{-1}+z^{-2}}$ is entered as an object as follows:

```
>>H=filt([-16.29 5.43],[-4 3 1]) ↵
```
← H holds whole $H(z)$, H is user-chosen

```
Transfer function:
16.29 - 5.43 z^-1
-----------------------
4 - 3 z^-1 - z^-2

Sampling time: unspecified
```

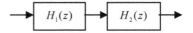

Figure 6.3(b) Two filter functions connected in series

In above return the **z^-1** means z^{-1}, **z^-2** means z^{-2}, and so on. Had you executed **H=filt([-16.29 5.43],[-4 3 1]);**, the functional popup would not have been displayed. Note that rationalization of the numerator and the denominator are done by -1 automatically to turn the first coefficient of the polynomial as positive.

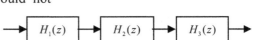

Figure 6.3(c) Three filter functions connected in series

⊟ **Cascade filters**

Filters placed side by side are called series or cascade filters. For example figures 6.3(b) and 6.3(c) show two and three filters in series respectively. If we are interested to find the equivalent filter function of the whole series combination, built-in function **series** is exercised provided that each filter is described as an object by the **filt**. The function needs the syntax **series**(filter 1 representing variable name, filter 2 representing variable name).

Figure 6.3(b) shows two filter functions $H_1(z) = \dfrac{5z^{-2} - z^{-1} + 1}{z^{-3} - 1}$ and

$H_2(z) = \dfrac{5.43(z^{-1} - 3)}{(z^{-1} - 1)(z^{-1} + 4)}$ connected in series. Their z domain equivalent system

function is given by $H_{eq}(z) = H_1(z)\, H_2(z) = \dfrac{27.15z^{-3} - 86.88z^{-2} + 21.72z^{-1} - 16.29}{z^{-5} + 3z^{-4} - 4z^{-3} - z^{-2} - 3z^{-1} + 4}$

which we wish to implement in MATLAB.

>>H1=filt([1 -1 5],[-1 0 0 1]); ↵   ← H1 is user-chosen variable and H1⇔ $H_1(z)$

>>H2=filt(5.43*[-3 1],conv([-1 1],[4 1])); ↵ ← H2 is user-chosen variable and H2
$\qquad\qquad\qquad\qquad\qquad\qquad\qquad$ ⇔ $H_2(z)$, see filters in factored form

>>Heq=series(H1,H2) ↵ ← $H_{eq}(z)$ is computed by series and assigned to Heq
$\qquad\qquad\qquad\qquad$ where Heq is user-chosen variable and Heq⇔ $H_{eq}(z)$

Transfer function:
-16.29 + 21.72 z^-1 - 86.88 z^-2 + 27.15 z^-3
---   ← the $H_{eq}(z)$ we expected

4 - 3 z^-1 - z^-2 - 4 z^-3 + 3 z^-4 + z^-5
Sampling time: unspecified

Filters in series like the figure 6.3(c) has three system functions where $H_3(z)$

$= \dfrac{3.2(z^{-1} - 2.2)}{(z^{-1} + 1)(z^{-1} + 1.5)}$ and $H_1(z)$ and $H_2(z)$ are the foregoing ones. With these

three filter functions, we should have the equivalent $H_{eq}(z) = H_1(z)\, H_2(z)\, H_3(z)$

$= \dfrac{114.7 - 205z^{-1} + 681.1z^{-2} - 469.2z^{-3} + 86.88z^{-4}}{6 + 5.5z^{-1} - 5z^{-2} - 11.54z^{-3} - 6.5z^{-4} + 5z^{-5} + 5.5z^{-6} + z^{-7}}$ by using the following:

>>H3=filt(3.2*[-2.2 1],conv([1 1],[1.5 1])); ↵   ← H3 is user-chosen variable and
$\qquad\qquad\qquad\qquad\qquad\qquad\qquad\qquad$ H3⇔ $H_3(z)$

>>Heq=series(Heq,H3) ↵ ← Assuming input argument Heq of series holds
$\qquad\qquad\qquad\qquad\qquad$ $H_1(z)\, H_2(z)$ equivalent from previous executions

Transfer function:
$\qquad$ 114.7 - 205 z^-1 + 681.1 z^-2 - 469.2 z^-3 + 86.88 z^-4
----
6 + 5.5 z^-1 - 5 z^-2 - 11.5 z^-3 - 6.5 z^-4 + 5 z^-5 + 5.5 z^-6 + z^-7

Sampling time: unspecified

In above execution the return from the **series** is again assigned to **Heq** which is a user-chosen variable. One can easily read off the three filter equivalent $H_{eq}(z)$.

One point has to be stressed out, the filter function available in **Heq** is in object form. For subsequent calculation the reader might need the numerical coefficients of $H_{eq}(z)$ which we get by the function **tfdata**. The function has a syntax [variable for the numerator of $H_{eq}(z)$, variable for the denominator of $H_{eq}(z)$]=**tfdata**(filter object variable, **'v'**) where the **v** is reserved to the function so then carry out the following:

>>[B,A]=tfdata(Heq,'v') ↲   ← B or A is user-chosen variable
                                                                                           and has ongoing meanings

B =

114.6816   -205.0368   681.1392   -469.1520   86.8800   0   0   0
                                                  ↑

B holds the numerator coefficients of $H_{eq}(z)$ as a row matrix

A =

6.0000   5.5000   -5.0000   -11.5000   -6.5000   5.0000   5.5000   1.0000
                                                  ↑

A holds the denominator coefficients of $H_{eq}(z)$ as a row matrix

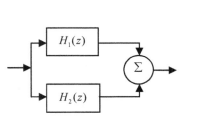

Figure 6.3(d) Two filter functions connected
in parallel

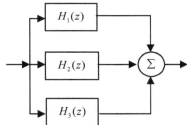

Figure 6.3(e) Three filter functions connected
in parallel

❖❖ Parallel filters

Cascade filter discussions are the prerequisite of this discourse. Figure 6.3(d) shows two filter functions connected in parallel. Taking the example filter functions $H_1(z)$ and $H_2(z)$ of the cascade connection into account, we should have the equivalent as $H_{eq}(z) = H_1(z) + H_2(z) = \dfrac{10.43z^{-4} - 2.29z^{-3} - 22z^{-2} + 1.57z^{-1} + 12.29}{z^{-5} + 3z^{-4} - 4z^{-3} - z^{-2} - 3z^{-1} + 4}$. MATLAB built-in function **parallel** determines the equivalent system function of two systems connected in parallel by applying the syntax **parallel**(filter 1 representing variable name, filter 2 representing variable name). We know that the $H_1(z)$ and $H_2(z)$ are stored in the workspace variables **H1** and **H2** respectively therefore we execute the following:

>>Heq=parallel(H1,H2) ↲   ← $H_{eq}(z)$ is computed and assigned to **Heq**, **Heq**
                                                    is user-chosen variable and **Heq**⇔ $H_{eq}(z)$

Transfer function:

$$\frac{12.29 + 1.57\ z^\wedge\text{-}1 - 22\ z^\wedge\text{-}2 - 2.29\ z^\wedge\text{-}3 + 10.43\ z^\wedge\text{-}4}{4 - 3\ z^\wedge\text{-}1 - z^\wedge\text{-}2 - 4\ z^\wedge\text{-}3 + 3\ z^\wedge\text{-}4 + z^\wedge\text{-}5}$$

Sampling time: unspecified

Figure 6.3(e) presents three filter functions connected in parallel whose equivalent is easily found by exercising **Heq=parallel(parallel (H1,H2),H3)** at the command prompt based on the foregoing $H_1(z)$, $H_2(z)$, and $H_3(z)$ where the symbols have the previously mentioned meanings and mathematically **Heq** returns $H_1(z) + H_1(z) + H_3(z)$. You may verify that $H_{eq}(z)$

$$= \frac{-9.725 + 67z^{-1} - 19.34z^{-2} - 31.9z^{-3} - 46z^{-4} + 26.34z^{-5} + 13.63z^{-6}}{6 + 5.5z^{-1} - 5z^{-2} - 11.5z^{-3} - 6.5z^{-4} + 5z^{-5} + 5.5z^{-6} + z^{-7}}$$ for the three filters.

Like the series counterpart you may also exercise **tfdata** on the **Heq** to extract the polynomial coefficients from **Heq**.

⊞ **Averaging filters**

An N-point causal average filter has the system function $H(z) =$ $\frac{1-z^{-N}}{N(1-z^{-1})}$ for example with $N = 5$ the $H(z)$ is $\frac{1-z^{-5}}{5(1-z^{-1})}$. As we have been discussing, the polynomial coefficients are [1   0   0   0   0   −1]/5 and [1 −1] for numerator and denominator respectively and enter them by:

```
>>B=[1 0 0 0 0 -1]/5; ↵      ← Assigning numerator coefficients to B
>>A=[1 -1]; ↵                ← Assigning denominator coefficients to A
```
where **B** and **A** have ongoing meanings.

⊞ **Comb filter**

A comb filter has the system function $H(z) = 1 - z^{-N}$ where N is the filter order for example with $N = 5$ the $H(z)$ is $1 - z^{-5}$. The magnitude frequency response of the filter looks much like the rounded teeth of a comb. Clearly the filter is FIR and has no denominator. As usual the polynomial coefficients are [1   0   0   0   0   −1]/5 and 1 for numerator and denominator respectively so entering is done by:

```
>>B=[1 0 0 0 0 -1]/5; ↵      ← Assigning numerator coefficients to B
>>A=1; ↵                     ← Assigning denominator coefficients to A
```
where **B** and **A** have usual meanings.

6.10 Application of digital filter on an audio

Before the application, the reader has to decide whether the operation be conducted in time or frequency domain. If the operation is in integer time index domain, filter impulse response $h[n]$ must be known and convolution provides the filtered audio as conducted in section 3.10. In the case of frequency or transform domain operation the filter system function $H(z)$ must be known regardless of filter type. In last two sections we illustrated a variety of filters and way to get their $H(z)$. Also in all filters the polynomial coefficients are made available in the variables **B** and **A** for numerator and denominator respectively. Section 6.4 mentioned **filter** is then applied on the **B** and **A** to get the filtered audio. Some filtering might change the audio amplitude range in that case section 3.2 cited function is applied to map the audio amplitudes linearly within [−1,1]. Two examples on audio filtering are illustrated in the following.

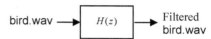

bird.wav ⟶ $H(z)$ ⟶ Filtered bird.wav

Figure 6.4(a) A filter function is applied to the digital audio bird.wav

⊟ Example 1

All pole filter $H(z) = \dfrac{1}{1+0.3z^{-1}+0.2z^{-2}+0.1z^{-3}+0.2z^{-4}}$ is to be applied on the digital audio **bird.wav** of section 2.7 which we have shown schematically in figure 6.4(a). Our objective is to play the original **bird.wav** and its filtered counterpart.

Get the audio and let the machine read (section 2.3) the audio as follows:

>>[v,fs]=wavread('bird.wav'); ↵

The workspace **v** and **fs** hold the digital audio samples as a column matrix and sampling frequency respectively. The **v** holds figure 6.1(b) shown $f[n]$ and we are looking for $y[n]$ after passing through $H(z)$. The $H(z)$ is entered by:

>>B=1; ↵   ← Entering numerator coefficient of $H(z)$ to **B**
>>A=[1 0.3 0.2 0.1 0.2]; ↵   ← Entering denominator coefficients of $H(z)$ to **A**

Call the function as:

>>y=filter(B,A,v); ↵   ← y holds $y[n]$ of figure 6.1(b), **y** is user-chosen variable

If we exercise **datastats** of section 2.6 on **v** and **y**, we find that the data range of the $y[n]$ is changed from that of $f[n]$ due to the filtering so apply the mapping on the last **y** as follows:

>>yn=2*mat2gray(y)-1; ↵   ← yn holds newly mapped audio data in [-1,1], yn is user-chosen

The **yn** in above execution then refers to the filtered audio of figure 6.4(a). Making the original and the filtered audio samples available in the workspace, we easily play them by:

>>wavplay(v,fs) ↵   ← Playing the original audio
>>wavplay(yn,fs) ↵   ← Playing the filtered audio

With concentration if you listen both audio several times, the filtered audio sounds slightly sharper meaning removal of some high frequencies from the original **bird.wav** due to the filtering.

⊟ Example 2

In this example we wish to apply the $H(z)$ of bandpass Chebyshev type I filter to the audio **bird.wav** with the specifications – order 4, 5 dB peak to peak ripple in the passband, and lower and upper crossover frequencies as 500 *Hz* and 2000 *Hz* respectively. The audio is to be played before and after the filtering.

From the example 1 the sampling frequency of the audio is available in **fs**. Last section mentioned **cheby1** helps us get the $H(z)$ filter coefficients as follows:

>>[B,A]=cheby1(4,5,[500 2000]/2/fs,'bandpass'); ↵ ← B or A has the same
                                                  meaning as that in example 1

Then we call (audio samples available in **v** from example 1) the **filter** for filtering operation:

>>y=filter(B,A,v); ↵   ← y holds the filtered audio
>>yn=2*mat2gray(y)-1; ↵   ← yn holds mapped filtered audio data in [-1,1]
>>wavplay(v,fs) ↵   ← Playing the original audio
>>wavplay(yn,fs) ↵   ← Playing the filtered audio

What do you infer from above two sounds? Definitely special audio effect is produced due to the filtering as if someone is trying to imitate the contents of the **bird.wav** audio.

With this special audio effect example we bring an end to the chapter.

Chapter 7

 ## Matrix Algebra and Digital Audio

Certain class of digital audio algorithms is largely based on linear algebra theory. A single digital audio can be treated as a vector which is the working element of matrix algebra. At times specially organized audio forms multiple vectors like frame by frame slicing of digital audio. Orthogonal matrix factorization techniques decompose an audio into subspace which helps in audio noise removal or enhancement. One advantage of the linear algebra based study is its direct signal dependence better than on the transform domain as happens in Fourier or Z analysis. Our linear algebra linking digital audio discussion highlights the following:

♦ ♦ Tools for matrix algebra operations like rank, inverse, etc
♦ ♦ Matrix factorization tools through eigenvalue-eigenvector
♦ ♦ Relevant matrix forming on discrete signal or digital audio
♦ ♦ Matrix covariance based study with supplied signal or audio

7.1 Rank of a matrix

The number of linearly independent rows or columns of a matrix A is called the rank of A. If the A is of order $M \times N$, then the rank of $A \leq$ minimum(M, N). The command **rank** determines the rank of a matrix when the matrix is its input argument. It is given that the matrix $A =$ $\begin{bmatrix} 1 & 2 & 4 & 3 \\ 3 & -1 & 2 & -2 \\ 5 & -4 & 0 & -7 \end{bmatrix}$ has the rank 2 which we wish to obtain.

Maintaining font equivalence for example $A \Leftrightarrow A$ we execute the following at the command prompt:

```
>>A=[1 2 4 3;3 -1 2 -2; 5 -4 0 -7]; ↵     ← Entering given matrix to A (section 1.3)
>>rank(A) ↵
```

ans =
    2

A row or column matrix has the rank 1. We know that the digital audio **bird.wav** is a column matrix and its rank should be 1 which we verify as follows:

```
>>[v,fs]=wavread('bird.wav'); ↵     ← Reading the audio (section 2.3), v holds the
                                       digital audio as a column matrix

>>rank(v) ↵
```

ans =
    1

7.2 Determinant of a square matrix

Determinant of a matrix is possible if the matrix is square. It is given that 124 and $2x^2 - 2x - 4x^3 + 4xy$ are the determinant values for the square

matrices $A = \begin{bmatrix} 2 & 4 & 3 \\ -4 & 6 & 9 \\ 3 & 7 & 10 \end{bmatrix}$ and $B = \begin{bmatrix} x & x^2 & y \\ 2x & x & 1 \\ 0 & 2 & 2 \end{bmatrix}$ respectively which we wish

to compute.

The command **det** computes the determinant of a square matrix when the matrix is its input argument. If any variable element like x is present in the given matrix, we declare that by using the command **syms** in advance. We attach both computations below:

For the square matrix A :
```
>>A=[2 4 3;-4 6 9;3 7 10]; ↵     ← Entering given matrix to A (section 1.3)
>>C=det(A) ↵
```

C =
    124

For the square matrix B :
```
>>syms x y ↵                      ← Declaring related variables
>>B=[x x^2 y;2*x x 1;0 2 2]; ↵    ← Entering given matrix to B
>>C=det(B); pretty(C) ↵
```

      2   3
  2 x - 2 x - 4 x + 4 x y

Font equivalence is maintained like $A \Leftrightarrow A$ in the last executions. In each case we assigned the computed value to the workspace **C** (any user-chosen variable). For the second matrix, the **pretty** (appendix B.13) is applied to see the readable form on the string stored in **C**.

7.3 Inverse of a square matrix

The function **inv** (abbreviation for the <u>inv</u>erse) computes the inverse of a square matrix when the matrix is its input argument but on the condition that the determinant of the square matrix is not zero. Rational form inverse is possible if we turn the given matrix elements to symbolic by using the command **sym** on the whole matrix.

Let us take the square matrix $A = \begin{bmatrix} 2 & 1 & 5 \\ 0 & 2 & 1 \\ 0 & 0 & 2 \end{bmatrix}$ as an example, whose

inverse is given by $B = \begin{bmatrix} \frac{1}{2} & -\frac{1}{4} & -\frac{9}{8} \\ 0 & \frac{1}{2} & -\frac{1}{4} \\ 0 & 0 & \frac{1}{2} \end{bmatrix} = \begin{bmatrix} 0.5 & -0.25 & -1.125 \\ 0 & 0.5 & -0.25 \\ 0 & 0 & 0.5 \end{bmatrix}$. Both the decimal

and rational form implementations are presented below:

for the inverse in decimal form,
```
>>A=[2 1 5;0 2 1;0 0 2]; ↵
>>B=inv(A) ↵
```

B =
```
    0.5000   -0.2500   -1.1250
         0    0.5000   -0.2500
         0         0    0.5000
```

for the inverse in rational form,
```
>>A=sym([2 1 5;0 2 1;0 0 2]); ↵
>>B=inv(A) ↵
```

B =
```
[ 1/2,   -1/4,   -9/8]
[   0,    1/2,   -1/4]
[   0,      0,    1/2]
```

Matrices of the symbolic variables are easy to deal with the **inv**. The only point is we have to declare the related variables by using the command **syms** in advance. Given that the inverse of $A = \begin{bmatrix} x & y \\ x+y & 1 \end{bmatrix}$ is $B =$

$\begin{bmatrix} -\dfrac{1}{y^2+yx-x} & \dfrac{y}{y^2+yx-x} \\ \dfrac{x+y}{y^2+yx-x} & -\dfrac{x}{y^2+yx-x} \end{bmatrix}$ whose implementation is also attached below:

```
>>syms x y ↵
>>A=[x y;x+y 1]; B=inv(A); pretty(B) ↵
         [              1                        y          ]
         [-  -------------------        -------------------  ]
         [              2                        2           ]
         [   -x + x y + y            -x + x y + y            ]
         [                                                   ]
         [       x + y                      x                ]
         [  -------------------      -  -------------------  ]
         [              2                        2           ]
         [  -x + x y + y            -x + x y + y             ]
```

We maintained font equivalence like A⇔A in just conducted executions (section 1.3 for matrix entering and appendix B.13 for **pretty**).

7.4 Characteristic polynomial of a square matrix

Characteristic polynomial of a square matrix A is defined by $|\lambda I - A|$ where I is the identity matrix of the same order as that of A, λ is the eigenvalue of A, and the modulus sign indicates the determinant of the square matrix. If any variable is present in the A, we declare that by using the command **syms** prior to computing. The built-in command **poly** computes the characteristic polynomial of a square matrix when the matrix is its input argument.

It is given that the $\lambda^3 + 6\lambda^2 - 23\lambda + 9$ and $\lambda^2 + 6\lambda - 7 + x$ are the characteristic polynomials of square matrices on numeric $A = \begin{bmatrix} 1 & 2 & -1 \\ -1 & -7 & 2 \\ 0 & 9 & 0 \end{bmatrix}$

and symbolic $B = \begin{bmatrix} 1 & x \\ -1 & -7 \end{bmatrix}$ respectively which we intend to compute. Both implementations are attached below:

For the coefficient form polynomial,
>>A=[1 2 -1;-1 -7 2;0 9 0]; P=poly(A) ↵

P =
        1.0000 6.0000 -23.0000   9.0000
For the polynomial with symbolic variables,
>>syms x, B=[1 x;-1 -7]; P=poly(B) ↵

P =
t^2+6*t-7+x
In each case we assigned the given matrix to corresponding variable like A to A and the output from the **poly** to P where P is a user-chosen variable. As a return from the **poly**, we obtain the polynomial coefficients that is [1 6 − 23 9] for the numeric case. For the symbolic case the output is the whole polynomial in terms of t (symbol λ is unavailable in the workspace and default $t \Leftrightarrow \lambda$).

7.5 Eigenvalues and eigenvectors of a square matrix

For a square matrix of order $N \times N$, the characteristic equation $|\lambda I - A| = 0$ is a polynomial of degree N and has N roots. These roots of the characteristic equation are called the eigenvalues of the matrix A.

For each eigenvalue, there is a matrix X of order $N \times 1$ which satisfies the matrix equation $A X = \lambda X$. If the equation is satisfied by the vector X, then the X is called an eigenvector of the matrix A. Matrix multiplication of A and X results the order of $A X$ as $N \times 1$ again the order of the matrix λX is also $N \times 1$. For the N eigenvalues, we must have N eigenvectors. The eigenvalues are unique but the eigenvectors are not. Multiplying an eigenvector by -1 does not change the concept.

The built-in function **eig** (from the <u>eig</u>envalue) helps us determine the eigenvalues or eigenvectors of a square matrix taking the matrix as its input argument.

It is given that the eigenvalues of $A = \begin{bmatrix} 2 & 2 & 0 \\ 2 & 1 & 1 \\ -7 & 2 & -3 \end{bmatrix}$ are -4, 1, and 3

and the eigenvectors are $\begin{bmatrix} 0.0747 \\ -0.2242 \\ 0.9717 \end{bmatrix}$, $\begin{bmatrix} -0.4364 \\ 0.2182 \\ 0.8729 \end{bmatrix}$, and $\begin{bmatrix} -0.6667 \\ -0.3333 \\ 0.6667 \end{bmatrix}$ which we intend to

$\quad\quad\quad\quad\quad\quad\;\; for\ \lambda = -4 \quad\quad for\ \lambda = 1 \quad\quad\quad for\ \lambda = 3$

obtain.

Eigenvalue implementation (font equivalence like A⟺A is maintained) is presented on the right side in this paragraph in which the first line is the matrix entering (section 1.3). The eigenvalues of A as a column matrix are returned to the workspace variable E (any user-chosen variable) in the

```
only for the eigenvalues,
>>A=[2 2 0;2 1 1;-7 2 -3]; ↵
>>E=eig(A) ↵

E =
    -4.0000
     3.0000
     1.0000
```

second line. First eigenvalue from the E is picked up by the E(1), second one by the E(2), and so on.

The **eig** keeps provision for returning the eigenvectors as well but with different number of output arguments. The syntax we apply is [user-supplied variable for eigenvectors, user-supplied variable for eigenvalues]=**eig**(square matrix). The implementation is on the right side of this paragraph. The Y and X correspond to the eigenvectors and eigenvalues respectively, both of which are our chosen variables. The matrix returned to Y is formed by placing the eigenvectors side by side. The X is a diagonal matrix whose diagonal elements are the eigenvalues of the A. The placement of the

```
for the eigenvalues and eigenvectors,
>>[Y X]=eig(A) ↵

Y =
     0.0747   -0.6667   -0.4364
    -0.2242   -0.3333    0.2182
     0.9717    0.6667    0.8729

X =
    -4.0000        0        0
         0    3.0000        0
         0        0    1.0000
only the second eigenvector,
>>T=Y(:,2) ↵

T =
    -0.6667
    -0.3333
     0.6667
```

eigenvectors and eigenvalues is in order for example the first column in the Y corresponds to the first diagonal element of the X or eigenvalue -4.

Once executed, one might need to access each eigenvalue and eigenvetor. For example the second eigenvector is picked up from the Y by using the command Y(:,2) and its eigenvalue from X by X(2,2). We picked up the second eigenvector and assigned that to T as shown above where the T is a user-chosen variable.

Again as a column matrix we can store the eigenvalues to the workspace D (user-chosen) by using the command **diag** as shown on the right side.

only the eigenvalues from X,
>>D=diag(X) ⏎

Depending on the nature of A, the eigenvalues can be real or complex however the **eig** is suitable for handling the complex eigenvalues in general.

D =
-4.0000
3.0000
1.0000

Every numeric eigenvector returned by the eig is normalized for example the first eigenvector $\begin{bmatrix} 0.0747 \\ -0.2242 \\ 0.9717 \end{bmatrix}$ has the magnitude $\sqrt{0.0747^2 + (-0.2242)^2 + 0.9717^2} = 1$.

Since eigenvectors are not unique, sometimes symbolic solution is appreciated. Ongoing matrix A has eigenvectors $\begin{bmatrix} 2 \\ 1 \\ -2 \end{bmatrix}$, $\begin{bmatrix} -2 \\ 1 \\ 4 \end{bmatrix}$, and $\begin{bmatrix} 1 \\ -3 \\ 13 \end{bmatrix}$ in integer form for 3, 1, and –4 respectively. It is implementable by declaring the elements of A as symbolic (with the help of the command **sym** on A) in that case we need the command [Y X]=eig(sym(A)).

The reader can verify that each eigenvalue-eigenvector set satisfies the matrix equation $A X = \lambda X$. To prove so, you may separately execute A*Y(:,1) and X(1,1)*Y(:,1) for the eigenvalue –4 at the command prompt.

7.6 Singular value decomposition and condition number

Definitions of the eigenvalues and eigenvectors hold true for square matrices. Spectral decomposition theorem helps any symmetric matrix be decomposed into the product of three matrices.

Let A be a matrix of order $M \times N$, A can be expressed as $A = U \times D \times V^T$ where the multiplication sign indicates the matrix multiplication and V^T is the transpose of V. The U and V are called the unitary matrices and their determinant should be 1. Orders of U, D, and V are $M \times M$, $M \times N$, and $N \times N$ respectively. This decomposition is termed as the singular value decomposition of A.

The decomposition exists for general matrices – square or rectangular. The D is a diagonal matrix (can be a rectangular one). Diagonal elements of D are called the singular values of A. Singular values are obtained by taking the positive square roots of the eigenvalues of $A \times A^T$ (A^T is the transpose of A).

Matrix $A^T \times A$ or $A \times A^T$ is symmetric and their eigenvalues are nonnegative. Columns of U and V are the eigenvectors of $A \times A^T$ and $A^T \times A$ respectively.

The decomposition $A = U \times D \times V^T$ for $A = \begin{bmatrix} 4 & 1 & 0 & 9 \\ 5 & 7 & -1 & 0 \\ 6 & 9 & 4 & 2 \end{bmatrix}$ is given

by $U = \begin{bmatrix} 0.4090 & 0.9062 & -0.1073 \\ 0.5168 & -0.3269 & -0.7912 \\ 0.7521 & -0.2682 & 0.6020 \end{bmatrix}$, $D = \begin{bmatrix} 15.0297 & 0 & 0 & 0 \\ 0 & 8.5537 & 0 & 0 \\ 0 & 0 & 3.3079 & 0 \end{bmatrix}$, and $V =$

$\begin{bmatrix} 0.5810 & -0.0445 & -0.2337 & -0.7783 \\ 0.7183 & 0.4438 & -0.0688 & 0.5314 \\ 0.1658 & 0.0872 & 0.9672 & -0.1717 \\ 0.3450 & -0.8908 & 0.0721 & 0.2869 \end{bmatrix}$. Our objective is to obtain U, D, and

V matrices starting from the A.

The built-in function **svd** (abbreviation for the <u>s</u>ingular <u>v</u>alue <u>d</u>ecomposition) performs the singular value decomposition of a matrix when the matrix is its input argument. Presented on the right side is the decomposition we are seeking for in which the first line is to assign the given matrix data to workspace A. In the second line we call the **svd** for the decomposition with three output arguments – **U**, **D**, and **V** where $U \Leftrightarrow U$, $D \Leftrightarrow D$, and $V \Leftrightarrow V$. The **U**, **D**, or **V** can be any user-chosen variables but in the shown order. We know that the

```
for the singular value decomposition:
>>A=[4 1 0 9;5 7 -1 0;6 9 4 2]; ↵
>>[U D V]=svd(A) ↵

U =
     0.4090  -0.9062  -0.1073
     0.5168   0.3269  -0.7912
     0.7521   0.2682   0.6020
D =
    15.0297        0        0   0
         0   8.5537        0   0
         0        0   3.3079   0
V =
     0.5810  -0.0445  -0.2337  -0.7783
     0.7183   0.4438  -0.0688   0.5314
     0.1658   0.0872   0.9672  -0.1717
     0.3450  -0.8908   0.0721   0.2869
```
for the condition number of A:
```
>>cond(A) ↵

ans =
     4.5436
```

eigenvectors are not unique. Multiplying an eigenvector by minus 1 does not alter the decomposition. This is obvious from the second column vectors of the U and the return to **U**.

The condition number of a matrix is defined as the ratio of its largest to smallest singular values. A large condition number indicates contiguous linear dependency among the columns of the matrix or in other words the matrix is nearly a singular one.

Just now we computed the singular values of the aforementioned A. Its condition number should be $\dfrac{15.0297}{3.3079} = 4.5436$. The function **cond** (abbreviation for <u>cond</u>ition number) determines the condition number of a matrix when the matrix is its input argument. Its implementation is presented in above too.

7.7 Singular value decomposition of a discrete signal

Without the loss of generality, a discrete signal is simply a row or column matrix but the definition of singular value stands for a rectangular matrix. This obstacle requires that we organize the given discrete signal in some special rectangular matrix for instance Toeplitz.

A Toeplitz matrix is one which has identical diagonal elements and the matrix is symmetric. Again another problem is associated with the Toeplitz. Due to symmetricity there will be redundant signal information in a perfect Toeplitz that is why in most audio literature a discrete audio signal $v[n]$ is arranged in a form which is slightly redundant rather than perfect

symmetric like $f[m,n] = \begin{bmatrix} v[L-1] & v[L-2] & \cdots & v[0] \\ v[L] & v[L-1] & \cdots & v[1] \\ \vdots & & & \\ v[N-1] & v[N-2] & \cdots & v[N-L] \end{bmatrix}$ where the discrete

signal $v[n]$ exists over $0 \le n \le N-1$ and has N samples. The L is a user-chosen integer with $L \le N$. Reference [4] reports the L to be $\dfrac{N}{3}$ or its closest integer.

♦ ♦ **Obtaining the** $f[m,n]$ **matrix from a given** $v[n]$

Let us consider the $v[n]$ as [9 6 −5 4 3] which indicates that the interval is $0 \le n \le 4$ because there are 5 samples in $v[n]$. If we choose $L=3$ which is less than 4 and should be, the $f[m,n]$ matrix should be $\begin{bmatrix} -5 & 6 & 9 \\ 4 & -5 & 6 \\ 3 & 4 & -5 \end{bmatrix}$.

Our objective is to obtain the $f[m,n]$ matrix starting from $v[n]$ and L.

There is no built-in function for this sort of matrix formation nevertheless author-written function **mtop** can be helpful in this regard (appendix C). Figure 7.1(a) shows the function file with two input arguments − **x** and **L**. The **x**

```
function X=mtop(x,L)
N=numel(x);
c=L:-1:1;
X=x(repmat([0:N-L]',1,L)+repmat(c,N-L+1,1));
```

Figure 7.1(a) Author-written function file to obtain $f[m,n]$ from $v[n]$

should be the $v[n]$ as a row matrix and the **L** is the user-chosen scalar L.

Open a new M-file, type the codes of the figure 7.1(a) in the M-file, save the file by the name **mtop**, and execute the following:

```
>>x=[9 6 -5 4 3]; ↵     ← v[n] is assigned to x as a row matrix, x is user-chosen
>>f=mtop(x,3) ↵          ← mtop return is assigned to f, f is user-chosen, f ⇔ f[m,n]

f =

   -5   6   9
    4  -5   6
    3   4  -5
```

❖❖ Singular value decomposition of a discrete signal $v[n]$

Singular value decomposition of a discrete signal $v[n]$ is basically the singular value decomposition of $f[m,n]$ in accordance with the section 7.6 quoted techniques. The rectangular matrix $f[m,n]$ utilizes the vector outer products of column and row matrices. The eigenvalue and eigenvector notions are only applicable for the square matrix whose discussions are found in section 7.5.

Obtaining the rectangular matrix $f[m,n]$ of order $M \times N$ from $v[n]$, it can be expressed by the summation formula as follows: $f[m,n] = \sum_{j=1}^{J} \sqrt{\lambda_j} u_j v_j$ where u_j and v_j are the eigenvectors of the matrices $g_1[m,n]$ and $g_2[m,n]$ ($g_1[m,n] = f[m,n] \times f[m,n]^T$ and $g_2[m,n] = f[m,n]^T \times f[m,n]$) respectively. The $g_1[m,n]$ and $g_2[m,n]$ are square matrix with the orders $M \times M$ and $N \times N$ and the u_j and v_j are the column and row matrices respectively. The J is the minimum between M and N and T is the matrix transposition operator. The λ_j is the j-th eigenvalue of the minimum order matrix between $g_1[m,n]$ and $g_2[m,n]$. The $\sqrt{\lambda_j}$ is called the j-th singular value of the decomposition.

Expanding the summation, one writes the $f[m,n]$ as the J term series like $f[m,n] = \sqrt{\lambda_1} u_1 v_1 + \sqrt{\lambda_2} u_2 v_2 + \sqrt{\lambda_3} u_3 v_3 + \ldots + \sqrt{\lambda_J} u_J v_J$, each of which is a component matrix of $f[m,n]$ and there are J such components. The u_j is a column matrix of length M but the v_j is a row matrix of length N on that account the order of any component matrix $\sqrt{\lambda_j} u_j v_j$ is $M \times N$. Both the $g_1[m,n]$ and $g_2[m,n]$ are symmetric matrices and share identical eigenvalues. The larger dimension of the two inherits 0 eigenvalues (apart from the identical ones) as well as corresponding eigenvectors but we ignore them for the decomposition.

The decomposition is illustrated by the modular discrete signal $v[n] =$ [9   6   −5   4   3]. If we choose $L=2$, the $f[m,n]$ becomes $\begin{bmatrix} 6 & 9 \\ -5 & 6 \\ 4 & -5 \\ 3 & 4 \end{bmatrix}$ from earlier discussion as follows:

```
>>x=[9 6 -5 4 3]; ↵          ← x holds v[n]
>>f=mtop(x,2); ↵             ← f holds f[m,n]
```

Based on the obtained $f[m,n]$, few intermediate computations are as follows:

$$f[m,n]^T = \begin{bmatrix} 6 & -5 & 4 & 3 \\ 9 & 6 & -5 & 4 \end{bmatrix}, \quad M=4, \quad N=2, \quad J=2, \quad g_1[m,n] = f[m,n] \times f[m,n]^T =$$

$$\begin{bmatrix} 117 & 24 & -21 & 54 \\ 24 & 61 & -50 & 9 \\ -21 & -50 & 41 & -8 \\ 54 & 9 & -8 & 25 \end{bmatrix}, \; g_2[m,n] = f[m,n]^T \times f[m,n] = \begin{bmatrix} 86 & 16 \\ 16 & 158 \end{bmatrix}, \text{ the eigenvalues of}$$

$$g_1[m,n] \text{ are } \begin{bmatrix} \lambda_1 \\ \lambda_2 \\ \lambda_3 \\ \lambda_4 \end{bmatrix} = \begin{bmatrix} 82.6046 \\ 161.3954 \\ 0 \\ 0 \end{bmatrix}, \text{ the eigenvectors of } g_1[m,n] \text{ are } u_1 = \begin{bmatrix} 0.4402 \\ -0.6752 \\ 0.5447 \\ 0.2315 \end{bmatrix}, u_2$$

$$= \begin{bmatrix} 0.7910 \\ 0.3803 \\ -0.3196 \\ 0.3570 \end{bmatrix}, u_3 = \begin{bmatrix} 0.4100 \\ 0.1572 \\ 0.2324 \\ -0.8679 \end{bmatrix}, \text{ and } u_4 = \begin{bmatrix} 0.1112 \\ -0.6122 \\ -0.7397 \\ -0.2564 \end{bmatrix} \text{ for } \lambda_1 = 82.6046, \; \lambda_2 = 161.3954,$$

$\lambda_3 = 0$, and $\lambda_4 = 0$ respectively, the eigenvalues of $g_2[m,n]$ are $\lambda_1 = 82.6046$ and $\lambda_2 = 161.3954$, and the eigenvectors of $g_2[m,n]$ are $v_1 = [-0.9782 \; 0.2076]$ and $v_2 = [0.2076 \; 0.9782]$ respectively.

There are two component matrices in $f[m,n]$, first of which is

$$\text{computed} \quad \text{as} \quad \sqrt{\lambda_1} u_1 v_1 = \sqrt{82.6046} \begin{bmatrix} 0.4402 \\ -0.6752 \\ 0.5447 \\ 0.2315 \end{bmatrix} \times [-0.9782 \qquad 0.2076] =$$

$$\begin{bmatrix} 2.0862 & 9.8306 \\ 1.0029 & 4.7261 \\ -0.8430 & -3.9723 \\ 0.9416 & 4.4368 \end{bmatrix} \text{ similarly the other becomes } \sqrt{\lambda_2} u_2 v_2 = \begin{bmatrix} 3.9138 & -0.8306 \\ -6.0029 & 1.2739 \\ 4.8430 & -1.0277 \\ 2.0584 & -0.4368 \end{bmatrix}.$$

If we add the last two component rectangular matrices, we end up with the obtained $f[m,n]$.

⊟⊟ How to obtain the component matrices in MATLAB?

Section 7.6 cited **svd** helps us get the component matrices in MATLAB. The **svd** takes the $f[m,n]$ matrix as its input argument and returns three matrices say **U**, **D**, and **V**. The **U** is composed of placing the eigenvectors of $g_1[m,n]$ side by side in columns that is U=$[u_1 \; u_2 \; u_3 \; ...]$. Again the **V** is composed of placing the eigenvectors of $g_2[m,n]$ side by side in columns that is V=$[v_1^T \; v_2^T \; v_3^T \; ...]$. The **D** is a diagonal matrix whose diagonal elements are the singular values but placed in descending order. The placement of the eigenvectors and singular values are in order. From earlier discussion the $f[m,n]$ is stored in the workspace **f**. Let us see the decomposition as follows:

>>[U D V]=svd(f) ↵   ← Applying the function on the f

U =

```
-0.7910    0.4402  -0.0145  -0.4246   ← u₁ is the second column of U
-0.3803   -0.6752   0.6319  -0.0131   ← u₂ is the first column of U
 0.3196    0.5447   0.7732  -0.0571
-0.3570    0.2315   0.0512   0.9035
```

D =

12.7042	0	$\leftarrow \sqrt{\lambda_2}$ =12.7042=D(1,1)
0	9.0887	$\leftarrow \sqrt{\lambda_1}$ =9.0887=D(2,2)
0	0	
0	0	

V =

-0.2076	0.9782	$\leftarrow v_1$ is transpose of the second column$\Leftrightarrow$V(:,2)'
-0.9782	-0.2076	$\leftarrow v_2$ is transpose of the first column$\Leftrightarrow$V(:,1)'

The command V(:,1) means the first column in V, so does the other. The eigenvalues are unique but not the eigenvectors so you may see the eigenvectors with a multiplication of −1. Multiplying a eigenvector by −1 does not change the concept of eigenvalue-eigenvector. The first diagonal element D(1,1) in D corresponds to the first columns of U and V, and so does the other.

The first matrix component is obtained by $\sqrt{\lambda_1} u_1 v_1$ whose code is D(2,2)*U(:,2)*V(:,2)' and the return is as follows:

>>l1=D(2,2)*U(:,2)*V(:,2)' ↵   ← The first component of $f[m,n]$ is assigned to

l1 or l1$\Leftrightarrow \sqrt{\lambda_1} u_1 v_1$, l1 is user-chosen

l1 =

3.9138	-0.8306
-6.0029	1.2739
4.8430	-1.0277
2.0584	-0.4368

Similarly the other matrix component is computed by the command l2= D(1,1)*U(:,1)*V(:,1)' where the l2 is user-chosen. We see whether these two components' sum retrieves $f[m,n]$ as follows:

>>l=l1+l2 ↵   ← The sum of l1 and l2 is assigned to l where the l is user-chosen

l =

6.0000	9.0000	
-5.0000	6.0000	← Exactly the obtained $f[m,n]$
4.0000	-5.0000	
3.0000	4.0000	

If we have many singular values, it is not feasible that we find each component matrix one at a time. Finding the component rectangular matrix one at a time can easily be implemented by a for-loop (appendix B.4) whose code is for k=1:2 D(k,k)*U(:,k)*V(:,k)'; end. We have two component matrices that is why the for-loop counter is 2. For the summation of the component rectangular matrices, a programming artifice is needed. Initially before the for-loop we assign 0 to l where the l is user-chosen. Each time we get one component matrix handled by the for-loop counter and add that to the

I successively until the last index of the for-loop is finished. However the retrieval of the $f[m,n]$ by using a for-loop takes place as follows:

```
>>I=0; ↵
>>for k=1:2 I=I+D(k,k)*U(:,k)*V(:,k)'; end ↵
```

If you call the I at the command prompt, you should see the retrieved $f[m,n]$. At the end we can say that the given discrete signal $v[n]$ is singular value decomposed and the decomposition result is available in the matrices U, D, and V.

7.8 Singular value decomposition of a digital audio

Last section mentioned **svd** works fine up to 3000 sample long digital audio or so in our laptop. We have to understand that MATLAB workspace keeps finite memory for data manipulation. The return matrices (i.e. U, D, and V) from **svd** for long digital audio become so large that MATLAB stops working or prints memory related error message. For this reason we should be careful about the sample number while working on a softcopy audio file.

As an example of singular value decomposition on a digital audio, let us consider the audio **bird.wav** of section 2.7. Our objective is to singular value decompose at least part of the audio and view the singular value plot.

Let the machine read the audio as follows:

```
>>[v,fs]=wavread('bird.wav'); ↵
```

Therefore the last section mentioned discrete signal $v[n]$ is stored in the workspace **v** from which we find the number of samples by:

```
>>N=numel(v) ↵
```

N =
  19456

So many samples can not be handled for earlier mentioned reason. We wish to choose the first 2500 samples of $v[n]$ which we obtain from **v** by the command **v(1:2500)**. Figure 2.2(a) shows the plot of the digital audio in which the selected portion of the audio corresponds approximately from 0 to 0.25×10^4 in the horizontal scale. However we assign the audio portion to some variable **y** as follows:

```
>>y=v(1:2500); ↵
```

One choice of section 7.7 cited L may be N/3. But the N should be for the y not for the v because we will be decomposing the audio portion rather than the whole audio. The number of samples held in N may not be multiple of 3 that is why we approximate the number to its nearest integer by the command **round** as follows:

```
>>N=numel(y); L=round(N/3); ↵
```

Now we need to obtain the special Toeplitz matrix $f[m,n]$ from **y** and based on chosen L by exercising the last section mentioned function as follows:
```
>>f=mtop(y,L); ⌐         ← f holds f[m,n], f is user-chosen
```
Making the $f[m,n]$ available in **f**, we call the **svd** of last section as shown below:
```
>>[U D V]=svd(f); ⌐
```
Therefore the first 2500 samples of the digital audio **bird.wav** are singular value decomposed and the decomposition result is stored in matrices **U**, **D**, and **V** where the symbols have section 7.6 quoted meanings. It took few seconds to obtain the decomposition although 2500 samples are chosen.

Referring to the workspace browser (section 1.4), we find that the matrix size of **D** is 1668×833 meaning $L=833$ or 833 singular values in the audio portion. The diagonal elements of **D** are the singular values (section 7.6) and stored in descending order. If we exercise the command **diag** on **D**, we obtain the diagonal values of **D** as a column matrix and do so as follows:
```
>>S=diag(D); ⌐
```

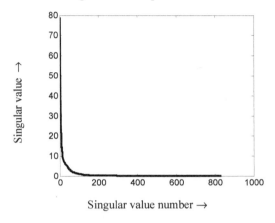

Figure 7.1(b) Singular value characteristic of first 2500 samples of the digital audio **bird.wav**

In the last execution we have the singular values assigned to **S** as a column matrix where the **S** is a user-chosen variable. Should the reader graph the singular values, appendix D cited **plot** is exercised as follows:
```
>>plot(S) ⌐
```
We see the graph of the figure 7.1(b) from the last execution. Vertical axis of the graph shows the singular values and the horizontal axis shows singular value number which is from 1 to 833. The graph says that the singular values are monotonically decreasing in nature. By inspection out of 833 singular values, we have negligible singular values approximately after 200 singular values.

⊟⊟ Some application of singular values

Singular value characteristic helps us detect the voice activity in a digital audio or sequence. The audio (section 2.7) **b.wav** is the speech signal for the English letter b and **nosound.wav** is just recorded sound in a room without any audio signal, both of which are of short duration. The two audio signals we treat as voiced and unvoiced respectively. In speech signal literature this sort of study is known as VAD – voice activity detection. Our objective is to graph the singular value characteristics of these two audio in a single plot so that the characteristic assists in VAD.

First you get the two audio through the link mentioned in section 2.7 and place the two files in your working path of MATLAB by using windows explorer. Let the machine read the two audio as follows:

>>[v1,fs]=wavread('b.wav'); ⏎   ← v1 holds $v[n]$ for b.wav, v1 is user-chosen
>>[v2,fs]=wavread('nosound.wav'); ⏎ ← v2 holds $v[n]$ for nosound.wav, v2 is
                                                                                                       user-chosen

If each sound has to be heard, the commands **wavplay(v1,fs)** and **wavplay(v2,fs)** can be exercised at the command prompt (section 2.3) respectively. Referring to the workspace browser (look for **v1** and **v2** matrix size), each audio contains 2800 samples. Then we obtain ongoing N and L for each audio as follows:

>>N=numel(v1); L=round(N/3); ⏎

Next we get the special Toeplitz $f[m,n]$ for each audio by:

>>f1=mtop(v1,L); ⏎   ← f1 holds $f[m,n]$ for b.wav, f1 is user-chosen
>>f2=mtop(v2,L); ⏎   ← f2 holds $f[m,n]$ for nosound.wav, f2 is user-chosen

Figure 7.1(c) Singular value characteristics of voiced and unvoiced digital audio – right side figure

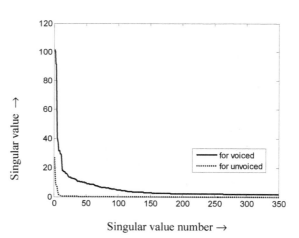

Singular value number →

After that singular value decomposition (each takes few seconds) is performed on each audio as follows:

>>[U1 D1 V1]=svd(f1); ⏎   ← Singular value decomposition of the first audio
>>[U2 D2 V2]=svd(f2); ⏎   ← Singular value decomposition of the second audio

In above executions the 1 and 2 refer to the first and second digital audio respectively where the **U1**, **D1**, etc are user-chosen variables. The singular values are stored as diagonal elements in the matrices **D1** and **D2** respectively. Only the diagonal from each is picked up by:

>>S1=diag(D1); ⏎ ← S1 holds singular values of **b.wav**, S1 is user-chosen
>>S2=diag(D2); ⏎ ← S2 holds singular values of **nosound.wav**, S2
is user-chosen

From the workspace browser there are 833 singular values in S1 or S2 wherefrom we intend to compare the first 350 for each. For the comparison, we generate a row matrix of integers ranging from 1 to 350 as follows:

>>n=1:350; ⏎   ← n holds the integers, n is user-chosen

Appendix D explained **plot** graphs the two characteristics together as shown in figure 7.1(c) by the following commands:

>>plot(n,S1(n),'k-',n,S2(n),'k-.') ⏎
>>legend('for voiced','for unvoiced') ⏎   ← Just to put a label in the graph

In the last executions, the command S1(n) inside the **plot** means the first 350 singular values out of S1, which holds 833 ones. Similar explanation goes for the S2(n). The command 'k-' means graphics curve line is black (understood by k) and continuous (understood by -). The graphics curve line indication is placed under a quote. The command **legend** adds label to the drawn graphics in which our chosen texts **for voiced** and **for unvoiced** differentiate the singular value characteristics respectively. The chosen texts are written under a quote.

What can we say about the characteristics of the figure 7.1(c)? For the unvoiced audio the singular value drops sharply compared to that of the voiced one. Starting singular value of the voiced audio is higher than that of the unvoiced one. We infer that starting singular value of the unvoiced one may serve as the threshold for VAD in a speech signal.

7.9 Covariance of random variables

In a rectangular matrix usually the columns represent the observations on random variables. The number of columns is taken as the number of random variables. Mean, median, or variance of a random variable describes the information about the variable itself. If we have two or more random variables placed in a rectangular matrix, covariance provides a relationship between the random variables or about their tendency to vary together rather than independently. The covariance of a matrix is another

matrix. Elements of the covariance matrix $V = \begin{bmatrix} V_{11} & V_{12} & \cdots & V_{1N} \\ V_{21} & V_{22} & \cdots & V_{2N} \\ \vdots & \vdots & \ddots & \vdots \\ V_{N1} & V_{N2} & \cdots & V_{NN} \end{bmatrix}$ of a

matrix $A = \begin{bmatrix} A_{11} & A_{12} & \cdots & A_{1N} \\ A_{21} & A_{22} & \cdots & A_{2N} \\ \vdots & \vdots & \ddots & \vdots \\ A_{M1} & A_{M2} & \cdots & A_{MN} \end{bmatrix}$ is defined as $V_{ij} = \dfrac{\sum_{k=1}^{M}(A_{ki} - \overline{A_i})(A_{kj} - \overline{A_j})}{M-1}$

where $A_1 = \begin{bmatrix} A_{11} \\ A_{21} \\ \vdots \\ A_{M1} \end{bmatrix}$, $A_2 = \begin{bmatrix} A_{12} \\ A_{22} \\ \vdots \\ A_{M2} \end{bmatrix}$ ......, and $A_N = \begin{bmatrix} A_{1N} \\ A_{2N} \\ \vdots \\ A_{MN} \end{bmatrix}$ and $\overline{A_j}$ is the mean of the

j^{th} column of A. The diagonal elements of V are the variances and the off-diagonal elements are the covariances that is why the V is called variance-covariance matrix. Another name of the V is dispersion matrix and it is a symmetric matrix. Orders of A and V are $M \times N$ and $N \times N$ respectively. To evaluate the covariance of a matrix, we use the command **cov**(matrix name) – (abbreviation for the <u>covariance</u>).

Let us consider $A = \begin{bmatrix} -2 & 6 & 6 \\ 4 & 30 & 1 \\ 1 & -4 & -4 \\ 5 & 0 & 5 \end{bmatrix}$ for elucidation where $M = 4$ and N

$= 3$. The order of V must be 3×3 and V prescribes the matrix form
$\begin{bmatrix} V_{11} & V_{12} & V_{13} \\ V_{21} & V_{22} & V_{23} \\ V_{31} & V_{32} & V_{33} \end{bmatrix}$. Since the V is a symmetric matrix, we have $V_{12} = V_{21}$,

$V_{13} = V_{31}$, and $V_{23} = V_{32}$. The means are $\overline{A_1} = 2$, $\overline{A_2} = 8$, and $\overline{A_3} = 2$. Some

intermediate computations are $V_{11} = \dfrac{\sum\limits_{k=1}^{4}(A_{k1} - \overline{A_1})^2}{4-1} = \dfrac{(-2-2)^2 + (4-2)^2 + (1-2)^2 + (5-2)^2}{4-1} =$

10, $V_{22} = \dfrac{\sum\limits_{k=1}^{4}(A_{k2} - \overline{A_2})^2}{4-1} = \dfrac{(6-8)^2 + (30-8)^2 + (-4-8)^2 + (0-8)^2}{4-1} = \dfrac{696}{3} = 232$, $V_{21} = V_{12} =$

$\dfrac{\sum\limits_{k=1}^{4}(A_{k1} - \overline{A_1})(A_{k2} - \overline{A_2})}{4-1} = \dfrac{(-2-2)(6-8) + (4-2)(30-8) + (1-2)(-4-8) + (5-2)(0-8)}{4-1} = \dfrac{40}{3} = 13.3333$, V_{31}

$= V_{13} = \dfrac{\sum\limits_{k=1}^{4}(A_{k1} - \overline{A_1})(A_{k3} - \overline{A_3})}{4-1} = \dfrac{(-2-2)(6-2) + (4-2)(1-2) + (1-2)(-4-2) + (5-2)(5-2)}{4-1} = -1$, and

eventually $V = \begin{bmatrix} 10 & 40/3 & -1 \\ 40/3 & 232 & 6 \\ -1 & 6 & 62/3 \end{bmatrix} = \begin{bmatrix} 10 & 13.3333 & -1 \\ 13.3333 & 232 & 6 \\ -1 & 6 & 20.6667 \end{bmatrix}$ which is what

we intend to find. The implementation is shown below:

for numeric form:
```
>>A=[-2 6 6;4 30 1;1 -4 -4;5 0 5]; ↵
>>V=cov(A) ↵
```

for rational form:
```
>>V=cov(sym(A)) ↵
```

V =

10.0000	13.3333	-1.0000
13.3333	232.0000	6.0000
-1.0000	6.0000	20.6667

V =

[10,	40/3,	-1]
[40/3,	232,	6]
[-1,	6,	62/3]

As you see the implementation can be of two types – numeric and symbolic. In numeric implementation we first assign the given matrix to workspace A which is a user-chosen variable. In the second line we call the function **cov** on A and assign the return to V which is also user-chosen. The rational form output is also achievable through the use of function **sym** on A which we also presented alongside the numeric computation. **Note that** later version of MATLAB may not work with the **sym**.

If the V_{ij} division takes place by M i.e. $V_{ij} = \frac{1}{M}\sum_{k=1}^{M}(A_{ki} - \overline{A_i})(A_{kj} - \overline{A_j})$, we

apply the command V=cov(A,1) where the second input argument 1 of **cov** is reserved and indicates the division by the M.

The **cov** is designed for handling many random variables. In the statistics literature, the random variable observations are given separately. For instance the two observations are given as $X = [-2 \quad 4 \quad 1 \quad 5]$ and $Y = [6$

$30 \quad -4 \quad 0]$ contrarily the covariance is defined as $c = \frac{1}{N}\sum_{k=1}^{N}(X_k - \overline{X})(Y_k - \overline{Y})$

where the c is covariance between the random variables X and Y and $\overline{X}$ and $\overline{Y}$ are their means respectively. For the given X and Y observation data, we should obtain $c = 10$.

Attached on the right side textbox is its implementation. In the implementation the first two lines are to assign the given X and Y data to the workspace X and Y respectively. In the third line there are three input arguments of the **cov** now, the given two vector assignees and division indicatory number respectively – another option implanted in **cov**. Therefore the

> Covariance of X and Y :
> >>X=[-2 4 1 5]; ↵
> >>Y=[6 30 -4 0]; ↵
> >>V=cov(X,Y,1); ↵
> >>c=V(1,2) ↵
>
> c =
>   10

outcome is a 2×2 matrix (because of two random variables) from which we pick up only the element indexed by (1,2) and which is assigned to c as far as our requirement is concerned.

7.10 Covariance of digital audio

In a digital audio the audio amplitudes become the random variable. There can be three situations – covariance between two audio, covariance of a single audio data arising from Toeplitz matrix, and covariance among multiple audio.

⊟⊟ Covariance between two audio

The two audio must be identical in length and if they are not, shorter one must be padded with zeroes. As an example our objective is to determine the covariance coefficients of the audio b.**wav** and nosound.**wav** of section 2.7.

Get the two audio and let the machine read (section 2.3) the two audio as follows:

```
>>[v1,fs]=wavread('b.wav'); ↵
>>[v2,fs]=wavread('nosound.wav'); ↵
```

The two audio are kept in the workspace **v1** and **v2** (user-chosen variables) respectively, both as a column matrix. These two variables refer to the X and Y of the last section. Fortunately the two audio have identical sample number (use **length** on each i.e. **length(v1)**) which is 2800 so we call the last section mentioned **cov** as follows:

```
>>V=cov(v1,v2,1); ↵
>>c=V(1,2) ↵
```

```
c =
          -1.3281e-004
```

Therefore the covariance between the two audio from above return is -1.3281×10^{-4}.

🗗🗗 Covariance of a single audio due to Toeplitz structure

In this sort of problem we have one audio and form a special Toeplitz matrix in accordance with the section 7.8. That Toeplitz matrix is section 7.7 mentioned $f[m,n]$ or last section mentioned A.

As an example we wish to find the variance-covariance matrix of audio **b.wav** by considering a Toeplitz with L=3 where L has section 7.7 cited meaning.

First machine reading of the audio is done by:

```
>>[v,fs]=wavread('b.wav'); ↵        ← v holds v[n] of section 7.7
>>f=mtop(v,3); ↵                    ← f holds f[m,n] of section 7.7
```

In above execution the **f** is a user-chosen variable and that is also A so call the **cov** as:

```
>>V=cov(f) ↵
```

```
V =
       0.0367    0.0337    0.0300
       0.0337    0.0367    0.0337
       0.0300    0.0337    0.0367
```

Above return says that the variance-covariance matrix V due to the Toeplitz matrix is $\begin{bmatrix} 0.0367 & 0.0337 & 0.03 \\ 0.0337 & 0.0367 & 0.0337 \\ 0.03 & 0.0337 & 0.0367 \end{bmatrix}$.

🗗🗗 Covariance of multiple audio

Here we have to have more than two audio and each audio has to be identical in length. Zero padding is performed on unequal length audio. For example we intend to determine the variance-covariance matrix of the audio **b.wav, nosound.wav,** and **pitch.wav**.

Obtain the three audio and let the machine read them as follows:

```
>>[v1,fs]=wavread('b.wav'); ↵
>>[v2,fs]=wavread('nosound.wav'); ↵
>>[v3,fs]=wavread('pitch.wav'); ↵
```

The three audio samples as a column matrix exist in workspace **v1, v2,** and **v3** respectively where the variables are user-chosen. Then we need the sample number of each audio which is obtained by the command **length:**

>>N1=length(v1); N2=length(v2); N3=length(v3); ⏎ ← N1, N2, N3 user-chosen

In last executions the variables **N1, N2,** and **N3** hold the sample number in the audio respectively. Just to see the sample numbers we form a row matrix as follows:

>>[N1 N2 N3] ⏎

ans =

     2800   2800   6616   ←Three sample numbers respectively

Obviously the third sample number is the largest so padding should be in accordance with the third one and we employ the **padarray** function of appendix B.12 for doing so (assuming padding is done at the end of the less sample number audio):

>>v1n=padarray(v1,[N3-N1,0],0,'post'); ⏎ ← v1n holds the padded v1, v1n is
                                                 user-chosen
>>v2n=padarray(v2,[N3-N2,0],0,'post'); ⏎ ← v2n holds the padded v2, v2n is
                                               user-chosen

Now the three column matrices **v1n, v2n,** and **v3** are identical in length and we form a three column matrix (appendix B.3) which will be A of the last section:

>>A=[v1n v2n v3]; ⏎   ← A holds A , A is user-chosen
>>V=cov(A) ⏎   ← Calling cov

V =

      0.0155 -0.0001 -0.0001
     -0.0001   0.0002 -0.0000
     -0.0001 -0.0000   0.5001

Easily we read off the variance-covariance matrix of the three audio as $V =$

$$\begin{bmatrix} 0.0155 & -0.0001 & -0.0001 \\ -0.0001 & 0.0002 & -0.0000 \\ -0.0001 & -0.0000 & 0.5001 \end{bmatrix}$$. Some element in V is seen as -0.0000 meaning

within four decimal accuracy that is 0 but that may not be zero in higher digit accuracy. The element location is (row,column)=(2,3) which we may call by:

>>V(2,3) ⏎

ans =

    -2.0879e-005   ← It indicates -0.0000 is -2.0879×10^{-5}

7.11 Principal component analysis

If N random variables form a data matrix A of order $M \times N$ (whose rows represent M observations on each variable), the matrix A can be transformed to N orthogonal random variables. The first few transformed variables (say P out of N) will bear nearly all the information possessed by the given N variables. The P transformed variables are termed as the principal components of A. Since $P < N$, the analysis is a dimension reduction approach. In most cases the given N random variables are correlated but the principal components hold orthogonality or uncorrelatedness.

The first step of the analysis is to have the variance-covariance matrix V from A (as introduced in section 7.9). The second step is to decompose the V as $E \times D \times E^T$ where the D is a diagonal matrix containing ordered eigenvalues (section 7.5 for the eigenvalues) of V and $E = [E_1 \ E_2 \ E_3 \ \cdots \ E_N]$. The E_1, E_2, E_3, $\cdots$ E_N are the normalized column eigenvectors of V.

The function **princomp** (abbreviation for the <u>prin</u>cipal <u>comp</u>onent) decomposes the given matrix data A to $E \times D \times E^T$ which is required from the analysis and with the four output argument syntax **[E Q D S]=princomp(A)**. The four output arguments indicated by **[E Q D S]**, out of which the second **(Q)** and the fourth **(S)** outputs are called the component scores and Hotelling's T^2 respectively. We excluded their descriptions. The first **(E)** and third **(D)** outputs correspond to the theory-mentioned E and D respectively. All variables like **E, Q**, etc are user-chosen.

Let us consider $A = \begin{bmatrix} 21 & 5 \\ 18 & 7 \\ 25 & 9 \\ 20 & 5 \\ 22 & 7 \end{bmatrix}$ where $M = 5$ and $N = 2$. Applying the

technique of section 7.9, we obtain $V = \begin{bmatrix} \frac{67}{10} & \frac{13}{5} \\ \frac{13}{5} & \frac{14}{5} \end{bmatrix}$. Concerning the section 7.5,

we have the eigenvalues and normalized eigenvectors of V as $\frac{3}{2}$, 8 and

$\begin{bmatrix} \frac{1}{\sqrt{5}} \\ -\frac{2}{\sqrt{5}} \end{bmatrix}$, $\begin{bmatrix} \frac{2}{\sqrt{5}} \\ \frac{1}{\sqrt{5}} \end{bmatrix}$ respectively. Having known the eigenvalues and eigenvectors,

one writes $D = \begin{bmatrix} 8 & 0 \\ 0 & \frac{3}{2} \end{bmatrix} = \begin{bmatrix} 8 & 0 \\ 0 & 1.5 \end{bmatrix}$ and $E = \begin{bmatrix} \frac{2}{\sqrt{5}} & \frac{1}{\sqrt{5}} \\ \frac{1}{\sqrt{5}} & \frac{2}{\sqrt{5}} \end{bmatrix} = \begin{bmatrix} 0.8944 & 0.4472 \\ 0.4472 & -0.8944 \end{bmatrix}$

hence the required decomposition is brought about. Its implementation is shown below:

```
>>A=[21 5;18 7;25 9;20 5;22 7]; ↵     ← Assigning A to A, A is user-chosen
>>[E Q D S]=princomp(A); ↵            ← Calling the function
>>E ↵                                 ← E⇔ E

E =
```

$$
\begin{matrix}
0.8944 & -0.4472 \\
0.4472 & 0.8944 \\
\uparrow & \uparrow \\
E_1 & E_2
\end{matrix}
$$

>>D ↲   ← D⇔ D

D =

$$
\begin{matrix}
8.0000 & \qquad\qquad & \leftarrow P=1 \\
1.5000 & \qquad\qquad & \leftarrow P=2
\end{matrix}
$$

Note that the eigenvectors are not unique. Multiplying an eigenvector with a negative sign does not make any difference. The variance is a measure of the information content of a random variable. Since the diagonal elements of D are the variances of the columns of A, it can be said that $\dfrac{8}{8+1.5}=84.21\%$ of the information contained in A is retained due to the transformation even if the second component is ignored (means setting 1.5 to 0).

⊟⊟ Principal components of a discrete signal

Suppose we have the discrete signal $f[n]=[21 \quad 18 \quad 25 \quad 20 \quad 22 \quad 5 \quad 7 \quad 9 \quad 5 \quad 7]$. We wish to analyze the signal based on principal components.

First the reader has to rearrange the $f[n]$ to a rectangular matrix form A. One option can be column by column arrangement. Another option can be semi Toeplitz structure like the section 7.7. Since most audio data storage takes place as a column matrix form, we consider the first option. Suppose the $f[n]$ is to be rearranged as earlier mentioned A which is $\begin{bmatrix} 21 & 5 \\ 18 & 7 \\ 25 & 9 \\ 20 & 5 \\ 22 & 7 \end{bmatrix}$ and where $M=5$ and $N=2$. The values of M and N are user-chosen.

The number of elements in $f[n]$ is 10 and the product of M and N is also so. Sometimes the product of M and N may not be equal to the number of elements in $f[n]$. In that case first we choose the M. Say the total number of elements in $f[n]$ is T (apply **length** function). Remainder after integer division (say R) of T by M is obtained by the **rem** of appendix B.8. If the R is 0, the T is divisible by M otherwise not. Integer division (say Q) of T/M is found by the **fix** of appendix B.8. Then we pad the $f[n]$ by $M-R$ zeroes by the **padarray** of appendix B.12. After that the padded $f[n]$ is rearranged to take the rectangular matrix size $M\times(Q+1)$ by using the **reshape** of appendix B.20.

As an example the $f[n]$ with $M=4$ should be $A=\begin{bmatrix} 21 & 22 & 5 \\ 18 & 5 & 7 \\ 25 & 7 & 0 \\ 20 & 9 & 0 \end{bmatrix}$ which we obtain by the following:

```
>>f=[21 18 25 20 22 5 7 9 5 7]; ↵      ← f[n] is assigned to f, f is user-chosen
>>M=4; T=length(f); ↵                  ← M⇔ M , T⇔ T
>>Q=fix(T/M); R=rem(T,M) ↵             ← Q⇔ Q , R⇔ R
```

R =
    2

Above return says that the R is not equal to 0 so the rest executions are the following:

```
>>fn=padarray(f,[0,M-R],0,'post'); ↵  ← fn holds the padded f, fn is user-chosen
>>A=reshape(fn,M,Q+1) ↵                ← reshape return is assigned to A, A⇔ A
```

A =

21	22	5
18	5	7	← The matrix we are after
25	7	0
20	9	0

Now you may call **princomp** on A as conducted in earlier executions.

⧉⧉ Principal components of a digital audio

We wish to compute the principal components of digital audio **b.wav** (section 2.7) considering M =600.

Let the machine read (section 2.3) the audio as follows:

```
>>[v,fs]=wavread('b.wav'); ↵
```

So the workspace **v** and **fs** hold the audio samples as a column matrix and sampling frequency respectively. The content of **v** is basically $f[n]$ so discrete signal mentioned executions follow immediately:

```
>>f=v; ↵                  ← Assigning v to f to exercise with the same variable
>>M=600; T=length(f); ↵   ← M⇔ M , T⇔ T
>>Q=fix(T/M); R=rem(T,M) ↵ ← Q⇔ Q , R⇔ R
```

R =
        400   ← R is not equal to 0 so 0 padding is essential

Earlier discrete signal happens to be as a row matrix but audio samples in **f** are as a column matrix so we carry out the following with the same symbology:

```
>>fn=padarray(f,[M-R,0],0,'post'); ↵ ← fn holds the padded f
>>A=reshape(fn,M,Q+1); ↵             ← reshape return is assigned to A, A⇔ A
>>[E Q D S]=princomp(A); ↵           ← Calling for principal component analysis
>>D ↵                                ← D⇔ D
```

D =

0.1187	← P =1	
0.0308	← P =2	
0.0109	← P =3	← Principal components
0.0031	← P =4	
0.0023	← P =5	

We bring an end to the chapter with this discussion.

Chapter 8

 ## Statistics and Digital Audio

The subject matter in this chapter is to address the very basics of statistical model based audio processing. Statistical estimation of a given audio needs many parameters, some of which is seen in chapter 2. Other elementary statistical parameters required either for feature extraction or enhancement of audio are addressed in this chapter. These parameters are termed as the properties of a digital audio and can be graphical or quantitative calculating. Our brief introduction aims at focusing the following:

- ❖ ❖ Histogram based computing and graphing for discrete random signals along with digital audio
- ❖ ❖ Probability and its density function with regard to digital audio
- ❖ ❖ Higher order statistics of a digital audio e.g. moment
- ❖ ❖ Correlation or autocorrelation of discrete random signals and digital audio

8.1 Histogram of an integer type discrete signal

By definition histogram is the frequency of occurrence of a random variable versus the random variable. Suppose we have the following integer data:

15   15   10   12   15   15   12   10   9   9   7   15   10   −1

The integers involved are −1, 7, 9, 10, 12, and 15 and their frequencies of occurrence are 1, 1, 2, 3, 2, and 5 respectively. The plot of frequencies 1, 1, 2, 3, 2, and 5 versus the integer data −1, 7, 9, 10, 12, and 15 is called the histogram. The total number of frequencies is 14 so as a percentage the elements are occupying 1/14, 1/14, 2/14, 3/14, 2/14, and 5/14 or 7.14%, 7.14%, 14.29%, 21.43%, 14.29%, and 35.71% respectively.

The built-in MATLAB function **tabulate** helps us find the frequencies related to a histogram. First we enter the given integer data as a row matrix as follows:

>>I=[15 15 10 12 15 15 12 10 9 9 7 15 10 -1]; ↵

The workspace I is a user-chosen variable and holds the given data. We call the function with I as the input argument:

>>tabulate(I) ↵

Value	Count	Percent
-1	1	7.14%
7	1	7.14%
9	2	14.29%
10	3	21.43%
12	2	14.29%
15	5	35.71%

As the return displays, there are three columns in the table – integers, frequencies of occurrence, and frequencies as a percentage respectively. We could have assigned above tabular data to some user-supplied variable T as follows:

>>T=tabulate(I); ↵

One selects the three columns in T by writing the commands T(:,1), T(:,2), and T(:,3) respectively. Let us select and assign the integer and the frequency columns as follows:

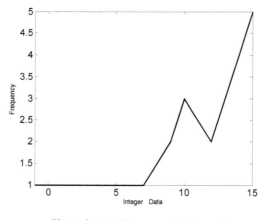

Figure 8.1(a) Histogram of integer data

>>x=T(:,1); ↵ ← x is a user-chosen variable, holds integers as a column matrix
>>y=T(:,2); ↵ ← y is a user-chosen variable, holds frequencies as a column matrix

Appendix D explained **plot** graphs the histogram like the figure 8.1(a) as follows:

>>plot(x,y) ↵

The horizontal and the vertical axes of the figure 8.1(a) refer to integer data and frequency respectively.

The reader might ask why we need to study the histogram of integer type data. Well in digital storage we store the audio data as fixed number of levels conventionally integer levels for example 256 levels ranging from 0 to 255. As implemented for integer data, a histogram reveals which integer level range shows more concentrations in a digital audio.

8.2 Histogram of a discrete signal with range

In section 8.1 we explained the histogram for integer type data which is usually applicable when we already have digitized data in fixed level numbers. Prior to level conversion often the audio data is in range form for what reason we need to study the histogram of a discrete signal which shows range type variations.

Let us consider the following data:

$$-0.9 \quad 0.8 \quad -0.65 \quad -0.6 \quad -0.53 \quad -0.5 \quad -0.1$$
$$-0.1 \quad -0.1 \quad 0.4 \quad 0.7 \quad -0.9 \quad 0.12$$

The reader decides the number of uniform data slots within the minimum and the maximum of the data. By inspection the minimum and the maximum in the data are −0.9 and 0.8 respectively. Let us consider 5 data slots so the data range in each slot should be $(0.8-(-0.9))/5=0.34$. Now if any given single data is called as X, the first slot variation should be between −0.9 and −0.9+0.34 or −0.56 i.e. $-0.9 \le X \le -0.56$, the second slot variation should be between −0.56 and −0.56+0.34 or −0.22 i.e. $-0.56 < X \le -0.22$, and so on. In a tabular form we present the histogram of the given data as follows:

Slot and slot range	Data in the slot	Frequency or number of data
First: $-0.9 \le X \le -0.56$	−0.9, −0.65, −0.6, −0.9	4
Second: $-0.56 < X \le -0.22$	−0.53, −0.5	2
Third: $-0.22 < X \le 0.12$	−0.1, −0.1, −0.1, 0.12	4
Fourth: $0.12 < X \le 0.46$	0.4	1
Fifth: $0.46 < X \le 0.8$	0.8, 0.7	2

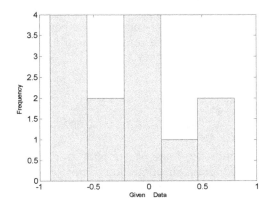

Figure 8.1(b)
Histogram of range type data – right side figure

MATLAB built-in function hist determines the histogram related frequency starting from the given data. One of the syntaxes we apply is user-supplied variable for frequency=hist(given data as a row or column matrix, user-supplied number of slot ranges or bins). The return from the hist is the slot range frequencies as a row matrix.

For the data we presented earlier, the number of slot ranges or bins is 5. Let us enter the given data to some variable I as follows:
>>I=[-0.9 0.8 -0.65 -0.6 -0.53 -0.5 -0.1 -0.1 -0.1 0.4 0.7 -0.9 0.12]; ↵

Then we call the hist with just quoted syntax as follows:
>>f=hist(I,5) ↵   ← f is user-chosen variable, holds frequencies as a row matrix

f =
    4   2   4   1   2

The usage of the command f=hist(I,5); does not display the frequencies. The command hist also keeps graphing option but in that case you do not need the output argument and we call it for the earlier data as follows:
>>hist(I,5) ↵

The result from above execution is the histogram plot of the figure 8.1(b). In the histogram now you find the horizontal and vertical axes as the data range and the frequency respectively. There are 5 range slots and you find 5 vertical bars in figure 8.1(b) accordingly. Horizontal axis of the figure covers $-0.9 \leq X \leq 0.8$.

The hist also keeps another option which is the center point of each slot range. In earlier mentioned data range table, the first range is $-0.9 \leq X \leq -0.56$ and the midpoint of the range is $(-0.9-0.56)/2=-0.73$. Continuing for the other slots we obtain the other four midpoints as -0.39, -0.05, 0.29, and 0.63 respectively.

This time the hist needs two output arguments, one for the frequency and the other for the midpoint i.e. the syntax is [user-supplied variable for frequency, user-supplied variable for midpoint]=hist(given data as a row or column matrix, user-supplied number of slot ranges or bins). Let us execute the following for the center point of the range slots:
>>[f,M]=hist(I,5) ↵   ← f or M is user-chosen variable

f =
    4   2   4   1   2   ← f holds frequencies
M =
    -0.7300 -0.3900 -0.0500 0.2900 0.6300   ← M holds midpoints

The returns to f and M are both as a row matrix. If the reader wishes to see the histogram in continuous line form rather than the bar form of figure 8.1(b), appendix D mentioned plot is exercised as plot(M,f) at the command prompt.

8.3 Histogram of a digital audio

The last two sections are the prerequisite for this section. Actually this section is the application of last section cited histogram.

As an example we wish to obtain the histogram of the digital audio **bird.wav** of section 2.7.

First let the machine read the audio (section 2.3) as follows:
>>[v,fs]=wavread('bird.wav'); ⏎

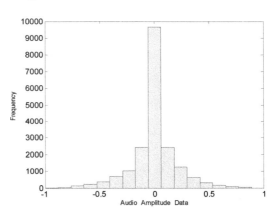

Figure 8.1(c)
Histogram of the
audio bird.wav
with 16 range slots
– right side figure

From sections 2.3-2.5 we know that the digital audio data as a column matrix and the sampling frequency are held in the variables **v** and **fs** respectively. Section 2.6 also says that the digital audio data lies in the range $-1 \le X \le 0.8906$. The range data exercise of section 8.2 is perfectly applicable here. Now we need to choose the number of slots or bins in the range say 16. Histogram frequencies of the audio **bird.wav** are then obtained by:
>>f=hist(v,16); ⏎   ← f is user-chosen variable, holds frequencies as a row matrix

By calling the **f** at the command prompt, you find the frequencies as 27, 47, 133, 206, 373, 680, 1048, 2401, 9671, 2404, 1274, 625, 301, 151, 83, and 32. Each slot spacing we then obtain as (0.8906–(–1))/16=0.1182. The range slots of the digital audio are as follows:

-1 to $-1+0.1182$ or -0.8818 i.e. $-1 \le X \le -0.8818$ for the 1st slot,

-0.8818 to $-0.8818+0.1182$ or -0.7636 i.e. $-0.8818 < X \le -0.7636$ for the 2nd slot,

$\vdots$

$0.8906-0.1182$ or 0.7724 to 0.8906 i.e. $0.7724 < X \le 0.8906$ for the last slot.

In order to see the histogram of the **bird.wav** we exercise the following:
>>hist(v,16) ⏎

Figure 8.1(c) depicts the histogram of the **bird.wav** with 16 range slots as a result of the last execution. Since we chose 16 slots, there are 16 vertical bars in the figure. If you inspect the figure, significant amplitudes of the audio are about 0 amplitude and concentrate one slot before and after the 0 amplitude slot.

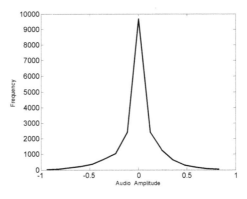

If the reader wishes to obtain the midpoints of the slots, following command is exercised:

```
>>[f,M]=hist(v,16); ↵
```

By calling the **M** at the command prompt, you find the midpoints as −0.9409, −0.8228, −0.7046, −0.5864, −0.4683, −0.3501, −0.2319, −0.1138, 0.0044, 0.1226, 0.2407, 0.3589, 0.4771, 0.5952, 0.7134, and 0.8315 respectively.

Figure 8.1(d) Continuous sense histogram of the audio bird.wav with 16 range slots

Also the command **plot(M,f)** shows the continuous sense histogram of the audio like the figure 8.1(d).

8.4 Probability and digital audio

A digital audio is a discrete signal with finite number of levels or variations. In audio coding applications we need the probability of digital audio data. Statistical modeling of audio signals also needs the probability of some particular level or range which we address in this section.

❖❖ When digital audio amplitudes are in integer level

In section 8.1 we explained that any discrete signal containing integer type data has the percentage of occurrence in the given data. That percentage is basically the probability when a digital audio is represented by integer type data.

The user has to decide first how many audio amplitude levels or integer type variations should be used. If the given audio data is in range form, one transforms the audio data to integer form by applying the section 3.2 mentioned techniques.

As an example we wish to determine the probability of amplitudes for the softcopy audio **bird.wav** considering 256 amplitude level variations.

Get the softcopy audio (section 2.7) in your working path of MATLAB and let the machine read (section 2.3) the audio as follows:

```
>>[v,fs]=wavread('bird.wav'); ↵
```

From sections 2.3-2.5 we know that the digital audio data as a column matrix and the sampling frequency are held in the variables **v** and **fs** respectively. Section 2.6 cited **datastats** shows that the audio amplitudes are in the range [−1,0.8906]. In 256 levels customarily we take the integer variation from 0 to 255 and which we obtain by the section 3.2 cited techniques as follows:

>>I=round(255*mat2gray(v)); ↵   ← I is a user-chosen variable

Therefore the **I** in above execution holds the converted audio amplitudes as a column matrix in integer range [0,255]. In order to know the probability, we exercise the **tabulate** of section 8.1 as follows:

>>T=tabulate(I); ↵   ← T is a user-chosen variable

From above execution, the **T** holds three columns − integer audio amplitude levels, frequency of occurrence, and percentage of occurrence or probability respectively. Let us pick up the integer audio amplitudes and the probability by following execution:

>>A=T(:,1); ↵ ← A is a user-chosen variable, holds integer audio amplitudes as a
column matrix

>>p=T(:,3); ↵ ← p is a user-chosen variable, holds percent probabilities of the
integer audio amplitudes as a column matrix

A problem appears in doing so which is the probability theoretically should be between 0 and 1. The data in the last **p** is out of 100 so we need to divide each data by 100 and assign to the same variable as follows:

>>p=p/100; ↵ ← This p holds probabilities between 0 and 1

Not necessarily all 256 audio amplitude levels will be present which is easily perceived by looking at the matrix size of the **A** or **p**. Workspace browser (section 1.4) says that the size is **<230×1>** meaning only 230 levels are present out of the 256.

The next question is how to see the probability of a particular amplitude level. The data in **A** and **p** is in order meaning first element in **p** for the first element in **A** and we may view that by **p(1)** or **A(1)**, similarly second element in **p** for the second element in **A** and we may view that by **p(2)** or **A(2)**, and so on.

As an example let us call the 56th element in both by the following:

>>A(56) ↵   >>p(56) ↵

ans =   ans =
66   0.0011

Above execution indicates that the probability of amplitude level 66 is 0.0011.

❖ ❖ When digital audio amplitudes are in range form

Often the given digital audio data is in range form which we elaborately explained in section 8.2. For the range type data the reader has to decide the number of range slots over the given audio data range. Usually we seek the probability for each range slot in such audio.

We apply the **hist** function for finding the frequencies of each range slot audio data. The **hist** returns only the frequencies. One divides the frequencies by the total number of audio samples to obtain the probability of each slot.

As an example we wish to determine the probability of range slots for the softcopy audio **bird.wav** considering 16 slots.

Let us carry out the following:
```
>>[v,fs]=wavread('bird.wav'); ↵      ← Machine reading of the audio with earlier
                                        mentioned symbol meanings
>>f=hist(v,16); ↵                    ← f is a user-chosen variable
```

In just conducted execution the **f** holds frequencies of the range slots as a row matrix. The command **length** determines the number of audio samples in **v** so the probabilities as a row matrix we obtain by:
```
>>p=f/length(v); ↵              ← p is a user-chosen variable
```

Therefore the **p** in last command line holds the probabilities. Recall that (section 8.3) with 16 slots the first range slot of the audio data is $-1 \leq X \leq -0.8818$ which refers to the first element in **p** or **p(1)**. Again the second range slot of the audio data is $-0.8818 < X \leq -0.7636$ which refers to the second element of **p** or **p(2)**, and so on. Just to see the first range slot probability we call the following at the command prompt:
```
>>p(1) ↵
```

ans =
   0.0014

i.e. The audio data in slot $-1 \leq X \leq -0.8818$ has the probability 0.0014.

8.5 PDF and digital audio

A probability density function (pdf) is basically for a continuous signal or random process. Our digital audio data does not fit to the definition due to discrete nature of the data yet discrete form of the pdf is helpful for probability based coding and feature extraction of audio signals.

By definition pdf is the frequency versus observation variable divided by the area of the curve so that the area under the pdf curve is always 1 obviously in continuous sense. In order to discretize the pdf we first determine the histogram and its area. We then divide each frequency by the area. By shape histogram and discretized pdf are similar. Horizontal axes of the two curves are identical but the vertical axis values are different.

Area finding of histogram curve is obtained by the function **trapz** which requires the syntax **trapz**(center points of the histogram bar as a row/column matrix, frequencies of the histogram as a row/column matrix). As always our audio data can be of two types – integer and range. The functions we need are explained in sections 8.1 and 8.2 respectively.

✦ ✦ When digital audio amplitudes are in integer level

Suppose we have the integer type audio data as illustrated in section 8.1. We wish to obtain the discrete version pdf of the data.

Reexecute the following four command lines which are taken from section 8.1:

```
>>I=[15 15 10 12 15 15 12 10 9 9 7 15 10 -1];  ↵
>>T=tabulate(I);  ↵
>>x=T(:,1);  ↵
>>y=T(:,2);  ↵
```

In the last execution workspace variables **x** and **y** hold the integers and the frequencies of the data as a column matrix respectively. Now we determine the area under the histogram by:

```
>>A=trapz(x,y);  ↵          ← A is a user-chosen variable, A holds the area
```
The horizontal and vertical axes data of the pdf will be in **x** and **y**/A respectively. Appendix D cited **plot** is exercised to graph the pdf as follows:

```
>>plot(x,y/A)  ↵
```

Figure 8.2(a) is the discretized pdf of the data, horizontal and vertical axes of which are the integer variable and area normalized frequency respectively.

Figure 8.2(a)
Discretized pdf of the integer data – right side figure

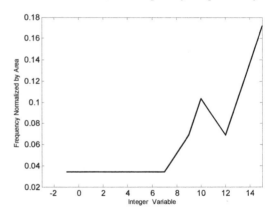

✦ ✦ When digital audio amplitudes are in range form

Concerning sections 8.2 and 8.3, the softcopy digital audio **bird.wav** data is in range form. We wish to obtain the discrete version pdf of the digital audio considering 16 range slots.

We reexecute the following two command lines from section 8.3:

```
>>[v,fs]=wavread('bird.wav');  ↵
>>[f,M]=hist(v,16);  ↵
```
From last executions, the workspace variables **f** and **M** hold the frequencies and range slot midpoints both as a row matrix respectively. Area under the histogram we obtain by:

```
>>A=trapz(M,f);  ↵          ← A is a user-chosen variable, A holds the area
```

Area normalized frequencies are given by f/A so the discretized pdf we graph by the command:

>>plot(M,f/A) ↵

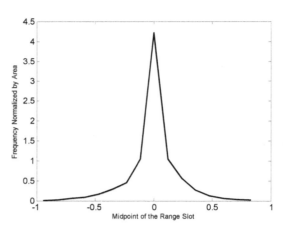

Figure 8.2(b)
Discretized pdf of the
audio bird.wav – right
side figure

Figure 8.2(b) depicts the discretized pdf of audio bird.wav. The horizontal and vertical axes of the figure 8.2(b) are the range slot midpoints and area normalized frequencies respectively.

8.6 Entropy and digital audio

Entropy gives a measure of information content by audio amplitudes in terms of the minimum number of bits per symbol required to encode the amplitude. Suppose a random variable X possesses N states given as $[x_1, x_2, x_3, \cdots, x_N]$ with state probabilities $[p_1, p_2, p_3, \cdots, p_N]$ where p_i is the probability of x_i, the entropy of X is defined as $H = -\sum_{i=1}^{N} p_i \log_2 p_i$ bits.

The p_i is in terms of a row or column matrix. The $p_i \log_2 p_i$ is obtained by the scalar code (.*) of appendix A and the summation ($\sum$) takes place by the command **sum** of appendix B.11.

For a digital audio, audio amplitude is the random variable and audio integer level or range slot is the state.

♦ ♦ When digital audio amplitudes are in integer level

As an example we wish to find the entropy of the integer amplitude levels of the section 8.1 mentioned data.

Let us execute section 8.1 mentioned commands as follows:

>>I=[15 15 10 12 15 15 12 10 9 9 7 15 10 -1]; ↵
>>T=tabulate(I); ↵

The T in above command line holds three columns, out of which the third column is the percentage probability (sections 8.1 and 8.4). Following the division by 100 we obtain the probabilities of the integers as follows:

>>p=T(:,3)/100; ⏎   ← p is a user-chosen variable, p holds p_i as a column matrix

>>H=-sum(p.*log2(p)) ⏎   ← H is a user-chosen variable,

-sum(p.*log2(p)) executes $-\sum_{i=1}^{N} p_i \log_2 p_i$

H =
　　　2.3527

The H in above implementation retains the entropy value which is 2.3527 bits.

✦✦ When digital audio amplitudes are in range form

We intend to calculate the entropy of audio **bird.wav** considering 16 range slots.

How probability of range data is calculated is illustrated in sections 8.2 and 8.3. Following three command lines are taken from the sections:

>>[v,fs]=wavread('bird.wav'); ⏎

>>[f,M]=hist(v,16); ⏎

>>p=f/length(v); ⏎

The workspace variable **p** in the last execution holds the probabilities of the range slots as a row matrix. Knowing so, we compute the entropy as follows:

>>H=-sum(p.*log2(p)) ⏎

H =
　　　2.5180

Therefore the softcopy audio **bird.wav** has the entropy 2.518 bits.

✦✦ Handling the 0 probability level

The function **tabulate** sometimes returns inbetween integer levels with 0 probability. The 0 probability attributes to computing with flaw or printing error message because logarithm of zero is minus infinity which machine can not handle. Under this circumstance we add a very small quantity called epsilon which has the MATLAB code **eps**. The **eps** should be added with the probability vector wherever the logarithm is applied. For example we could have written H=-sum(p.*log2(eps+p)) to determine the entropy for the integer level probabilities.

Not only with the integer level probabilities, 0 probabilities may occur in the case of range data like in audio **bird.wav**. Even for the range type data we apply the command H=-sum(p.*log2(eps+p)) too.

8.7 Moment and digital audio

The Pth central moment of a discrete random process or signal $X[n]$ is defined as $M_X^P = \frac{1}{N} \sum_{n=0}^{N-1} (X[n] - m)^P$ where the $X[n]$ is assumed to exist over $0 \le n \le N-1$, N is the total sample number in $X[n]$, m is the average of $X[n]$ i.e. $m = \frac{1}{N} \sum_{n=0}^{N-1} X[n]$, and P is the order of the moment.

The built-in function **moment** computes the Pth moment with the syntax **moment**(discrete random process or signal as a row/column matrix, P).

❖❖ Moment example on a discrete signal

For example let us compute the 4th moment of $X[n] = [-1 \quad 2 \quad 90 \quad 34]$. Relevant calculations are $m = \frac{-1+2+90+34}{4} = 31.25$ and $M_X^4 =$ $\frac{(-1-31.25)^4 + (2-31.25)^4 + (90-31.25)^4 + (34-31.25)^4}{4} = 3431764.52$ which we wish to obtain.

Its straightforward computation is as follows:
```
>>X=[-1 2 90 34]; ↵        ← X is a user-chosen variable that holds X[n]
>>M=moment(X,4) ↵          ← M is a user-chosen variable that holds M_X^4

M =
        3.4318e+006        ← It indicates the moment is 3.4318×10^6
```

❖❖ Moment example on a digital audio

We intend to calculate the 4th moment of **bird.wav** audio samples (section 2.7).

Let the machine read (section 2.3) the digital audio samples as follows:
```
>>[v,fs]=wavread('bird.wav'); ↵
```

From sections 2.4-2.5 we know that the digital audio data as a column matrix and the sampling frequency are held in the variables **v** and **fs** respectively. The **v** basically holds the moment definition cited discrete signal $X[n]$. The moment computation then takes place by:
```
>>M=moment(v,4) ↵          ← M is a user-chosen variable that holds M_X^4

M =
        0.0109
```
Above return says that the 4th moment of audio **bird.wav** (i.e. M_X^4) is 0.0109.

❖❖ Moment with sampling frequency consideration

Last two illustrations assume that the sampling frequency has no relationship with the moment. If the sampling frequency sensitivity is required, the moment expression is modified as $M_X^P = \frac{f_s}{N} \sum_{n=0}^{N-1} \left(\frac{X[n]}{f_s} - m \right)^P$ where the f_s is the sampling frequency.

For the **bird.wav** example, we would have executed the command as **M=fs*moment(v/fs,4)** and which would have returned the 4th order moment of the audio as 1.0159×10^{-15} obviously the sampling frequency information should be available to **fs** followed by the reading of the audio.

8.8 Correlation and digital audio

Correlation of two discrete signals gives a measure of the similarity or dependency of their variations in time or space that is how much the two signals are different or alike. Mostly the measure of correlation is through coefficients which we address in the following.

♦ ♦ Correlation coefficient of two discrete signals

The correlation coefficient ρ of two discrete signals or random processes $X[n]$ and $Y[n]$ is defined as
$$\rho = \frac{\frac{1}{N}\sum_{k=0}^{N-1}(X[k]-\overline{X})(Y[k]-\overline{Y})}{\sqrt{\frac{1}{N}\sum_{i=0}^{N-1}(X[i]-\overline{X})^2}\sqrt{\frac{1}{N}\sum_{i=0}^{N-1}(Y[i]-\overline{Y})^2}}$$

where the $\overline{X}$ is average value of $X[n]$ samples, N is the number of samples in $X[n]$, and the integer n variation covers $0 \le n \le N-1$ for N given samples. Similar symbol meanings also hold true for the $Y[n]$. It needs to be stressed that both signals are identical in sample number i.e. exist over $0 \le n \le N-1$.

MATLAB built-in function **corrcoef** computes the correlation coefficient with the syntax **corrcoef(** $X[n]$ as a row or column matrix, $Y[n]$ as a row or column matrix) but the function is devised to handle multiple signals so the outcome from the **corrcoef** is in general a square symmetric matrix. For 2 signals, 3 signals, etc, the return matrix size is 2×2, 3×3, etc respectively. Suppose the return matrix is R then for the 2 signals' case, we take only the element whose position indexes or coordinates in the matrix is (1,2). In other words we call the element R(1,2) for the ρ.

As an example, two discrete signals are given as $X[n]=[-2 \quad 4 \quad 1 \quad 5]$ and $Y[n]=[6 \quad 30 \quad -4 \quad 0]$. Based on these two signals earlier defined ρ should be 0.2768 which we wish to compute.

The step by step execution is the following:

```
>>X=[-2 4 1 5]; ↵    ← X is a user-chosen variable that holds X[n] as a row matrix
>>Y=[6 30 -4 0]; ↵   ← Y is a user-chosen variable that holds Y[n] as a row matrix
>>R=corrcoef(X,Y); ↵    ← R is a user-chosen variable that holds the return
                                       matrix of size 2×2
>>r=R(1,2) ↵            ← r is a user-chosen variable, r⇔ρ


r =
        0.2768
```

When the two signals $X[n]$ and $Y[n]$ are unequal in length, we pad the shorter length one by zeroes. For example $X[n]=[-2 \quad 4 \quad 1 \quad 5]$ and $Y[n]=[6 \quad 30]$ should be handled by considering $Y[n]=[6 \quad 30 \quad 0 \quad 0]$. Given that the correlation coefficient of the $X[n]$ and the padded $Y[n]$ is 0.2657 which we intend to calculate.

The zero padding is carried out by the **padarray** of appendix B.12 for many sample signals. The sample number in each signal is found by the command **length** so for the unequal length signal we exercise the following:

>>X=[-2 4 1 5]; ↵   ← X is a user-chosen variable that holds $X[n]$ as a row matrix
>>Y=[6 30]; ↵   ← Y is a user-chosen variable that holds $Y[n]$ as a row matrix
>>N1=length(X); ↵ ← N1 is a user-chosen variable that holds length of $X[n]$
                                which is 4
>>N2=length(Y); ↵ ← N2 is a user-chosen variable that holds length of $Y[n]$
                                which is 2

Now the padding should be by 2 zeroes or **N1-N2**. Since the entered signals are in row matrix form and we wish to pad afterwards, the required row index in the padding should be 0 and the post padding is done by:

>>Yn=padarray(Y,[0,N1-N2],'post'); ↵ ← Yn is a user-chosen variable that
                                holds padded Y
>>R=corrcoef(X,Yn); ↵   ← R is a user-chosen variable that holds the return
                                matrix of size 2×2
>>r=R(1,2) ↵   ← r is a user-chosen variable, r⇔ ρ

r =
        0.2657

♦ ♦ Correlation coefficient of two digital audio

Starting from two softcopy audio we may need to find the correlation coefficient ρ of the digital signals. Given that the ρ is -0.0031 for the audio **bird.wav** and **hello.wav** (section 2.7), we wish to verify the correlation coefficient.

As a first step, obtain the two softcopies and let the machine read (section 2.3) the audio as follows:

>>[v1,fs]=wavread('bird.wav'); ↵ ← v1 is a user-chosen variable
>>[v2,fs]=wavread('hello.wav'); ↵ ← v2 is a user-chosen variable

From the last executions, the two audio each as column matrix are held in the workspace variables **v1** and **v2** respectively. Obviously the **v1** and **v2** are the correlation definition mentioned $X[n]$ and $Y[n]$ respectively. Next what we need is the sample number in each audio. We may see the workspace browser (section 1.4) or apply the following commands in order to know the sample number:

>>N1=length(v1) ↵   ← N1 is a user-chosen variable that holds length of $X[n]$

N1 =
        19456
>>N2=length(v2) ↵   ← N2 is a user-chosen variable that holds length of $Y[n]$

N2 =
        50388

From the last two executions, clearly the **N2** is greater than the **N1** so the padding would be by **N2-N1** zeroes and to the signal stored in **v1**. Since the held signals in column matrix form, we need padding only in the row direction consequently the column index of the **padarray** function should be 0 as exercised in the following:

>>vn=padarray(v1,[N2-N1,0],'post'); ↲ ← vn is a user-chosen variable that
                                                            holds padded v1
>>R=corrcoef(vn,v2); ↲   ← R is a user-chosen variable that holds the
                                            correlation matrix return of size 2×2
>>r=R(1,2) ↲   ← r is a user-chosen variable, r⇔ ρ

r =
            -0.0031

8.9 Autocorrelation and digital audio

Autocorrelation $r[k]$ of a finite random process or discrete signal $X[n]$ is defined as

$r[k]=\sum_{n=-(N-1+k)}^{N-1-k} X[n]X[n+k]$ without scaling,

$r[k]=\frac{1}{N}\sum_{n=-(N-1+k)}^{N-1-k} X[n]X[n+k]$ with biased scaling, and

$r[k]=\frac{1}{N-|k|}\sum_{n=-(N-1+k)}^{N-1-k} X[n]X[n+k]$ with unbiased scaling

where the integer interval of $X[n]$ covers $0 \leq n \leq N-1$ (i.e. signal sample existence), the number of samples in $X[n]$ is N, and the variation of k is $-(N-1) \leq k \leq (N-1)$. There are $2N-1$ samples in $r[k]$ with length of $X[n]$ being N.

MATLAB built-in function **xcorr** provides the autocorrelation with the syntax **xcorr**($X[n]$ as a row or column matrix, indicatory reserve word under quote for autocorrelation type). The autocorrelation return from the **xcorr** is a row or column matrix depending on the input matrix type. The reserve words are **biased** and **unbiased** for the biased and the unbiased scalings of $r[k]$ respectively. The default type is without scaling for which no reserve word is necessary. Let us see the following examples in this regard.

◆ ◆ **Example 1**

The discrete signal $X[n]=[-2\ \ 4\ \ 1\ \ 5]$ provides the autocorrelation $r[k]=\sum_{n=-3-k}^{3-k} X[n]X[n+k]=[-10\ \ 18\ \ 1\ \ 46\ \ 1\ \ 18\ \ -10]$ without scaling which we wish to compute.

The autocorrelation is computed as follows:
>>X=[-2 4 1 5]; ↲ ← X is a user-chosen variable that holds $X[n]$ as a row matrix
>>r=xcorr(X) ↲   ← r is a user-chosen variable that holds $r[k]$ as a row matrix

r =
        -10.0000 18.0000 1.0000 46.0000 1.0000 18.0000 -10.0000

MATLAB just returns $r[k]$ values, the k information is completely lost. It is the user who keeps a mark on the k. There are four samples in $X[n]$ (i.e. $N=4$) so the n and the k intervals are $0 \le n \le 3$ and $-3 \le k \le 3$ respectively.

✦✦ Example 2

One property of the autocorrelation $r[k]$ is that it is symmetric about the $k=0$ over $-(N-1) \le k \le N-1$. The $r[k]$ so obtained is also called two sided autocorrelation. Sometimes in order to avoid redundancy, we consider only one side or the right half of the autocorrelation i.e. considering the interval $0 \le k \le N-1$.

Obviously example 1 mentioned autocorrelation is two sided. We wish to obtain one sided autocorrelation of the $r[k]$ i.e. $r[k] = \sum\limits_{n=0}^{3-k} X[n]X[n+k]$

$=[46 \quad 1 \quad 18 \quad -10]$.

In example 1 the two sided $r[k]$ is stored in the workspace **r**. Position indexes in **r** are all integers and start from 1 so the indexes from N to $2N-1$ in **r** refer to the one sided $r[k]$ which we obtain as follows:

      >>r1=r(4:7) ↵   ← r1 is a user-chosen variable that holds one sided
                                                     $r[k]$ as a row matrix

      r1 =
              46.0000   1.0000 18.0000 -10.0000

The 4 samples in $X[n]$ indicate $N=4$ so the position indexes needed are 4 to 7 which we call by **4:7** (section 1.3).

✦✦ Example 3

Two sided autocorrelation with unbiased scaling of the example 1 mentioned $X[n]$ is $r[k] = \dfrac{1}{4-|k|} \sum\limits_{n=-3-k}^{3-k} X[n]X[n+k] = [-10 \quad 9 \quad 0.3333 \quad 11.5$

$0.3333 \quad 9 \quad -10]$ which we easily calculate by ongoing function and symbology as follows:

      >>X=[-2 4 1 5]; ↵
      >>r=xcorr(X,'unbiased') ↵

      r =
              -10.0000 9.0000 0.3333 11.5000 0.3333 9.0000 -10.0000

✦✦ Example 4

Autocorrelation plotting may take place in discrete sense for which we apply the **stem** of appendix D. Example 3 mentioned autocorrelation we wish to plot.

As explained earlier, MATLAB returns only $r[k]$ not k. The k variation should be over $-3 \le k \le 3$. In order to graph the autocorrelation, we execute the following:

      >>k=-3:3; ↵   ← k is a user-chosen variable that holds integers as a row matrix
                                      over $-3 \le k \le 3$

      >>stem(k,r) ↵   ← Calling stem for the graph where the r is taken from the
                                      example 3

Figure 8.3(a) shows the outcome from the last execution. The horizontal and vertical axes of the graph correspond to k and $r[k]$ respectively. By inspection the $r[k]$ value at $k = -3$ coincides with the vertical axis. The graph is better perceived if we change the axis setting for example over $-4 \leq k \leq 4$ and $-15 \leq r[k] \leq 15$. The command **axis** helps us do so with the syntax **axis**(bounds as a four element row matrix) where the row matrix is formed by [minimum k maximum k minimum $r[k]$ maximum $r[k]$]] therefore we exercise the following for the example intervals:

>>axis([-4 4 -15 15]) ↵

The graph is not shown for the space reason. Certainly the graph will be perceived better.

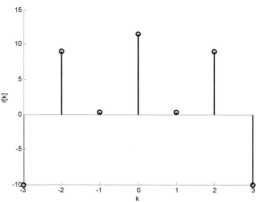

Figure 8.3(a) Plot of two sided autocorrelation of the example 3 mentioned $X[n]$

✦✦ Example 5

We may choose many samples of a discrete signal and assume the signal as a continuous one without the loss of generality. In this example we generate a discrete sine wave which has the equation $X[n] = 3\sin\dfrac{n\pi}{10}$ over $0 \leq n \leq 20$. The autocorrelation of the sine wave without scaling is to be computed and both the sine wave and its autocorrelation are to be graphed in continuous sense in the same window.

Let us generate the n values over $0 \leq n \leq 20$ as follows:

>>n=0:20; ↵   ← n is a user-chosen variable that holds integers as a row matrix

$X[n]$ values for different n are calculated by the scalar code (appendix A):

>>x=3*sin(pi*n/10); ↵   ← x is a user-chosen variable that holds $X[n]$ values
as a row matrix

The autocorrelation $r[k]$ of $X[n]$ without scaling is computed by:

>>r=xcorr(x); ↵   ← r is a user-chosen variable that holds $r[k]$ values
as a row matrix

Over $0 \leq n \leq 20$ there are 21 samples so $r[k]$ domain will be over $-20 \leq k \leq 20$ hence we generate the k values as a row matrix by:

```
>>kn=-20:20; ↵
```
← kn is a user-chosen variable that holds k values
as a row matrix

The main graphing here is $X[n]$ versus $0 \le n \le 20$ and $r[k]$ versus $-20 \le k \le 20$ in the same window. One of many solutions can be the use of **subplot** of appendix D. Let us say we plot the first graph on top of the other. Treating each graph as an element we have a matrix of graphs of size 2×1 so the **subplot** window control command should be 211 and 212.

Continuous sense graph is drawn by the **plot** of appendix D. Following executions we have to conduct for viewing the outcome as shown in figure 8.3(b):

```
>>subplot(211),plot(n,x) ↵
>>subplot(212),plot(kn,r) ↵
```

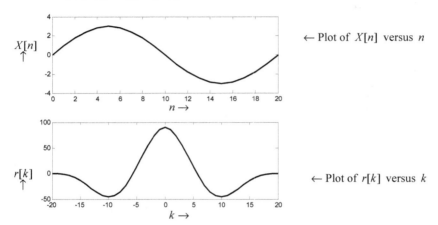

← Plot of $X[n]$ versus n

← Plot of $r[k]$ versus k

Figure 8.3(b) Plot of the discrete sine function and its autocorrelation in continuous sense

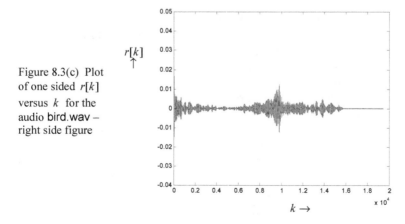

Figure 8.3(c) Plot of one sided $r[k]$ versus k for the audio bird.wav – right side figure

-162-

♦ ♦ Autocorrelation of a digital audio

We wish to compute and graph the autocorrelation of audio **bird.wav** (section 2.7).

Obtain the audio in your working path of MATLAB and let the machine read the audio as follows:

>>[v,fs]=wavread('bird.wav'); ⏎   ← v is a user-chosen variable, $v \Leftrightarrow X[n]$

From sections 2.3-2.5 we know that the digital audio data as a column matrix and the sampling frequency are held in the variables **v** and **fs** respectively. This **v** holds the data of autocorrelation theory mentioned $X[n]$. Now we have to decide the type of autocorrelation for example without scaling and let us find it as follows:

>>r=xcorr(v); ⏎   ← r is a user-chosen variable which holds $r[k]$ of the audio
bird.wav

From the sample number information the N is 19456 so the n interval is $0 \le n \le 19455$ and the k interval is $-19455 \le k \le 19455$, involvement of so many integers in the intervals.

If we are interested about one sided autocorrelation of the audio, we may select the elements from the **r** whose indexes are from 19456 to 2×19456−1 or 38911 i.e. **r(19456:38911)**.

If we were interested about one sided autocorrelation with unbiased scaling, the command would have been:

>>r=xcorr(v,'unbiased'); ⏎
>>r1=r(19456:38911); ⏎   ← r1 is a user-chosen variable that holds one sided
                              unbiasedly scaled $r[k]$ of bird.wav over
                              $0 \le k \le 19455$

If we graph the discrete plot like example 4, the graph may not be perceived better for so many samples. Continuous sense graph like example 5 is perceived better so we exercise the following:

>>k=0:19455; ⏎
>>plot(k,r1) ⏎

Figure 8.3(c) depicts the graph of one sided $r[k]$ versus k for the audio **bird.wav**.

That brings an end to this chapter.

Mohammad Nuruzzaman

Chapter 9

 ## Linear Prediction and Digital Audio

Quantitatively linear prediction model forecasts the future value of a discrete signal. The prediction is viewed as the linearly weighted superposition of the past signal samples. The model so obtained serves the purpose of having the spectral envelop of the time domain data. This theory plays an important role in the model-based spectral analysis of the digital audio signals. MATLAB embedded functional tools are very handy in the analysis. However considered topics for linear prediction demonstration are the following:

- ❖ ❖ Linear prediction basics along with the computing tools
- ❖ ❖ Modeling of the linear prediction in the Z transform domain
- ❖ ❖ Frequency response based on the linear prediction model
- ❖ ❖ Digital audio linkage with the linear prediction model whether in time or transform domain

9.1 Linear prediction basics in time index domain

Given a discrete signal $f[n]$, linear prediction models or forecasts the future sample values of the signal employing linear combination of the past signal sample values. There are several types of linear prediction, we address here the very basic one.

Concerning the figure 9.1(a), a sample of the discrete signal $f[n]$ is predicted from the past samples. The number of past samples is counted for the prediction order P which is completely user-chosen and an integer. The $f[n]$ samples we assume start from $n=0$ and continue up to $n=N-1$ where N is the number of samples in $f[n]$ i.e. $0 \le n \le N-1$ with integer n. Usually the P is less than the N. If the predicted sample is $y[n]$, then the $y[n]$ is formed from the weighted linear combination of $f[n-1]$, $f[n-2]$, $f[n-3]$, $\cdots$ and $f[n-P]$ that is

$$y[n]=a_1 \; f[n-1]+a_2 \; f[n-2]+a_3 \; f[n-3]+\cdots \cdots +a_P \; f[n-P]$$

or in summation form $y[n]=\sum_{k=1}^{P} a_k f[n-k]$.

The main problem here is to determine the linear prediction coefficients $[\,a_1,a_2,a_3,\cdots,a_P\,]$ starting form $f[n]$.

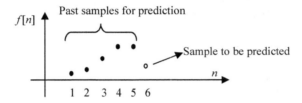

Figure 9.1(a) Linear prediction strategy in a discrete signal

In order to determine the prediction coefficients, we apply the autocorrelation vector $r[n]$ which is given by $r[n]=\dfrac{1}{N}\sum_{k=0}^{N-1} f[k]f[k-n]$. The Toeplitz matrix R and a column vector V are obtained from the $r[n]$ as follows: $R=$

$$\begin{bmatrix} r[0] & r[1] & r[2] & r[P-1] \\ r[1] & r[0] & r[1] & r[P-2] \\ r[2] & r[1] & r[0] \; \cdots\cdots & r[P-3] \\ & & \vdots & \\ r[P-1] & r[P-2] & r[P-3] & r[0] \end{bmatrix} \text{ and } V=\begin{bmatrix} r[1] \\ r[2] \\ r[3] \\ \vdots \\ r[P] \end{bmatrix}.$$ Once the R and the V

are found, the prediction coefficients as a column matrix are computed by

$$\begin{bmatrix} a_1 \\ a_2 \\ a_3 \\ \vdots \\ a_P \end{bmatrix} =-R^{-1}\,V$$ where the R^{-1} is the matrix inverse of the R.

Having found the linear prediction coefficients, the error signal due to prediction is given by $e[n]=f[n]-y[n]$. The power in the error signal $e[n]$ attributed to prediction is given by $E=r[0]+\sum_{k=1}^{P} a_k r[k]$.

❖❖ Illustration by a numerical example

Let us say we have the discrete signal $f[n]=[2 \quad 3 \quad -7 \quad 9 \quad 0 \quad 11]$. There are 6 samples in $f[n]$ so $N=6$ and the n interval is $0 \le n \le 5$. The reader has to decide the linear prediction order P say $P=4$.

With this P, the predicted signal $y[n]=a_1 f[n-1]+a_2 f[n-2]+a_3 f[n-3]+a_4 f[n-4]$ or in summation form $y[n]=\sum_{k=1}^{4} a_k f[n-k]$.

The autocorrelation vector $r[n]$ we obtain by $\frac{1}{6}\sum_{k=0}^{5} f[k]f[k-n]$ and which becomes $[44 \quad -13 \quad 18.6667 \quad -9.8333 \quad 5.5 \quad 3.6667]$ over the interval $0 \le n \le 5$ so $r[0]=44$, $r[1]=-13$, $r[2]=18.6667$, $r[3]=-9.8333$, $r[4]=5.5$, and $r[5]=3.6667$.

Our chosen prediction order P is 4 so $R = \begin{bmatrix} r[0] & r[1] & r[2] & r[3] \\ r[1] & r[0] & r[1] & r[2] \\ r[2] & r[1] & r[0] & r[1] \\ r[3] & r[2] & r[1] & r[0] \end{bmatrix} =$

$\begin{bmatrix} 44 & -13 & 18.6667 & -9.8333 \\ -13 & 44 & -13 & 18.6667 \\ 18.6667 & -13 & 44 & -13 \\ -9.8333 & 18.6667 & -13 & 44 \end{bmatrix}$ and $V = \begin{bmatrix} r[1] \\ r[2] \\ r[3] \\ r[4] \end{bmatrix} = \begin{bmatrix} -13 \\ 18.6667 \\ -9.8333 \\ 5.5 \end{bmatrix}$.

The linear prediction coefficient then becomes $\begin{bmatrix} a_1 \\ a_2 \\ a_3 \\ a_4 \end{bmatrix} = -R^{-1}V =$

$\begin{bmatrix} 0.1743 \\ -0.3972 \\ 0.0619 \\ 0.1008 \end{bmatrix}$.

The power E in the error signal is calculated as $r[0]+\sum_{k=1}^{4} a_k r[k] = 34.2646$.

9.2 Implementation of linear prediction

Depending on digital audio study, you may need $r[n]$, R, V, or prediction coefficients of section 9.1. These quantities can be implemented easily by applying built-in MATLAB functions in conjunction with little programming. Here we implement the numerical illustration on linear prediction of the last section.

The given signal $f[n]=[2 \quad 3 \quad -7 \quad 9 \quad 0 \quad 11]$ is entered as a row matrix as follows:

```
>>f=[2 3 -7 9 0 11]; ↵       ← f holds f[n] where f is user-chosen variable
```

The command **length** finds the number of samples N in $f[n]$. Also the chosen prediction order 4 is assigned to P by the following:

```
>>N=length(f); P=4; ↵       ← N⇔ N and P⇔ P
```

Mohammad Nuruzzaman

In section 8.9 we explained the two sided autocorrelation exercise with the help of the function **xcorr**. The $r[n]$ of section 9.1 is basically the one sided biased autocorrelation of section 8.9. First we find the two sided biased autocorrelation by **xcorr** and then obtain the $r[n] = \frac{1}{N}\sum_{k=0}^{N-1} f[k]f[k-n]$ from the xcorr return by choosing samples indexed from N to the last one (command N:end does so) so we execute the following:

```
>>r2=xcorr(f,'biased');  ↵    ← r2 is a user-chosen variable which holds the two
                                  sided autocorrelation vector as a row matrix
>>r=r2(N:end) ↵               ← r is a user-chosen variable which holds the one
                                  sided r[n] vector of section 9.1
```

r =

    44.0000 -13.0000 18.6667 -9.8333   5.5000   3.6667

The Toeplitz matrix R of the last section is obtained from $r[n]$ but not from all elements of the matrix and that is decided by the linear prediction order P. In time index term the n is from 0 to $P-1$ or in MATLAB index term the variation is from 1 to P because MATLAB array indexing starts from 1 instead of 0. We need to apply r(1:P) to extract $r[n]$ elements necessary for R. Appendix B.14 cited function **toeplitz** easily helps us obtain the R from the existing r as follows:

```
>>R=toeplitz(r(1:P)) ↵    ← R is a user-chosen variable which holds R
```

R =

    44.0000 -13.0000   18.6667   -9.8333
   -13.0000   44.0000 -13.0000   18.6667
    18.6667 -13.0000   44.0000 -13.0000
    -9.8333   18.6667 -13.0000   44.0000

The vector V of the section 9.1 is also obtained from the one sided $r[n]$ but from the n index 1 through P or in MATLAB index from 2 to P+1. Knowing so we get the V by:

```
>>V=r(2:P+1)';  ↵         ← V is a user-chosen variable which holds the V
```

The V has to be a column matrix but the r is a row matrix. For this reason the transposition operator is added at the end in last command line (section 1.3). Section 7.3 cited **inv** computes the inverse of the R hence the linear prediction coefficients are calculated as:

```
>>a=-inv(R)*V ↵           ← a is a user-chosen variable which holds − R^{-1} V
```

a =

    0.1743
   -0.3972
    0.0619
    0.1008

The power term $r[0]+\sum_{k=1}^{4} a_k r[k]$ is implemented by r(1)+r(2:P+1)*a where r and a are row and column matrices respectively. Also remembering the fact

-168-

that MATLAB indexing starts from 1. The part $\sum_{k=1}^{4} a_k r[k]$ is equivalent to the matrix multiplication of $r[k]$ (must be a row matrix) and a_k (must be a column matrix). The r(1) and r(2:P+1) mean the $r[0]$ and $r[1]$ through $r[P]$ respectively. However the execution is as follows:

>>E=r(1)+r(2:P+1)*a ⏎   ← E is a user-chosen variable which holds the E

    E =
            34.2646

◆ ◆ Readymade tool of MATLAB

If intermediate quantities are needed in linear prediction, just mentioned commands are exercised. If the reader says only the linear prediction coefficients are needed, then built-in function **lpc** is exercised with the syntax **lpc(** $f[n]$ as a row or column matrix, prediction order P) and the return from the function are the prediction coefficients as a row matrix. The $f[n]$ is stored in **f** so we call the function as follows:

>>c=lpc(f,4) ⏎   ← c is a user-chosen variable which holds linear
                              prediction coefficients

    c =
        1.0000   0.1743   -0.3972   0.0619   0.1008

By default the **lpc** returns the first coefficient as 1 because of normalization. If you need the coefficients other than the first one, execute **c(2:end)** and assign the picked up coefficients to some other variable as shown below:

>>b=c(2:end) ⏎   ← b is a user-chosen variable that keeps the earlier mentioned
                      coefficients as a row matrix

    b =
        0.1743   -0.3972   0.0619   0.1008

Again from the above **b**, the a_1, a_2, a_3, and a_4 are picked up by **b(1)**, **b(2)**, **b(3)**, and **b(4)** respectively.

The **lpc** also keeps provision for returning the error signal power E in that case we need two output (section 1.3) arguments – one for prediction coefficient and the other for the error power thereby requiring the syntax $[a_k, E]=$**lpc(** $f[n]$, P) and let us execute that as follows:

>>[b,E]=lpc(f,P) ⏎   ← b,E bears earlier mentioned meanings

    b =
        1.0000   0.1743   -0.3972   0.0619   0.1008
    E =
        34.2646

9.3 Linear prediction and Z transform

Linear prediction models the time domain discrete data to Z transform domain. With the symbology of section 9.1, the linear prediction coefficients $[a_1, a_2, a_3, \cdots, a_p]$ along with the error signal power E give birth of the Z transform system function $H(z)$ as:

$$H(z) = \frac{\sqrt{E}}{1 + a_1 z^{-1} + a_2 z^{-2} + a_3 z^{-3} + \ldots\ldots + a_p z^{-p}}$$

Figure 9.1(b) shows the filter structure of the $f[n]$ discrete signal modeling starting from $u[n]$ where $u[n]$ is a zero mean and unity variance random signal. The system function $H(z)$ can be written as:

$$H(z) = \frac{F(z)}{U(z)}$$

where $F(z)$ and $U(z)$ are the Z transforms of the $f[n]$ and the $u[n]$ respectively.

The system function $H(z)$ attributing to linear prediction needs $[a_1, a_2, a_3, \cdots, a_p]$ and E starting from $f[n]$.

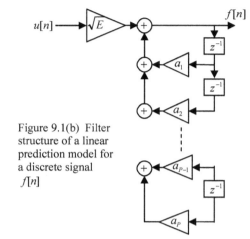

Figure 9.1(b) Filter structure of a linear prediction model for a discrete signal $f[n]$

❖❖ **An example of the model** $H(z)$

From sections 9.1 and 9.2, we obtained $a_1 = 0.1743$, $a_2 = -0.3972$, $a_3 = 0.0619$, and $a_4 = 0.1008$ and $E = 34.2646$ considering the 4th order prediction on $f[n] = [2\ \ 3\ \ -7\ \ 9\ \ 0\ \ 11]$ so the $H(z)$ can be written as

$$\frac{\sqrt{34.2646}}{1 + 0.1743z^{-1} - 0.3972z^{-2} + 0.0619z^{-3} + 0.1008z^{-4}}.$$ Let us reexecute the necessary commands of section 9.2 as follows:

```
>>f=[2 3 -7 9 0 11]; ↵     ← f holds f[n] where f is a user-chosen variable
>>P=4; ↵                    ← P holds prediction order 4, P is user-chosen
>>[b,E]=lpc(f,P); ↵         ← b holds denominator coefficients of H(z) as a row
                              matrix and E holds E
```

❖❖ **Poles and zeroes of system function** $H(z)$

Just discussed Z transform system function $H(z)$ is rearranged (multiplying the numerator and the denominator by z^4) to be written as

$$\frac{z^4 \sqrt{34.2646}}{z^4 + 0.1743z^3 - 0.3972z^2 + 0.0619z + 0.1008} =$$

$$\frac{z^4 \sqrt{34.2646}}{(z - 0.4662 - j0.3214)(z - 0.4662 + j0.3214)(z + 0.5534 + j0.0892)(z + 0.5534 - j0.0892)}$$

from which the zeroes (roots of the numerator polynomial) and the poles (roots of the denominator polynomial) of $H(z)$ are 0 of order 4 and $\{0.4662 + j\,0.3214,\ 0.4662 - j\,0.3214,\ -0.5534 - j\,0.0892,\ \text{and}\ -0.5534 + j\,0.0892\}$ respectively. We wish to obtain these poles or zeroes from $H(z)$.

Appendix B.15 cited **roots** helps us get the roots of the denominator of $H(z)$ as follows:

>>q=roots(b) ↵   ← q holds poles of $H(z)$ as a column matrix, q is user-chosen

q =
    0.4662 + 0.3214i
    0.4662 - 0.3214i
    -0.5534 + 0.0892i
    -0.5534 - 0.0892i

The zeroes of $H(z)$ are determined by the prediction order P and not returned by the **roots**. The **roots** operated on **b** only returns the poles of the $H(z)$ which are available in **q**. Consecutive poles in **q** are accessed by the commands q(1), q(2), q(3), and q(4) respectively.

It is given that the polar form poles are $0.5663\angle 34.5819^{\circ}$, $0.5663\angle -34.5819^{\circ}$, $0.5606\angle 170.8451^{\circ}$, and $0.5606\angle -170.8451^{\circ}$ respectively which we also wish to obtain.

Appendix B.6 mentioned **abs** and **angle** compute the magnitude and the phase angle in radian of the poles respectively. For viewing reason, we may form a two element row matrix by placing the two quantities side by side as follows:

>>[abs(q) rad2deg(angle(q))] ↵

ans =
    0.5663   34.5819   ← 1st pole
    0.5663   -34.5819   ← 2nd pole
    0.5606   170.8451   ← 3rd pole
    0.5606   -170.8451   ← 4th pole

If the phase angle in degree is needed, we apply the **rad2deg** command on the return from the **angle** and did so in above command line.

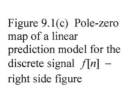

Figure 9.1(c) Pole-zero map of a linear prediction model for the discrete signal $f[n]$ – right side figure

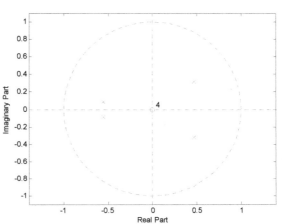

Graphical display of the poles and the zeroes is conducted by the **zplane** command of section 6.6 as follows:

```
>>zplane(sqrt(E),b) ⏎
```

As a result of the last execution, figure 9.1(c) shows the pole-zero map of the linear prediction model for the discrete signal $f[n]$. The number 4 at the center of the figure 9.1(c) indicates linear prediction order as well as order of the zeroes located at the origin.

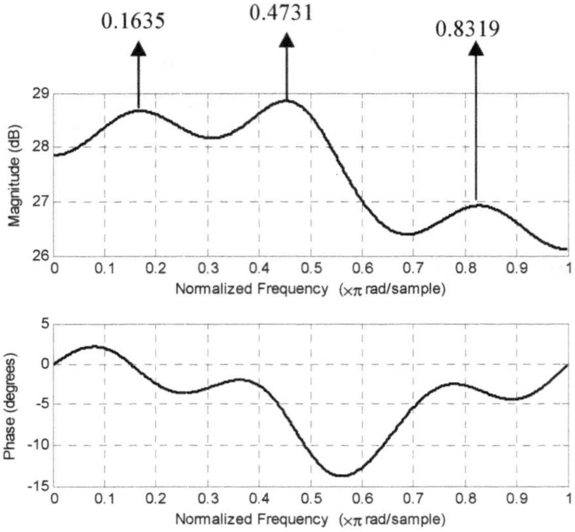

Figure 9.2(a)
Frequency
response of a linear
prediction model
$H(z)$ – right side
figure

9.4 Frequency response of linear prediction model

From the last three sections we know that the linear prediction system $H(z)$ is obtained starting from a discrete signal $f[n]$. Having found the $H(z)$, the frequency response of the model is carried out by the **freqz** of section 6.7. The frequency response means the variation of $H(e^{j\omega})$ versus ω.

✦✦ Frequency response of $H(z)$

As an example we wish to obtain the frequency response of the linear prediction model based on a prediction order 6 for the signal $f[n]=[2\ 3\ -7\ 9\ -10\ 11\ -23\ 45\ 49]$.

Following the symbology and the technique of section 9.2, we execute the following at the command prompt:

```
>>f=[2 3 -7 9 -10 11 -23 45 49]; ⏎    ← f holds f[n] as a row matrix
>>P=6; ⏎                              ← P holds the prediction order
>>[b,E]=lpc(f,P) ⏎                    ← Calling lpc of section 9.2, symbols have same
                                        meanings of the section
b =
        1.0000  -0.1262   0.0506   0.0314  -0.0196  -0.0132   0.0508
```

E =
578.3372

From the last return, one writes the linear prediction system function as $H(z)$

$$= \frac{\sqrt{578.3372}}{1-0.1262z^{-1}+0.0506z^{-2}+0.0314z^{-3}-0.0196z^{-4}-0.0132z^{-5}+0.0508z^{-6}}$$

or in ω domain $H(e^{j\omega})=$

$$\frac{\sqrt{578.3372}}{1-0.1262e^{-j\omega}+0.0506e^{-j2\omega}+0.0314e^{-3j\omega}-0.0196e^{-4j\omega}-0.0132e^{-5j\omega}+0.0508e^{-6j\omega}} \ .$$

Frequency spectrum of the $H(e^{j\omega})$ is seen by:
>>freqz(sqrt(E),b) ↵
Figure 9.2(a) shows the frequency spectrum of the linear prediction model. Basically the figure shows the variations of $20\log_{10}|H(e^{j\omega})|$ in dB and $\angle H(e^{j\omega})$ in degree against the normalized ω frequency between 0 and 1.

The $20\log_{10}|H(e^{j\omega})|$ has even symmetry about π where $0\le\omega\le 2\pi$. The interval $0\le\omega\le 2\pi$ translates to $0\le f\le f_s$ where the f and the f_s are discrete frequency and sampling frequency respectively, both in Hz. Also the ω is in radian/sec. Figure 9.2(a) depicted frequency axis covers $0\le\omega\le\pi$ or $0\le f\le\frac{f_s}{2}$ equally.

❖❖ Poles of $H(z)$ as a resonator

A complex conjugate pole pair of the linear prediction system $H(z)$ shows a local peak at the phase angle of the pole. In audio applications it has very important implication. We wish to illustrate the pole resonance phenomenon of linear prediction quantitatively. In order to do so, first we need the poles and obtain them by the last section mentioned commands as follows:
>>q=roots(b); ↵   ← q holds poles of $H(z)$ as a column matrix, q is user-chosen
>>[abs(q) rad2deg(angle(q))] ↵

ans =
```
        0.5978     29.4372
        0.5978    -29.4372
        0.6368     85.1577
        0.6368    -85.1577
        0.5919    149.7417
        0.5919   -149.7417
```
From above return we are sure that the pole pairs of the linear prediction system $H(z)$ are $0.5978\angle\pm29.4372°$, $0.6368\angle\pm85.1577°$, and $0.5919\angle\pm149.7417°$. Each of the three pairs exhibits a local peak in the magnitude spectrum in figure 9.2(a). The pole phase angle $0-180°$ is also proportional to 0-1, $0\le\omega\le\pi$, or $0\le f\le\frac{f_s}{2}$. The first pole pair phase angle is $\pm29.4372°$ which

translates to 0-1 scale as $\frac{29.4372}{180}$ =0.1635. The other two pole pairs similarly relate the normalized scale as 0.4731 and 0.8319 respectively. In figure 9.2(a) we have shown their positions in conjunction with the local peaks by an uparrow indication.

♦♦ Resonator frequencies from $H(z)$

Having known the poles of $H(z)$, the phase angles of the poles relate to the resonator frequencies when compared to the sampling frequency. Suppose the sampling frequency is f_s=16 KHz. Just mentioned pole phase angles ±29.4372°, ±85.1577°, and ±149.7417° when compared to f_s become 1.3083 KHz, 3.7848 KHz, and 6.6552 KHz respectively. We wish to obtain these resonator frequencies by programming means.

Let us get the phase angles in degrees from the poles stored in q to some user-chosen variable A:

>>A=rad2deg(angle(q)); ⏎

If we take absolute value of each angle in A, all angles in A turn to positive. The operation **abs(A)-A** turns one of the conjugate angles doubled while the other to 0. Appendix B.16 cited **nonzeros** only takes the nonzero elements from the resulting matrix due to the operation:

>>U=nonzeros(abs(A)-A)/2 ⏎   ← U holds the nonzero elements, U is a user-chosen variable

U =
        29.4372
        85.1577   ← Positive phase angle of each pole pair
       149.7417

Next we assign the sampling frequency to **fs** where **fs** is a user-chosen variable:

>>fs=16e3; ⏎

From earlier discussion the returned phase angles are normalized to 0-1 thereby making it proportional to 0-$\frac{f_s}{2}$:

>>fr=U*fs/2/180 ⏎   ← fr holds the resonance frequencies, fr is a user-chosen variable

fr =
    1.0e+003 *
        1.3083
        3.7848
        6.6552

In above return the **fr** holds the sampling frequency proportionate resonance frequencies as a column matrix. The first frequency **1.0e+003 * 1.3083** means 1.3083×10^3 Hz and so does the other. The three frequencies are accessed by the commands **fr(1)**, **fr(2)**, and **fr(3)** respectively.

9.5 Linear prediction on a digital audio

A digital audio contains numerous samples. If the length of the discrete signal $f[n]$ as introduced in previous sections increases, so does the computing burden on the machine. For this reason it is a standard practice that we take some segment of the digital audio and work on that. The reader may go through section 3.3 for frame by frame selection of a digital audio.

As an example we wish to obtain the linear prediction model parameters or related quantities for the first 5000 samples of the digital audio **bird.wav**.

Get the audio (section 2.7) and let the machine read (section 2.3) the audio as follows:

>>[v,fs]=wavread('bird.wav'); ↵

The workspace **v** and **fs** hold the digital audio samples as a column matrix and sampling frequency respectively. The first 5000 samples from the **v** as a column matrix are obtained by **v(1:5000)** and that is basically section 9.1 mentioned $f[n]$:

>>f=v(1:5000); ↵   ← f holds $f[n]$ as a column matrix

Next we have to decide linear prediction order say $P=8$ so execute the following:

>>P=8; ↵   ← P holds the prediction order
>>[b,E]=lpc(f,P) ↵   ← Calling **lpc** of section 9.2, symbols have same
                              meanings of the section

b =
    1.0000 -1.6594 0.7916 -0.0868 0.2987 -0.0938 0.0073 -0.2487   0.2616
E =
    7.3958e-004

From the above return the linear prediction model function can be written as $H(z) =$

$$\frac{\sqrt{7.3958\times10^{-4}}}{1-1.6594z^{-1}+0.7916z^{-2}-0.0868z^{-3}+0.2987z^{-4}-0.0938z^{-5}+0.0073z^{-6}-}$$

$$\overline{0.2487z^{-7}+0.2616z^{-8}} \;.$$

Whatever operation is conducted on **b** and **E** in the last three sections can be conducted on just found **b** and **E**. Not only that, you may exercise other functions like **xcorr** on **f** for obtaining $r[n]$, R, or V with identical commands if it is necessary. For space reason we did not include further examples pertaining to the digital audio prediction.

9.6 Spectrum envelope of digital audio by linear prediction

Linear prediction coefficients help us obtain the approximate magnitude spectrum shape which we plan to illustrate in this section. This property is used to extract the feature from a digital audio. There are two key elements here for obtaining magnitude spectrum – by discrete Fourier transform (section 4.1) and by linear prediction modeling (section 9.4). One important criterion is that the two spectra have to vary commonly i.e. magnitude spectrum amplitudes should be normalized between 0 and 1 for each case. Preferably the comparison of the two spectra takes place in dB scale. The horizontal axis of the spectrum may be in integer frequency index, discrete frequency in Hz, or in angular frequency in radian/sec. Due to symmetry it is also customary to view the comparison up to half variation on the horizontal axis – in any of the three forms. For graphical display in a single window appendix D mentioned **subplot** is useful in which three graphs will be displayed on top of the other – magnitude spectrum from discrete Fourier transform, magnitude spectrum attributing to linear prediction model, and both together from top to bottom respectively. Each of the three graphs uses appendix D cited command **plot** for graphing – a discrete graph in continuous sense.

As an example we wish to demonstrate the comparison for the digital audio **bird.wav** considering first 3000 samples.

Get the audio (section 2.7) and let the machine read (section 2.3) the audio as follows:

>>[v,fs]=wavread('bird.wav'); ⏎

The workspace **v** and **fs** hold the digital audio samples as a column matrix and sampling frequency respectively. The first 3000 samples from the **v** as a column matrix is obtained by **v(1:3000)** and that is basically section 4.1 mentioned $f[n]$:

>>f=v(1:3000); ⏎   ← f holds $f[n]$ as a column matrix, f is user-chosen

✦ ✦ Magnitude spectrum from discrete Fourier transform

Section 4.1 mentioned **fft** helps us get the complex discrete Fourier transform from which the magnitude spectrum is obtained by the command **abs** and section 4.8 cited technique is used to get the spectrum in decibel as follows:

>>F=abs(fft(f)); M=max(F); F1=20*log10(eps+F/M); ⏎

In above execution the **F**, **M**, and **F1** are user-chosen variables and hold the absolute magnitude spectrum $|F[k]|$, maximum value of the magnitude spectrum $|F[k]|_{max}$, and decibel magnitude spectrum $20\log_{10}\dfrac{|F[k]|}{|F[k]|_{max}}$ respectively. There are 3000 samples in the spectrum and the first half is going to be from 1 to 1500 and as a row matrix we generate the frequency index k (section 1.3) by:

>>k=1:1500; ⏎   ← k holds the frequency index integers as a row matrix where k is user-chosen

>>F2=F1(k); ⏎   ← F2 holds the first half of the spectrum where F2 is user-chosen

Treating each of the three graphs as a matrix element, we have a matrix of graphs of the size 3×1 and control to each graph is accessed by 311, 312, and

313 respectively so the half magnitude discrete Fourier transform is graphed by:

```
>>subplot(311), plot(k,F2) ↵
```

MATLAB responds with the top figure and we did not include the figure for space reason. The function **plot** keeps its own setting for axes but for better perception user-defined axis setting is entered by the command **axis** with the syntax **axis**([minimum horizontal index, maximum horizontal index, lowest dB, highest dB]) obviously the input argument as a four element row matrix. In most audio study the vertical axis dB variation from −60 to 0 is exercised and of coarse the horizontal integer frequency index varies form 0 to 1500 so we execute the following:

```
>>axis([0 1500 -60 0]) ↵
```

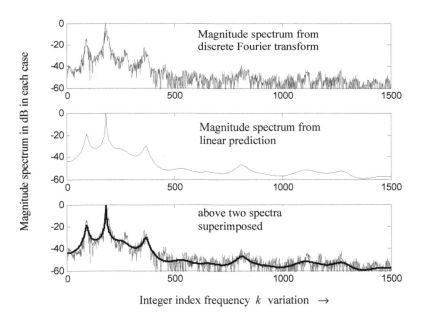

Figure 9.2(b) Magnitude spectrum envelop detection property of the linear prediction model

♦ ♦ Magnitude spectrum from linear prediction model

In order to get the magnitude spectrum due to linear prediction, first we need the system function $H(z)$. Prior to that, prediction order selection is necessary so let it be 24 then obtain the $H(z)$ parameters as follows:

```
>>[b,E]=lpc(f,24); ↵          ← Symbols have section 9.2 cited meanings
```

System function $H(z)$ denominator coefficients are available in **b** so call the function **freqz** of section 6.7 to get the $H(e^{j\omega})$ complex values for earlier decided 1500 points:

>>H=freqz(sqrt(E),b,1500); ⏎ ← H holds complex $H(e^{j\omega})$ values as a
column matrix where H is user-chosen variable
>>H1=abs(H); ⏎ ← H1 holds | $H(e^{j\omega})$ | values where H1 is user-chosen variable
>>Hd=20*log10(eps+H1/max(H1)); ⏎ ← Converting H1 held values to dB to be
in Hd where Hd is user-chosen variable
>>subplot(312), plot(k,Hd) ⏎ ← Calling the grapher for plotting, graph
response from MATLAB not shown for space reason
>>axis([0 1500 -60 0]) ⏎ ← User-defined axes setting for better perception

✦ ✦ Plotting both spectra together

Magnitude spectrum values arising from discrete Fourier transform and linear prediction are stored in the workspace variables **F2** and **Hd** respectively. We graph them together by the same **plot** function as follows:
>>subplot(313), plot(k,F2,k,Hd) ⏎
>>axis([0 1500 -60 0]) ⏎ ← User-defined axes setting for better perception

Figure 9.2(b) shows the resulting graph from all these executions. Vertical and horizontal axes of each of the three traces in figure 9.2(b) refer to the magnitude spectrum in dB (which is selected to vary between −60 and 0) and integer frequency index (which is selected to vary between 0 and 1500) respectively.

The superimposed third figure needs extra graphical manipulation. The default thickness of the linear prediction curve does not appear distinctive in the first appearance. Click the Edit plot icon (figure 1.2(c)), select the linear prediction curve by the mouse in the superimposed trace, rightclick the mouse, find the option for the line width, change the width to some larger value for instance 2, and click the Edit plot icon again to deactivate the selection.

What can we say from above analysis? Studying the first 3000 samples of the audio **bird.wav**, linear prediction model magnitude spectrum follows the trajectory of magnitude discrete Fourier transform of the audio.

9.7 Linear prediction error versus order

In section 9.1 we introduced the linear prediction error E and order P. The dependency of E on P helps us decide the prediction order. In section 9.2 we also illustrated the computing of E employing **lpc**. In this section our objective is to address the programming technique for graphing the E versus P.

As an example we wish to demonstrate the variation for the digital audio **bird.wav** considering first 5000 samples.

Get the audio (section 2.7) and let the machine read (section 2.3) the audio as follows:
>>[v,fs]=wavread('bird.wav'); ⏎

The workspace **v** and **fs** hold the digital audio samples as a column matrix and sampling frequency respectively. The first 5000 samples from the **v** as a column matrix are obtained by **v(1:5000)** and that is basically section 9.1 mentioned $f[n]$:

>>f=v(1:5000); ↵   ← f holds $f[n]$ as a column matrix

The P variation is conducted by a for-loop (appendix B.4). Let us choose the prediction order P variation from 2 to 200 with increment 1. The complete for-loop command is the following:

>>for P=2:200, [b,E]=lpc(f,P); Ev(P-1)=E; end ↵

All symbols in above execution have section 9.2 cited meanings. The for-loop counter **P** gives the control on the P. For each **P**, we call the **lpc**. Out of the two returns from the **lpc** – **b** and **E**, only the **E** is used. Each **E** is assigned as the **(P-1)**-th element to **Ev** where the **Ev** is a user-chosen variable. At the end of the for-loop, the **Ev** becomes a row matrix containing errors for P ranging from 2 to 200. Since array indexing starts from 1, **P-1** is used rather than **P**. Nevertheless the P is not stored in any variable so as a row matrix we generate that by:

>>O=2:200; ↵   ← O holds the P variation as a row matrix, O is user-chosen

Appendix D cited **plot** graphs the E versus P as follows:

>>plot(O,Ev) ↵

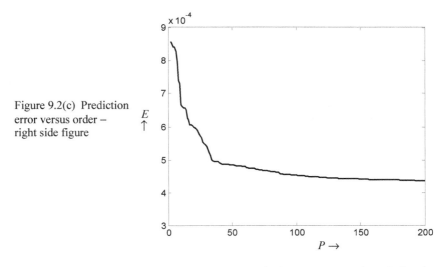

Figure 9.2(c) Prediction error versus order – right side figure

E
↑

$P \rightarrow$

Figure 9.2(c) depicts the E versus P variation as a result of the last execution. Besides normalized error is often feasible rather than absolute error which needs $\dfrac{E}{E_{max}}$ operation and which we carry out by the command **max** on (appendix B.5) Ev i.e. **Ev/max(Ev)**. Knowing so, we execute the following command for the normalized error graph:

Mohammad Nuruzzaman

>>figure,plot(O,Ev/max(Ev)) ↵

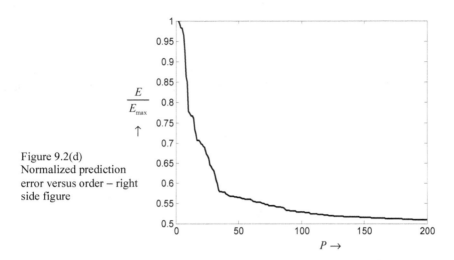

Figure 9.2(d)
Normalized prediction
error versus order – right
side figure

Figure 9.2(d) is the response of the normalized error from the last command line. The command **figure** opens a new window for the **plot**. We did so because already one graph has been conducted by the **plot**. We have chosen 5000 samples so prediction order might go up to 5000 yet only the order 100 causes the prediction error between 50% and 55% which is obvious from above normalized graph.

9.8 Linear prediction model and inverse filter

Recall that the linear prediction model (section 9.3) is given as

$H(z) = \dfrac{\sqrt{E}}{1 + a_1 z^{-1} + a_2 z^{-2} + a_3 z^{-3} + \dots\dots + a_p z^{-P}}$. An inverse filter has the system

function $G(z) = 1 + a_1 z^{-1} + a_2 z^{-2} + a_3 z^{-3} + \dots\dots + a_p z^{-P}$ which is taken from the denominator of $H(z)$. In this section our aim is to implement one important property linking linear prediction model to inverse filter which is magnitude spectra of $H(z)$ and $G(z)$ are mutually reflective. In other words the $|H(e^{j\omega})|$ is the reflection of $|G(e^{j\omega})|$ and vice versa.

Best way to implement the property is to graph the $|H(e^{j\omega})|$ and the $|G(e^{j\omega})|$ preferably in dB scale over a common ω interval.

As an example we wish to consider the first 5000 samples of the **bird.wav** of section 2.7 and the prediction order being 8.

In section 9.5 we addressed how to get the $H(z)$ for the audio. Also from the same section the denominator polynomial coefficients of the $H(z)$ are stored in **b** so the system function for the inverse filter is going to be

$$G(z) = 1 - 1.6594z^{-1} + 0.7916z^{-2} - 0.0868z^{-3} + 0.2987z^{-4} - 0.0938z^{-5} + 0.0073z^{-6} - 0.2487z^{-7} + 0.2616z^{-8}.$$

The polynomial coefficients of the $G(z)$ are also the same **b**. In section 6.7 we explained how the function **freqz** graphs and computes $H(e^{j\omega})$ but that is suitable for a single system. For more than one system we have to have the system function values and graph them on our own by the **plot** of appendix D. To do so, we need to quantify the frequency interval and to normalize each of the $|H(e^{j\omega})|$ and the $|G(e^{j\omega})|$.

Let us say the common angular frequency interval is $0 \le \omega \le \pi$. The **freqz** needs frequency step selection and let it be 0.01 rad/sec. The ω values we easily generate (section 1.3) by:

>>w=0:0.01:pi; ↵

At the ω values stored in **w**, $H(e^{j\omega})$ and $G(e^{j\omega})$ values are computed by:

>>H=freqz(sqrt(E),b,w); ↵   ← H holds the complex $H(e^{j\omega})$ values as a row
matrix, H is user-chosen variable

>>G=freqz(b,1,w); ↵   ← G holds the complex $G(e^{j\omega})$ values as a row
matrix, G is user-chosen variable

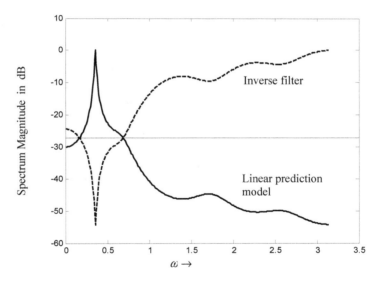

Figure 9.3(a) Image property of the inverse filter and linear prediction model

For the normalization between 0 and 1 of each spectrum we apply the section 4.8 cited techniques:

>>Hn=abs(H)/max(abs(H)); ↵   ← Hn holds $\dfrac{|H(e^{j\omega})|}{|H(e^{j\omega})|_{max}}$ values as a row
matrix, Hn is user-chosen variable

>>Gn=abs(G)/max(abs(G)); ↵   ← Gn holds $\dfrac{|G(e^{j\omega})|}{|G(e^{j\omega})|_{max}}$ values as a row

matrix, Gn is user-chosen variable

>>Hd=20*log10(eps+Hn); ↵   ← Hd holds dB values based on Hn stored
normalized values, Hd is user-chosen variable

>>Gd=20*log10(eps+Gn); ↵   ← Gd holds dB values based on Gd stored
normalized values, Gd is user-chosen variable

Having the normalized spectrum dB values available, one exercises the **plot** as follows:

>>plot(w,Hd,w,Gd) ↵

Figure 9.3(a) depicts the image property of inverse filter and linear prediction model as a result of the last line execution. Horizontal and vertical axes of the graph refer to $0 \le \omega \le \pi$ and normalized dB spectrum values respectively.

The reference line about which the image property is demonstrated one might be interested in. The average of the first elements in normalized magnitude spectrum values is the line about which the reflection occurs. The first two elements we access by the commands **Hd(1)** and **Gd(1)** respectively so the dB corresponding to the line we see by:

>>(Hd(1)+Gd(1))/2 ↵

ans =
         -27.1415

Above return says that the dB related to the reference line is −27.1415. By inspection one can easily verify the line's occurring at −27.1415 dB in the vertical axis of the figure 9.3(a).

The property so illustrated is implemented over $0 \le \omega \le \pi$. If the reader says the horizontal axis variation should be in Hertz frequency for example over $0 \le f \le 8KHz$ with a sampling frequency $16\ KHz$ and frequency step $0.05\ KHz$, we modify the computations as follows:

>>fr=0:0.05e3:8e3; ↵   ← fr holds the chosen frequency values as a row
matrix, fr is user-chosen

>>H=freqz(sqrt(E),b,fr,16e3); ↵
>>G=freqz(b,1,fr,16e3); ↵

The rest commands are the same as those conducted earlier with the exception in the **plot** and which would be **plot(fr,Hd,fr,Gd)**. The graph is not shown for the space reason.

With the above image property, we bring an end to the chapter.

Chapter 10

 Noise and Elementary
Speech Enhancement

Speech in environments such as vehicles, aircrafts, and public spaces are often mixed with noises. Even indoor generated speech may contain noise. In order to improve the quality or intelligibility of speech, enhancing algorithms are developed. Speech enhancement scope is vast owing to unpredictability and variability of noise characteristics and sophisticated algorithm needs rigorous discourse which is beyond the scope of the text. Since the text basis is hand-on computing, some fundamental speech enhancing algorithms are selected just to render a flavor in the topic. Our elementary illustration addresses the following:

❖ ❖ The concept of noise and its additive behavior in audio
❖ ❖ Basic spectral subtraction algorithm and its application to audio
❖ ❖ Wiener adaptive filtering of audio in integer time index domain
❖ ❖ Enhancing digital audio by singular value decomposition

10.1 What is noise?

In audio engineering term any unwanted audio is noise. The noise issue is profoundly related with a speech signal because of its widespread applications. Audio noise may appear from many sources. Suppose you are talking by a mobile and beside you somebody is drilling a hole so on the other side of the mobile the listener will hear your speech as well as drilling sound. For this example your voice is audio signal and drilling sound is the noise. The concepts of clean audio signal and noise are relative. To the mobile talker the drilling sound is the noise while to the person who is drilling, the mobile talking is noise. The problem is your voice and the drilling sound are mixed during transmission, storage, or propagation – in audio literature this sort of mixing is called additive noise. To improve the quality and intelligibility of speech and to reduce the communication fatigue, noise simulation, noise estimation, and noise reduction techniques are studied.

Theoretically we illustrate the additive noisy signal model by the equation $f[n] = y[n] + w[n]$ considering the discrete form where $y[n]$ is the clean audio signal, $w[n]$ is the noise signal, and $f[n]$ is the noisy signal. Going back to speech corrupted by drilling sound – speech signal without the drilling noise is $y[n]$ and only the drilling sound is $w[n]$. The problem is when we talk by the mobile, we listen $f[n]$ – neither $y[n]$ nor $w[n]$.

The challenge is how to get the $y[n]$ from $f[n]$. The difficulty arises from the fact that the noise signal is not unique and its characteristics change with place and time. Sources of noise can be many. Certainly the task is not easy that is why dozens of noise estimation and speech enhancement algorithms are devised. We will try to implement few of which and show how the algorithms improve the noisy signal quality at least slightly in the following sections – obviously the basic ones as far as the text scope is concerned.

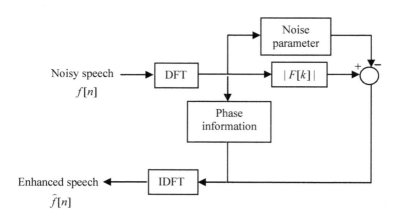

Figure 10.1(a) Block diagram of basic spectral subtraction algorithm

10.2 Basic spectral subtraction algorithm

In order to reduce the additive noise as explained in section 10.1, the first algorithm we consider is the spectral subtraction algorithm whose block diagram is seen in figure 10.1(a). The DFT and IDFT of the figure correspond to the forward discrete Fourier transform (performed by the built-in function **fft**) and inverse discrete Fourier transform (performed by the function **ifft**) respectively, details of which are seen in section 4.1.

The DFT turns the noisy $f[n]$ to complex $F[k]$ which has two components symbolically $|F[k]|$ and $\angle F[k]$. From the $F[k]$ we obtain the $|F[k]|$ and the $\angle F[k]$ by the functions **abs** and **angle** respectively (appendix B.6). The phase information $\angle F[k]$ is kept as it is returned by the **angle** in accordance with the block diagram.

The noise parameter (let us call it $D[k]$) of the figure 10.1(a) is usually some scalar and changes with time or audio frame (section 3.3). For implementation we assume some scalar value of $D[k]$. The magnitude spectrum of the recovered signal happens to be $|F[k]| - D[k]$ and the complex spectrum is formed by the expression with the magnitude $|F[k]| - D[k]$ and phase $\angle F[k]$ or mathematically $(|F[k]| - D[k])e^{j\angle F[k]}$. Inverse transform returns the $\hat{f}[n]$ from $(|F[k]| - D[k])e^{j\angle F[k]}$. The $\hat{f}[n]$ may contain some imaginary part which we ignore by taking the real part of $\hat{f}[n]$ – this is the spectral subtraction algorithm. The last obtained $\hat{f}[n]$ is the enhanced audio.

In audio system books you find the $F[k]$ as $F(\omega)$ which is the continuous spectrum. Since our demonstration is in discrete form, we prefer the $F[k]$ to $F(\omega)$. A computer always works on discrete data never with continuous one that is why our notation is so.

As a modular example, it is given that $\hat{f}[n]$=[4.9342   5.9526 −0.0342   −8.9526] is the enhanced discrete signal by applying basic spectral subtraction algorithm on the noisy signal $f[n]$=[5   6   0   −9] with $D[k]$=0.1. We wish to implement this computation.

As a procedure, first we enter the noisy signal $f[n]$ as a row matrix as follows:
>>f=[5 6 0 -9]; ↵   ← f holds $f[n]$, f is user-chosen variable
Then the DFT is computed for determining the $F[k]$ by **fft** as follows:
>>F=fft(f); ↵   ← F holds $F[k]$, F is user-chosen variable
We pick up the phase information in radian from **F** by the command **angle** as follows:
>>P=angle(F); ↵   ← P holds $\angle F[k]$, P is user-chosen variable
Next the $|F[k]|$ is computed by the command **abs**:
>>A=abs(F); ↵   ← A holds $|F[k]|$, A is user-chosen variable
The computation $|F[k]| - D[k]$ of the subtractive algorithm occurs by:
>>An=A-0.1; ↵   ← An holds $|F[k]| - D[k]$, An is user-chosen variable
The complex exponential $e^{j\angle F[k]}$ is formed by the command **exp(i*P)**. All involved variables so far are as a row matrix. Element by element multiplication takes place by the scalar code of appendix A which is simply

the operator .*. Knowing so, the term $(|F[k]|-D[k])e^{j\angle F[k]}$ of the algorithm we form by:

>>Fn=An.*exp(i*P); ↵   ← Fn holds $(|F[k]|-D[k])e^{j\angle F[k]}$, Fn is
                                user-chosen variable

After that IDFT of the $(|F[k]|-D[k])e^{j\angle F[k]}$ is performed by the section 4.1 mentioned ifft as follows:

>>f1=ifft(Fn); ↵   ← f1 holds IDFT of the $(|F[k]|-D[k])e^{j\angle F[k]}$, f1 is user-
                            chosen variable

Finally considering only the real part of elements in the f1 provides the $\hat{f}[n]$ we are seeking for:

>>fn=real(f1) ↵   ← fn holds the $\hat{f}[n]$, fn is user-chosen variable

fn =
          4.9342   5.9526   -0.0342   -8.9526

At least half a dozen varieties of the spectral subtraction algorithms are reported in [4]. We hope the introductory one as exercised just now will pave the way for those of the others.

❖ ❖ **Devising a function file**

    Having gone through the step by step approach, one might be interested to put all the codes in a file. We wrote a function file (appendix C) by the name **basspec** with two input arguments f and D – the modular noisy signal $f[n]$ as a row matrix and the user-

```
function fn=basspec(f,D)
F=fft(f);
P=angle(F);
A=abs(F);
An=A-D;
Fn=An.*exp(i*P);
f1=ifft(Fn);
fn=real(f1);
```

Figure 10.1(b) Function file
for the basic spectral
subtraction algorithm

supplied $D[k]$ scalar respectively as shown in figure 10.1(b). Type the codes of the figure 10.1(b) in a new M-file (section 1.3) and save the file by the name **basspec** in your working path of MATLAB where the **basspec** is a user-chosen word. Verify that you are able to get the previously obtained $\hat{f}[n]$ as follows:

>>f=[5 6 0 -9]; ↵
>>fn=basspec(f,0.1) ↵

fn =
          4.9342   5.9526   -0.0342   -8.9526   ← what found before

10.3 Application of the spectral subtraction algorithm

    In section 10.2 we just exercised the basic subtraction algorithm considering a prototype discrete signal. We wish to implement the algorithm now in this section to a practical noisy softcopy audio.

    Since a practical softcopy audio holds millions of samples, we have to apply the subtraction algorithm on frame by frame basis whose discussion is found in section 3.3. Figure 3.1(c) presents the codes for frame by frame selection of a digital audio. From the section we know that the **g** and **fn** hold each audio frame and retrieved audio respectively so whatever computation we perform should be between the **g** and the **fn** of the last for-loop in figure 3.1(c).

In section 10.2 we also demonstrated the function file derivation of the basic subtraction algorithm through the MATLAB function **basspec**. The functional calling should be between the **g** and **fn** of the for-loop in figure 3.1(c).

As an example let us consider the **test.wav** of section 2.7. We intend to apply the subtraction algorithm to this softcopy where the $D[k]$ of section 10.2 is chosen to be 0.05 and frame length of the audio is chosen as 300 samples.

```
[v,fs]=wavread('test.wav');
f=v;
L=300;
N=numel(f);
R=rem(N,L);
if R~=0
    f1=[f;zeros(L-R,1)];
end
fn=[ ];
for k=1:L:numel(f1)
    g=f1(k:k+L-1);
    gn=basspec(g,0.05);
    fn=[fn;gn];
end
```

Figure 10.1(c) M-file for applying the subtraction algorithm on test.wav

The modification done on the program codes of the figure 3.1(c) is presented in figure 10.1(c). The first line of figure 10.1(c) is to read the audio. The second line is to assign the audio samples available in **v** to **f**. In the for-loop of figure 10.1(c) the **basspec** is called on each sequential frame **g** with chosen value of $D[k]$. The return from the **basspec** is assigned to some variable **gn** which is user chosen. The noise removed frame **gn** is put in order to form the **fn** or enhanced audio.

The next question is how to verify the audio quality improvement? Well we can first play the original audio by:
>>wavplay(v,fs) ↵
Then the enhanced one we play by:
>>wavplay(fn,fs) ↵
Aurally we hear the quality difference certainly not so much. The reason is basic subtraction algorithm has some drawbacks. If you apply other variants of the algorithm [4], the audio quality may improve.

10.4 Wiener filter in integer time index domain

Wiener or minimum mean square error filter utilizes the local statistical knowledge in filtering a discrete signal. The filter is also called the Wiener adaptive filter. The filter becomes adaptive in the sense that it works on discrete signal frame characteristics. The discrete signal filtering happens through the following algorithm and formulation:

⇒ The computation occurs in the time or index domain i.e. for every sample in the signal

⇒ The frame $f_b[n]$ is chosen from given discrete signal row or column matrix $f[n]$ where n is the integer index variable

⇒ The frame $f_b[n]$ sample number is user-chosen say P

⇒ The local mean and variance for the signal frame $f_b[n]$ are given by $\mu = \frac{1}{P}\sum_{n=0}^{P-1} f_b[n]$ and $\sigma^2 = \frac{1}{P}\sum_{n=0}^{P-1}(f_b^2[n]-\mu^2)$ respectively

⇒ The filtered signal $r[n]$ of the $f[n]$ is given by the expression $r[n]$ $= \mu + \frac{\sigma^2 - v^2}{\sigma^2}(f[n]-\mu)$ where v^2 is user defined noise variance

⇒ The expression $\sum_{n=0}^{P-1} f_b[n]$ is equivalent to the sum of all elements in the frame $f_b[n]$

⇒ The filtered signal $r[n]$ has the same size as that of the original signal $f[n]$

⇒ At the time or integer index bound of the signal, we pad the undefined element within the frame by zeroes

⇒ In any of the discrete signals $f[n]$, $r[n]$, and $f_b[n]$, the integer index n starts from 0

❖❖ Modular example

Let us consider the prototype discrete signal $f[n] = [6\ 7\ -9\ 8\ 4]$ for the Wiener filtering with frame length 3 (meaning $P = 3$) and variance $v^2 = 3$. We wish to illustrate how Wiener filtered $r[n]$ discrete signal is obtained by computing every given sample.

Let us calculate two sample points' filtered values. For the first sample in $f[n]$, the $f_b[n]$ with frame length 3 samples becomes $f_b[n] = [0\ 6\ 7]$ after padding by zeroes. We have mean of $f_b[n] = \mu = 4.3333$ and $\sigma^2 = \frac{1}{P}\sum_{n=0}^{P-1}(f_b^2[n]-\mu^2) = 9.5556$ and the first filtered sample is $r[0] = \mu +$

$\frac{\sigma^2 - v^2}{\sigma^2}(f[0]-\mu) = 4.3333 + \frac{9.5556-3}{9.5556}(6-4.3333) = 5.4767$.

Similarly for the third filtered sample in $f[n]$ we get $r[2] = -8.456$ where $f_b[n] = [7\ -9\ 8]$, mean of $f_b[n] = \mu = 2$, $\sigma^2 = \frac{1}{P}\sum_{n=0}^{P-1}(f_b^2[n]-\mu^2) = 60.6667$,

and $\mu + \frac{\sigma^2 - v^2}{\sigma^2}(f[2]-\mu) = 2 + \frac{60.6667-3}{60.6667}(-9-2) = -8.456$. Also let us not forget that the index n is 0, 1, 2, etc for the first, second, third, etc sample in $f_b[n]$, $f[n]$, or $r[n]$ respectively. Thus applying Wiener filtering for every sample we obtain $r[n] = [5.4767\ \ 6.6826\ \ -8.456\ \ 7.6013\ \ 4]$.

❖ ❖ MATLAB approach to obtain the Wiener filtered signal

Now focus needs to be given in MATLAB implementation for which we utilize the built-in function **wiener2** with the syntax **wiener2**(discrete signal $f[n]$ as a row or column matrix, frame $f_b[n]$ matrix dimension as a two element row matrix, user-supplied variance v^2). If you choose row matrix for $f[n]$, the $f_b[n]$ frame matrix dimension is $1 \times P$ whereas for the column matrix the dimension will be $P \times 1$. The second input argument of **wiener2** will be $[1 \quad P]$ or $[P \quad 1]$ accordingly. Of coarse the frame type, row or column, follows that of $f[n]$. The return from the **wiener2** is the Wiener filtered signal $r[n]$ whose matrix size is the same as that of the $f[n]$.

Just now we demonstrated the Weiner filtered $r[n] = [5.4767 \quad 6.6826 \quad -8.4560 \quad 7.6013 \quad 4]$ from the discrete signal $f[n] = [6 \quad 7 \quad -9 \quad 8 \quad 4]$ with frame $f_b[n]$ of length 3 and variance $v^2 = 3$. By dint of **wiener2**, we have the computation done as follows:

```
>>f=[6 7 -9 8 4];  ⏎ ← Entering f[n] as a row matrix to f, f is user-chosen variable
>>r=wiener2(f,[1 3],3) ⏎   ← r holds the r[n], r is user-chosen variable

r =
        5.4767   6.6826   -8.4560   7.6013   4.0000
```

In MATLAB array indexing starts from 1 but the n starts from 0 so there will be one sample lag between theory mentioned $r[n]$ and MATLAB implemented r i.e. $r[0]$, $r[1]$, etc correspond to r(1), r(2), etc respectively.

10.5 Wiener filter on a softcopy audio in time index domain

In this section we plan to apply the section 10.4 explained Wiener filtering to a softcopy audio. Let us consider the audio **test.wav** of section 2.7 and the machine read (section 2.3) the audio as follows:

```
>>[v,fs]=wavread('test.wav'); ⏎
```

So the variables **v** and **fs** hold the discrete signal $f[n]$ as a column matrix and sampling frequency respectively. Let us choose the frame $f_b[n]$ length as 400 so the matrix size of $f_b[n]$ as a two element row matrix is 400×1. The next point is the variance v^2 selection say 0.00001 therefore we apply the filtering on the discrete signal as follows:

```
>>vn=wiener2(v,[400 1],0.00001); ⏎
```

In the last execution the **vn** holds the Wiener filtered audio signal $r[n]$ where the **vn** is a user-chosen variable. How are we certain about the audio improvement? First play the original audio by:

```
>>wavplay(v,fs) ⏎
```

The Wiener filtered audio you play by:

>>wavplay(vn,fs) ↵

You may not hear any audible difference between the two audio. The reason is we applied the basic Wiener filtering and did not consider other variants [4],[11]. Only one arbitrary frame number and one variance are exercised. You may try with other frame number and variance to hear some improvement in the audio quality.

10.6 Artificial noise generation

Any unwanted audio signal is noise which we explained in section 10.1. Sometimes we generate artificial noise and add the noise to a clean audio signal specially when noise removal algorithm has to be devised and tested. In this section we plan to demonstrate mainly two types of noise – uniform and Gaussian.

✦ ✦ Uniformly distributed random variable

The function **unifrnd** (abbreviation for un̲iformly f̲ractional ra̲n̲d̲om) generates samples of uniformly distributed continuous random variable X from user-defined range $A \leq X \leq B$ where $B > A$. One of the syntaxes of the function is **unifrnd(A , B)** for example any fractional value over $-4 \leq X \leq 5$ is generated by:

>>X=unifrnd(-4,5) ↵   ← X is user-chosen variable that holds the value

X =
         -2.2373

The random numbers in row, column, or rectangular matrix are generated by four input arguments with the syntax **unifrnd(A , B ,user-required row number,user-required column number)**. Let us see the following generation:

Generation of a column matrix X of length 4 in which every element is between −4 and 5:

>>X=unifrnd(-4,5,4,1) ↵

X =
         4.0428
        -3.4790
        -0.8242
         1.3456

Generation of a row matrix X of length 4 in which every element is between −4 and 5:

>>X=unifrnd(-4,5,1,4) ↵

X =
         -0.3486   4.4192   4.2521   -0.3076

Generation of a rectangular matrix X of order 2×3 in which every element is between −4 and 5:

>>X=unifrnd(-4,5,2,3) ↵

X =
$$3.3185 \quad -2.7500 \quad -2.2115$$
$$-3.9112 \quad -2.1751 \quad 1.4341$$

When you execute the commands, these output numbers may not appear in the command window due to the randomness of generation but the numbers will be between the defined range that is for sure. A row and a column matrices of length 4 have the rectangular matrix dimensions 1×4 and 4×1 which we used in the generation of the random numbers in the row and column matrices respectively. The fractional value means the generation is in the continuous sense. The term uniform means any value between -4 and 5 is equally likely.

✦ ✦ Gaussian random variable

Samples of Gaussian random variable from user-supplied mean μ and standard deviation σ we implement by the function **normrnd** (abbreviation for normal random number). The function in general provides matrix based Gaussian random numbers for which the common syntax is **normrnd**(μ, σ,user-required row number,user-required column number). It should be noted that the random number might appear from $-\infty$ to $+\infty$ theoretically but with more likelihood towards the user-given mean. However following is the example on sample generation of the random numbers:

Generation of a single Gaussian number X of mean $\mu = 2$ and standard deviation $\sigma = 4$:
>>X=normrnd(2,4,1,1) ↵

X=
6.2671

Generation of a row matrix X of length 4 in which every element is a Gaussian random number with $\mu = 2$ and $\sigma = 4$:
>>X=normrnd(2,4,1,4) ↵

X =
$$-3.3447 \quad 4.8573 \quad 8.4942 \quad -0.7671$$

Generation of a column matrix X of length 3 in which every element is a Gaussian random number with $\mu = 2$ and $\sigma = 4$:
>>X=normrnd(2,4,3,1) ↵

X =
7.1610
4.6744
6.7634

Generation of a rectangular matrix X of order 3×3 in which every element is a Gaussian random number with $\mu = 2$ and $\sigma = 4$:
>>X=normrnd(2,4,3,3) ↵

$$X =$$

5.4320	-3.7639	4.7600
7.0160	4.2846	5.2625
-4.3749	0.4005	4.8476

The sample generated is for the continuous random variable.

A normal random number has $\mu = 0$ and $\sigma = 1$ which is just a special case of the Gaussian random number.

✦ ✦ Corrupting an audio by noise addition

Suppose we have the audio bird.wav of section 2.7. Our objective is to add uniformly distributed noise between −0.05 and 0.05 to the audio and to play the original audio, noise signal, and noisy audio.

Get the audio and let the machine read (section 2.3) the audio as follows:

>>[v,fs]=wavread('bird.wav'); ↵

Therefore the audio samples as a column matrix and sampling frequency are stored in v (v indicates $y[n]$ of the section 10.1) and fs respectively. The command length finds the number of samples in v by:

>>N=length(v); ↵   ← N holds the total sample number in the audio

If any noise has to be added, that should corrupt every sample in the audio and we generate the same sample number noise as an identical size column matrix by earlier mentioned unifrnd:

>>X=unifrnd(-0.05,0.05,N,1); ↵   ← X holds the noise samples and represents
                                    $w[n]$ of the section 10.1, X is user-chosen

Simply addition of v and X provides the corrupted signal $f[n]$ of the section 10.1:

>>A=v+X; ↵   ← A holds the noisy $f[n]$, A is user-chosen

In doing so the $f[n]$ range might be other than $[-1,1]$ which is essential for audio playing:

>>f=2*mat2gray(A)-1; ↵   ← Mapped noisy signal is assigned to f in accordance
                            with section 3.2, f is user-chosen

Now all three signals are available in the workspace so we play them by:

>>wavplay(v,fs) ↵   ← Playing the original audio
>>wavplay(X,fs) ↵   ← Playing the noise only
>>wavplay(f,fs) ↵   ← Playing the noisy signal

In a similar fashion you can add Gaussian noise to the audio.

10.7 Noise reduction by singular value decomposition

The reader needs to go through sections 7.6 and 7.7 which are the prerequisite of this section. Maintaining the symbology of just mentioned sections, we wish to singular value decompose the signal $v[n]=[9 \quad 6 \quad -5 \quad 4 \quad 3 \quad -2 \quad 7 \quad 12]$ with $L=5$ so carry out the following:

```
>>x=[9 6 -5 4 3 -2 7 12]; ↵   ← x holds v[n]
>>f=mtop(x,5); ↵              ← f holds f[m,n]
>>[U D V]=svd(f); ↵           ← Applying the svd function on the f
```

Our concern here is on the diagonal matrix D or D. Let us see its content by:
```
>>D ↵
```

```
D =
    18.1493        0        0        0    0
         0   13.1293        0        0    0
         0        0   8.2058        0    0
         0        0        0   3.5905    0
```

We know that the diagonal elements of the D are the singular values and extract those by the command **diag** as follows:
```
>>s=diag(D) ↵                 ← s holds the singular values, s is user-chosen variable
```

```
s =
    18.1493
    13.1293
     8.2058
     3.5905
```

❖❖ Understanding only noise in terms of singular value

Noise is also a signal. If we singular value decompose only noise, we also have noise's U, D, and V representations. If the noise is uncorrelated and white, then the D takes the form λI where I is an identity matrix and λ is the singular value for example $D = \begin{bmatrix} 4 & 0 & 0 & 0 \\ 0 & 4 & 0 & 0 \\ 0 & 0 & 4 & 0 \\ 0 & 0 & 0 & 4 \end{bmatrix}$ with $I = \begin{bmatrix} 1 & 0 & 0 & 0 \\ 0 & 1 & 0 & 0 \\ 0 & 0 & 1 & 0 \\ 0 & 0 & 0 & 1 \end{bmatrix}$ and $\lambda = 4$.

When $D \neq \lambda I$, the noise is called colored noise which is often the case for example $D = \begin{bmatrix} 7 & 0 & 0 & 0 \\ 0 & 6 & 0 & 0 \\ 0 & 0 & 5 & 0 \\ 0 & 0 & 0 & 4 \end{bmatrix}$.

❖❖ Understanding noisy signal and SNR in terms of singular values

The diagonal elements in D or elements in **s** are a measure of the signal $v[n]$ energy. But the problem is clean signal and noise both energies present in the diagonal are in mixed form. Besides noise is an unknown enemy whose energy is not known beforehand that is why noise reduction algorithm frequently works on the approach "check and see". Earlier illustrated $v[n] = [9 \ 6 \ -5 \ 4 \ 3 \ -2 \ 7 \ 12]$ with $L=5$ has the singular values 18.1493, 13.1293, 8.2058, and 3.5905. Usually noise is related with the lowest singular values for instance with 3.5905 and 8.2058.

We simply discard the lowest singular values but the user has to decide how many singular values should be discarded. For the $v[n]$ we have 4 singular values namely $\lambda_1 = 18.1493$, $\lambda_2 = 13.1293$, $\lambda_3 = 8.2058$, and $\lambda_4 =$

3.5905. Depending on the discarding of λ's, the signal to noise power ration (SNR) is defined that is

when λ_4 is discarded, the SNR is $\dfrac{\lambda_1^2+\lambda_2^2+\lambda_3^2}{\lambda_4^2}$,

when λ_3 and λ_4 are discarded, the SNR is $\dfrac{\lambda_1^2+\lambda_2^2}{\lambda_3^2+\lambda_4^2}$, and

when λ_2, λ_3, and λ_4 are discarded, the SNR is $\dfrac{\lambda_1^2}{\lambda_2^2+\lambda_3^2+\lambda_4^2}$.

As an example when λ_3 and λ_4 are discarded, the SNR is $\dfrac{\lambda_1^2+\lambda_2^2}{\lambda_3^2+\lambda_4^2}=$

$\dfrac{18.1493^2+13.1293^2}{8.2058^2+3.5905^2}=6.2545.$

The decibel (dB) SNR is also widely used in noise reduction algorithms which just needs the $10\log_{10}x$ operation. For the numerical example we are going to have $10\log_{10}6.2545=7.9619\,dB$.

Note that if there are 4 singular values, we may have 3 SNRs maximum so a programming tactic is needed to calculate the SNRs starting from the singular value column matrix **s**. Following is the implementation:

```
>>P=length(s); ↵      ← P holds the number of singular values, P is user-chosen
>>SN=[ ]; for k=1:P-1, SN=[SN;sum(s(1:k).^2)/sum(s(k+1:P).^2)]; end ↵
>>SND=10*log10(SN); ↵
```

Data accumulation technique of appendix B.3 is applied in conjunction with the for-loop of appendix B.4. The for-loop counter **k** changes from 1 to P-1 where the **P** is the number of singular values. The numerator of the SNR is the summation of λ's from 1 to **k** whereas the denominator of the SNR is that of from **k+1** to **P** consequently **s(1:k)** and **s(k+1:P)** are used to select only those singular values. The **.^** operator just squares every element selected. The **SN** and **SND** in above execution are user-chosen variables and hold the SNR in absolute and in dB respectively. You may see them side by side by:

```
>>[SN SND] ↵
```

ans =

1.3040	1.1528
6.2545	7.9619
44.1460	16.4489
↑	↑
ratio SNR	dB SNR

❖❖ **Decision making from given SNR**

Just now we finished the discussion on how one gets the SNR if singular values are consecutively discarded. In noise reduction usually we select some SNR and try to find which singular value should be discarded so another programming needs to be conducted. In the last execution the **SND** holds the SNR in *dB*. Suppose we wish to discard the singular values which correspond to SNR>10 *dB*. Obviously the answer is the 4th one as in earlier discussion. Appendix B.9 cited **find** we apply for that as follows:

```
>>I=find(SND>10); ↵        ← I holds the integer index values, I is user-chosen
```

We need to add 1 to I because singular value discarding takes place from 2 to P but in SND the values are indexed from 1 to P-1. The calling then occurs by:

```
>>1+I ↵
```

```
ans =
        4        ← Meaning the 4th singular value should be discarded
```

As another example the SNR>7 *dB* is achieved by discarding the 3rd and the 4th singular values which we verify similarly by:

```
>>I=find(SND>7); ↵
>>1+I ↵
```

```
ans =
        3        ← Meaning the 3rd singular value should be discarded
        4        ← Meaning the 4th singular value should be discarded
```

Note that the return is as a column matrix and we could have assigned the return to c by writing c=1+I; where c is a user-chosen variable.

✦ ✦ Obtaining approximate signal after singular value discarding

After discarding the lowest singular values, one needs to know how to recover the approximated signal. From earlier discussions, singular values are stored in the diagonal D and we need to set 0 to the position which is decided by the SNR study. Let us say SNR>7 *dB* singular values are to be discarded. We know that the workspace c holds the singular value position indexes. Making the c available, we carry out the following:

```
>>for k=1:length(c), D(c(k),c(k))=0; end ↵
```

In the diagonal D basically (3,3) and (4,4) indexed elements should be 0 as far as the SNR is concerned. In the last execution the for-loop provides control over the elements in c staring from 1 to **length(c)** or the number of elements in the c. For this reason we used the indexes as (c(k),c(k)). You may call the variable to see its contents:

```
>>D ↵
```

```
D =
    18.1493         0    0    0    0
         0    13.1293    0    0    0
         0         0    0    0    0
         0         0    0    0    0
```

Compare the above D to the last D, what is seen? Clearly 0 is set to the mentioned positions. In section 7.6 we quoted that the starting matrix *A* (special Toeplitz $f[m,n]$ is A) is calculated by the matrix multiplication $U \times D \times V^T$ which needs the code:

>>fn=U*D*V' ⏎   ← fn holds the approximated $f[m,n]$ after singular value
                                                            discarding, fn is user-chosen

fn =

```
 4.7101   5.9241  -3.1269   2.4861   8.4230
-4.0124   1.8646  -0.7940  -2.2297   5.3266
 6.6334  -1.5088   0.5253   3.6608  -5.9590
10.7225   5.4261  -3.0858   5.7901   4.5948
```

Appendix B.19 mentioned **top2m** recovers the approximated signal from **fn** and which we call by:

>>xn=top2m(fn) ⏎   ← xn holds the approximated signal, xn is user-chosen

xn =

```
10.7225   6.0297  -2.8690   3.2225   3.3464  -3.7719   3.9063   8.4230
```

Still there is a problem. When original $f[m,n]$ was formed from the $v[n]$, the element placing was having the following order: first element of $v[n]$ refers to the upper rightmost element in $f[m,n]$ (which is 9), second element of $v[n]$ refers to second super diagonal from the right (which is 6). You may verify that by calling:

>>f ⏎

f =

```
 3   4  -5   6   9
-2   3   4  -5   6
 7  -2   3   4  -5
12   7  -2   3   4
```

```
function xn=svd_noise(x,L,SNR)
f=mtop(x,L);
[U D V]=svd(f);
s=diag(D);
P=length(s);
SN=[ ];
for k=1:P-1
  SN=[SN;sum(s(1:k).^2)/sum(s(k+1:P).^2)];
end
SND=10*log10(SN);
I=find(SND>SNR);
c=1+I;
for k=1:length(c)
  D(c(k),c(k))=0;
end
fn=U*D*V';
xn=top2m(fn);
xn=fliplr(xn);
```

Figure 10.2(a) Function file for the singular value based noise reduction

All we need is flip the elements in **xn** from the left to right and which takes place by appendix B.7 mentioned **fliplr** as follows:

>>xn=fliplr(xn) ⏎   ← Flipped signal is again assigned to xn

xn =

```
8.4230   3.9063  -3.7719   3.3464   3.2225  -2.8690   6.0297   10.7225
```

Finally the last **xn** contains the approximated signal of earlier starting $v[n]$ after discarding more than $7\,dB$ singular values.

❖ ❖ Devising a function file

Whatever is conducted so far can be placed in a file. Figure 10.2(a) shows the code as a function file (appendix C). We named the function file as **svd_noise** which can be any name of your choice. The function has three input arguments namely **x**, **L**, and **SNR** which correspond to earlier mentioned $v[n]$, L, and SNR in *dB* respectively. The return from the function is the approximated signal **xn** following the singular value discarding. Both the **x** and **xn** in the function file are as a row matrix. However type the codes of the figure 10.2(a) in a new M-file (section 1.3) and save the file by the name **svd_noise** or get the softcopy from the link of section 2.7. Making the file available, you may verify the following:

```
>>x=[9 6 -5 4 3 -2 7 12]; ↵        ← x holds v[n]
>>xn=svd_noise(x,5,7) ↵            ← Calling the file with L=5 and 7 dB SNR

xn =
        8.4230   3.9063   -3.7719   3.3464   3.2225   -2.8690   6.0297   10.7225
```

Above is the approximated signal $\hat{v}[n]$ what we found before.

❖ ❖ Applying the algorithm to a softcopy audio file

When we apply the algorithm to a softcopy, we do not apply the singular value discarding technique to whole digital audio because intensive calculation is involved. Instead we take some samples and apply the algorithm to the samples. If the digital audio is 3000 or 4000 samples long, we may apply the algorithm to the audio without taking frames. For long audio, we apply the algorithm on frame by frame approach as mentioned in section 3.3.

For example we consider the audio **b.wav** (section 2.7) where the algorithm is to be applied with L=60 and 11 *dB* SNR.

Since the audio is 2800 samples long, we consider the whole audio rather than frame by frame selection. Let the machine read (section 2.3) the audio as follows:

```
>>[v,fs]=wavread('b.wav'); ↵
```

So the workspace **v** and **fs** hold the audio data as a column matrix and sampling frequency respectively. The **v** is synonym of earlier illustrated **x**. The execution might take few seconds depending on your PC even with 2800 samples so please be patient. We call the function then:

```
>>xn=svd_noise(v,60,11); ↵
>>wavplay(v,fs) ↵          ← Playing the original audio
>>wavplay(xn,fs) ↵         ← Playing the algorithm applied audio
```

The **b.wav** has some high frequency buzzing noise associated. Certainly that noise is removed which you verify aurally by listening the above two playings.

We addressed three elementary speech enhancing algorithms. We hope the elaborate demonstrations on the algorithms will pave the way to implement other speech enhancing algorithms. Keeping this hope, we closed the chapter.

Chapter 11

Miscellaneous Topics

As the name implies, a mixture of digital audio based topics are chosen to implement in this chapter. While devising a particular chapter, all topics covering the chapter may not receive evenhanded attention yet the material may bear special importance to the course. Furthermore a certain digital audio problem may also need the exercise of various chapter cited functions. This sort of problems is tackled here too and the medley includes the following:

✦ ✦ Telephone dial pad signal properties through graphical means
✦ ✦ Source-filter model of engineering speech generation
✦ ✦ Power-energy based computing of discrete signal/digital audio
✦ ✦ Most common forms of coding and softcopy writing of audio

11.1 Telephone dial pad signals

Built-in function **phone** of MATLAB displays a graphical user interface window which interactively shows the commonly dialed number signals both in the time and in the frequency domains. All you have to do is execute the following:

```
>>phone ↵
```

Figure 11.1(a) shows the graphical user interface as a result of the last line execution. By inspection the upper and the lower left figures correspond to time and frequency domain responses respectively. The check button **Sound** in the window if checked, clicking the numbers by mouse sounds like a push button telephone. The shown one in figure 11.1(a) is for the number 1. If you click other number in the window, the response automatically updates.

Figure 11.1(a) Phone
Pad Graphical User
Interface

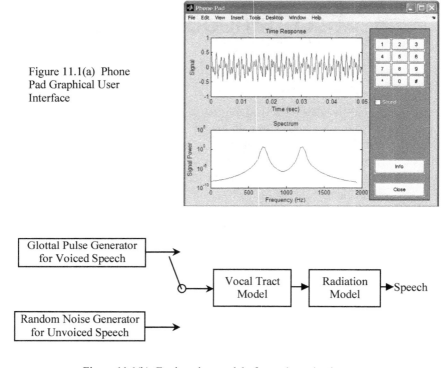

Figure 11.1(b) Engineering model of speech production

11.2 Engineering model of speech

In audio processing literature speech signal occupies a substantial portion because of its practical implication. Basic statistics (section 2.6) of a speech signal change over time. Speech production involves a number of organs and muscles. Biological production of speech is quite complex and not easy to implement. However audio researchers found some model to imitate the speech production which is depicted in figure 11.1(b). There are three basic elements in the engineering production process – glottal pulse or noise generator, vocal tract model, and radiation model, each of which needs rigorous discussion. In this elementary text we wish to cite some hand-on generation of speech in MATLAB. Let us see what we need for a speech production.

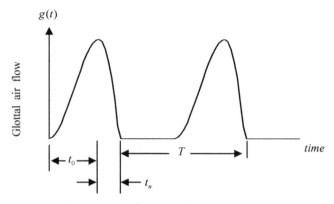

$g(t)$

Glottal air flow

t_0

T

time

t_n

Figure 11.1(c) Glottal airflow waveform of speech

Glottal pulse and random noise generator:

A mathematical polynomial relationship for one cycle of the glottal flow waveform is given by:

$$g(t) = \begin{cases} 3\left(\dfrac{t}{t_0}\right)^2 - 2\left(\dfrac{t}{t_0}\right)^3 & 0 \le t \le t_0 \\[3mm] 1 - \left(\dfrac{t-t_0}{t_n}\right)^2 & t_0 < t \le t_0 + t_n \end{cases}$$

Figure 11.1(c) depicts the glottal airflow waveform of speech over a time period T. The T is called the pitch period or in other words $\dfrac{1}{T}$ is the pitch or fundamental frequency of the speaker. The pitch frequency $\dfrac{1}{T}$ varies from speaker to speaker. The frequency range is about 60-150 Hz for male speakers and 200-400 Hz for children and female speakers. Glottal pulse $g(t)$ is usually generated in

```
function y=g(t,to,tn)
    y=[ ];
for k=1:length(t)
    n=t(k);
    if 0<=n&n<=to
        y=[y 3*(n/to)^2-2*(n/to)^3];
    elseif to<=n&n<=(to+tn)
        y=[y 1-((n-to)/tn)^2];
    else
        y=[y 0];
    end
end
```

Figure 11.1(d) Author written function file for the generation of one cycle samples of glottal airflow waveform

time domain meaning we have some sample values of the $g(t)$ as a row or column matrix. The parameters t_0 and t_n of glottal pulse in figure 11.1(c) approximately follow some fixed ratio in relation to T for instance $t_0 = 0.4T$ and $t_n = 0.16T$. The number of glottal pulse depends on the duration of speech of the speaker.

Single cycle glottal pulse:

Author written function file **y=g(t,to,tn)** as seen in figure 11.1(d) returns one cycle $g(t)$ functional values in which the input arguments – **t, to,** and **tn** refer to t, t_0, and t_n respectively and **y** to $g(t)$. The **t** is a row or column matrix and each of **to** and **tn** is a scalar. The reader may type the codes of the figure 11.1(d) in a new M-file (section 1.3 and appendix C) and save the file by the name **g** or get the softcopy from the link mentioned in section 2.7.

The $g(t)$ value say for $t_0 = 4\,m\sec$ and $t_n = 1.6\,m\sec$ is 0.1563 at $t = 1\,m\sec$ which we wish to obtain.

Just call the function as follows:

>>y=g(1e-3,4e-3,1.6e-3) ⏎ ← Appendix A for engineering
scale factor

y =
          0.1563

Multiple cycle glottal pulse:

If you wish to generate many cycle samples of $g(t)$ and which is required for engineering speech production, some programming is necessary. With the limitation that the cycle number is integer well for a thousand data audio that is not a serious issue.

```
function yn=gn(y,N)
yn=[ ];
for k=1:N-1
    yn=[yn y(2:end)];
end
yn=[yn y];
```

Figure 11.1(e) Author written function file for the generation of multi cycle samples from one cycle samples

For example with pitch period $T = 10\,m\sec$, $t_0 = 0.4T$, and $t_n = 0.16T$, we wish to get the samples of $g(t)$ for four cycles and plot them.

The four cycles mean the interval $0 \le t \le 4T$ or $40\,m\sec$. We need sampling frequency to generate the data say $f_s = 4\,KHz$ (also from sampling theorem $f_s > \dfrac{2}{T}$). The $\dfrac{1}{f_s}$ becomes the t step over the required interval. First we get one cycle samples of $g(t)$ as follows:

```
>>T=10e-3; to=0.4*T; ⏎
>>tn=0.16*T; fs=4e3; ⏎
```

In the last two line executions we entered given T, t_0, and t_n to like name variables e.g. t_n to tn. The t values over one period or $0 \leq t \leq T$ is generated by a row matrix:

```
>>t=0:1/fs:T; ⏎     ← t is a user-chosen variable
>>y=g(t,to,tn); ⏎   ← Calling the g for one cycle samples, y
                      holds the samples where y is user-chosen variable
```

Author written another function file gn(y,N) (shown in figure 11.1(e)) generates multiple cycle glottal pulse samples by taking one cycle samples as a row or column matrix and repetition number which are indicated by the input arguments y and N respectively. The return from the gn(y,N) is the multiple cycle samples. Either you type the code of the figure 11.1(e) in a new M-file and save the file by the name gn or get the softcopy from the link of section 2.7. For the 4 required cycles we call the gn as follows:

```
>>yn=gn(y,4); ⏎   ← yn is a user-chosen variable that holds
                     the 4 cycle samples of g(t)
```

For the whole four cycles over $0 \leq t \leq 4T$, the t points as a row matrix we generate by:

```
>>tn=0:1/fs:40e-3; ⏎ ← tn is a user-chosen variable that holds
                        the t points of the 4 cycles
```

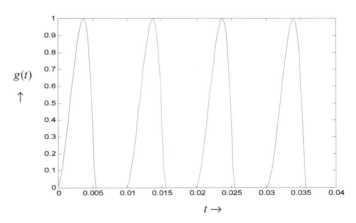

Figure 11.1(f) Four cycles of the glottal pulse

Graphing the glottal pulse:

Continuous sense $g(t)$ versus t is graphed by the appendix D mentioned plot as follows:

>>plot(tn,yn) ↵

Figure 11.1(f) shows the plot of the 4 cycle glottal pulse as a result of the last line execution.

Random noise generator is addressed in chapter 10.

Vocal tract model:

Vocal tract model is considered as an acoustic resonator and similar to the pole resonator of the linear prediction model as addressed in chapter 9. Vocal tract resonances are termed as the formants and corresponding frequencies are called formant frequencies. Vocal tract modeling takes place in Z transform domain e.g. a vocal tract system function $V(z)$ can be $\dfrac{1}{1+0.17z^{-1}-0.39z^{-2}+0.09z^{-3}+0.1008z^{-4}}$. Vowel and consonant of speech have different $V(z)$ expressions or in other words different denominator polynomials of $V(z)$. How we get the $V(z)$ is not straightforward. You need to do some research on that.

Radiation model:

The radiation model $R(z)$ is usually in Z transform domain for example $R(z)=1+z^{-1}$.

Way to implement:

Obviously there is some algorithm for generating a speech in computer which is explained in the following:

(a) obtain the vocal tract polynomial system function $V(z)$ mainly the denominator polynomial coefficients of $V(z)$.

(b) form a filter object (say V) of $V(z)$ using **filt** function of section 6.9.

(c) form a radiation model object (say R) $R(z)$ using **filt**.

(d) obtain the series equivalent (say **Heq**) of V and R using **series** of section 6.9 because $V(z)$ and $R(z)$ of the speech model in figure 11.1(b) are in cascade form.

(e) use **tfdata** of section 6.9 on the **Heq** to extract the numerator (say **n1**) and the denominator (say **d1**) polynomial coefficients as a row matrix because **Heq** is an object in Z transform domain.

(f) generate one cycle glottal pulse $g(t)$ samples (say **y**) as a row or column matrix in time domain by using author written **g(t,to,tn)** from given T, t_0, and t_n.

(g) decide the number of cycles (say **N**) of the glottal pulse from speech duration and generate so cycles (say **yn**) by using author written **gn(y,N)**.

(h) apply the **filter** function of section 6.4 to obtain the speech signal samples (say **bn**) in time domain by using **filter(n1,d1,yn)**.

(i) to be consistent with the MATLAB player, the **bn** samples should be in $[-1,1]$ and do so by section 3.2 mentioned technique, say the mapped samples are in **b**.

(j) finally play the generated speech by using **wavplay** of section 2.3 employing the **b** and provided sampling frequency.

All variables **V**, **R**, etc in above procedure are user-chosen.

Table 11.A Vocal tract model coefficients of the english letter b

Coefficient of	Coefficient	Coefficient of	Coefficient	Coefficient of	Coefficient
constant	1	z^{-11}	-0.0706	z^{-21}	0.0108
z^{-1}	-0.6533	z^{-12}	-0.0357	z^{-22}	-0.0242
z^{-2}	-0.0422	z^{-13}	0.0461	z^{-23}	-0.0218
z^{-3}	0.1423	z^{-14}	0.1095	z^{-24}	0.0077
z^{-4}	0.0683	z^{-15}	0.1198	z^{-25}	0.0245
z^{-5}	-0.0798	z^{-16}	0.0734	z^{-26}	0.0193
z^{-6}	-0.1973	z^{-17}	0.0260	z^{-27}	0.0089
z^{-7}	-0.2213	z^{-18}	0.0397	z^{-28}	-0.0090
z^{-8}	-0.1368	z^{-19}	0.0588	z^{-29}	-0.0305
z^{-9}	-0.0704	z^{-20}	0.0464	z^{-30}	-0.0436
z^{-10}	-0.0608				

A practical example:

Recall that in section 2.7 the softcopy audio **b.wav** is the speech signal of english letter b and that was obtained from the voice recording of a direct speaker. We wish to generate the same accent by the model of the figure 11.1(b).

The given parameters are the following:

numerator polynomial coefficient of $V(z)$ is 1,

denominator polynomial coefficients of $V(z)$ are presented in table 11.A,

radiation model $R(z)$ is $0.9 - 0.87z^{-1}$,

window sampling frequency f_s is $22050\ Hz$,

single cycle glottal pulse parameters are: glottal pulse pitch period $T = 6.3\ m\sec$, $t_0 = 0.4T$, and $t_n = 0.16T$, and

cycle repetition of glottal pulse is 20.

Let us enter the given parameters as follows:

>>d=[1 -0.6533 -0.0422 0.1423 0.0683 -0.0798 -0.1973 ... ↵

-0.2213 -0.1368 -0.0704 -0.0608 -0.0706 -0.0357 0.0461 ... ┘
0.1095 0.1198 0.0734 0.0260 0.0397 0.0588 0.0464 0.0108 ... ┘
-0.0242 -0.0218 0.0077 0.0245 0.0193 0.0089 -0.0090 ... ┘
-0.0305 -0.0436]; ┘

In above execution the **d** holds the denominator polynomial coefficients of $V(z)$ which is taken from the table 11.A. The ellipsis (...) reference is seen in section 1.4. Form the $V(z)$ object by (step b of the procedure):

>>V=filt(1,d); ┘   ← V holds $V(z)$

If you do not like typing, get the coefficients file from the link of section 2.7 in your working path of MATLAB (stored in the file **bcoeff.mat**) and conduct first the command **load bcoeff** and then the V=filt(1,d);. The radiation model object $R(z)$ is formed by (step c of the procedure):

>>R=filt([0.9 -0.87],1); ┘   ← R holds $R(z)$

Equivalent of $V(z)$ and $R(z)$ is obtained by (step d of the procedure):

>>Heq=series(V,R); ┘   ← Heq holds the series equivalent

From the **Heq** the polynomial coefficients are extracted by (step e of the procedure):

>>[n1,d1]=tfdata(Heq,'v'); ┘

Single glottal pulse parameters are entered by (step f of the procedure):

>>T=6.3e-3; to=0.4*T; tn=0.16*T; ┘ ← T⇔T , to⇔t_0 , tn⇔t_n

From sampling frequency f_s one cycle t samples of the glottal pulse as a row matrix are generated by:

>>fs=22050; t=0:1/fs:T; ┘ ← fs⇔f_s , t holds one cycle t samples

>>y=g(t,to,tn); ┘   ← y holds one cycle $g(t)$ samples

>>yn=gn(y,20); ┘   ← yn holds 20 cycle $g(t)$ samples (step g
                            in the procedure)

Generated glottal pulses are filtered by the equivalent of $V(z)$ and $R(z)$ by (step h of the procedure):

>>bn=filter(n1,d1,yn); ┘ ← bn holds the filtered glottal pulses

Mapping the samples in **bn** to [−1,1] takes place by (step i of the procedure):

>>b=2*mat2gray(bn)-1; ┘ ← b holds the playable samples

At last we call the player by (step j of the procedure):

>>wavplay(b,fs) ┘

You must be hearing the sound of the letter b generated by your computer.

Advantage of the engineering speech model:

Recall that we have 2800 samples in the softcopy **b.wav** of section 2.7. Apart from the glottal pulse we have only 31 coefficients of the vocal tract model along with the two coefficients of the radiation model. For a particular speaker the glottal pulse parameters are fixed so lot of computer memory space is saved by engineering speech model above all physical presence of the speaker is not necessary and which is frequently applied in TTS (text to speech) system.

11.3 Power-energy of discrete signal and digital audio
We are going to address power and energy related subject matter of discrete signal in conjunction with a digital audio.

♦ ♦ Power and energy of a discrete signal
When we have a discrete signal $x[n]$ in integer time index domain, it is assumed that the $x[n]$ may be a voltage or a current signal in standard unit (meaning voltage in Volt and current in Ampere) besides the signal is taken through a one Ohm resistor. If we square every sample in $x[n]$, that signal turns to power $p[n]$ in Watt.

As an example say $x[n] = [4 \quad 8 \quad -9 \quad 0 \quad 2]$ in V or A
then $p[n] = x[n]^2 = [16 \quad 64 \quad 81 \quad 0 \quad 4]$ in W

There are five samples in $x[n]$ so we write the n interval variation as $0 \le n \le 4$.

Now the operation $T_s \sum_{n=0}^{4} p[n]$ provides the total energy E in the $x[n]$ in Joule where T_s is the sampling period in sec i.e.
$$E = 16 + 64 + \cdots = 165 \, J \text{ when } T_s = 1 \text{sec.}$$

Next query might be how to implement signal power-energy in MATLAB. The $x[n]$ is usually entered (section 1.3) as a row or column matrix so enter the example $x[n]$ as follows:
>>x=[4 8 -9 0 2]; ↵   ← x holds the $x[n]$ as a row matrix, x is user-chosen
Scalar code of $x[n]^2$ as mentioned in appendix A computes the $p[n]$ as follows:
>>p=x.^2 ↵   ← p holds the $p[n]$, p is user-chosen variable

p =
    16   64   81   0   4
With $T_s = 1$sec, the total energy E in Joule we calculate by the command **sum** of appendix B.11:
>>E=sum(p) ↵   ← E holds the E, E is user-chosen variable

E =
    165

When the T_s is not equal to 1sec, we must multiply the E with the T_s value. For instance a sampling frequency $f_s = 2\,KHz$ is used in $x[n]$ so the T_s is $1/f_s$ that is the E would have been 165/2000 or 0.0825 Joules. You may implement that as follows:

```
>>fs=2e3; ↵            ← fs holds the f_s, fs is user-chosen variable
>>E=sum(p)/fs ↵

E =
      0.0825
```

❖ ❖ Power and energy of a digital audio

As an example we wish to compute the power and energy of the digital audio **bird.wav** of section 2.7.

Let the machine read (section 2.3) the audio as follows:

```
>>[x,fs]=wavread('bird.wav'); ↵
```

The workspace **x** and **fs** hold the discrete audio signal $x[n]$ as a column matrix and the sampling frequency f_s respectively. Earlier mentioned command we then exercise in the following:

```
>>p=x.^2; ↵            ← p holds the p[n] of the audio
```

The **p** is also a column matrix of the same size as that of **x**. Computer screen is not enough to fit all 19456 samples of $x[n]$ or $p[n]$ so suppression is used (i.e. ;) in the last command. Energy in the audio is then computed as:

```
>>E=sum(p)/fs ↵

E =
      0.0367
```

Above return says that the total energy in the audio is $0.0367\,J$.

❖ ❖ Power as a function of frequency for discrete signal

We just mentioned how to obtain power $p[n]$ from $x[n]$ certainly in integer time index domain. The integer time index has link with the sampling frequency. As quoted in chapter 6, maximum frequency in $x[n]$ can be $\frac{f_s}{2}$ or the frequency interval had better be written as $0 \le f < \frac{f_s}{2}$. We may distribute (in accordance with the sampling theorem) the N samples of $p[n]$ over $0 \le f < \frac{f_s}{2}$ with the frequency increment $\frac{f_s}{2(N-1)}$. Then power $p[n]$ as a function of discrete frequency becomes $p[f]$.

Let us consider the earlier 5 sample $p[n] = [16\ \ 64\ \ 81\ \ 0\ \ 4]$ with f_s $=200\,Hz$. Obviously the N is 5 so the 5 frequencies are $0\,Hz$, $25\,Hz$, $50\,Hz$, $75\,Hz$, and $100\,Hz$ respectively which we implement as follows:

```
>>N=length(p); ↵            ← N holds the N, N is user-chosen
>>fs=200; ↵                 ← fs holds the f_s, fs is user-chosen
```

```
>>f=0:fs/2/(N-1):fs/2 ↵          ← f holds the 5 frequencies as a row matrix
                                    (section 1.3), f is user-chosen
f =

      0    25    50    75    100
```

The same **p** now holds the power for different frequencies for example the third element in the **p** which is 81 refers to 50 *Hz* .

✦ ✦ Digital audio power as a function of frequency

Just mentioned power variation equally applies to a digital audio e.g. the earlier **bird.wav** as (with the same symbology):

```
>>[x,fs]=wavread('bird.wav'); ↵
>>p=x.^2; ↵                      ← p holds the power samples of the audio
>>N=length(p); ↵
>>f=0:fs/2/(N-1):fs/2; ↵         ← f holds the frequency samples
```

The graph of *p*[*f*] versus discrete frequency *f* is called power spectrum which you may graph by the **stem** of appendix D but continuous sense graph is better for so many samples and that happens by the command **plot** as follows:

```
>>plot(f,p) ↵
```

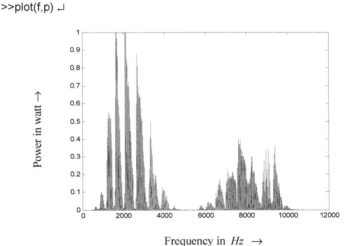

Figure 11.2(a) Power spectrum of the audio bird.wav

Figure 11.2(a) depicts the power spectrum property of the audio. If the reader is willing to see the dB watt variation, the command **plot(f,10*log10(p+eps))** is executed (section 4.8). The graph is not shown though, unnecessary dB (for example more than −50 dB) is avoided by applying the same section cited technique.

❖ ❖ Power spectral density (PSD)

PSD is a continuous term which is the Fourier transform of the autocorrelation. It is also a power spectrum (more appropriately energy spectrum) and has the unit Watt/Hertz. Discrete version of the PSD (say $P[f]$) is defined through the discrete autocorrelation of section 8.9 where f is the frequency in Hertz. The $P[f]$ of a discrete signal $x[n]$ is given by:

$$P[f] = r[0] + \sum_{k=1}^{N-1} 2r[k]\cos\left(2\pi\frac{f}{f_s}k\right) \text{ where}$$

f_s is the sampling frequency,

$r[n]$ is the one sided autocorrelation function, and

$0 \le n \le N-1$ is the interval over which the discrete signal $x[n]$ or $r[n]$ exists.

Since the $r[n]$ does not have a single definition, the $P[f]$ also has many forms moreover large sample number $x[n]$ is handled differently thus various algorithmic approaches are developed to compute the power spectrum like Burg, Welch, etc.

```
function P=pwsd(x,f,fs)
N=length(x);
M=length(f);
rr=xcorr(x);
r=rr(N:end);
P=[ ];
for m=1:M
    P=[P r(1)+sum(2*r(2:end).*cos(2*pi*f(m)*[1:N-1]/fs))];
end
```

Figure 11.2(b) Author written function file for discrete PSD computing based on without scaling autocorrelation

We introduce the PSD computing through unscaled autocorrelation $r[n]$ by function **P=pwsd(x,f,fs)** as shown in figure 11.2(b). Type the codes of the figure 11.2(b) in a new M-file (section 1.3) and save the file by the name **pwsd** in your working path of MATLAB or get the softcopy from the link of section 2.7. The input arguments of the function which are **x**, **f**, and **fs** refer to the signal $x[n]$ as a row or column matrix, given frequency f in Hertz where PSD is to be computed at, and sampling frequency f_s in Hertz respectively. The **P** return is simply the PSD value in Watt/Hertz. The f variation can be over $0 \le f \le f_s$.

❖ ❖ Power spectral density at a single frequency

Suppose $P[f]$ is $170.799\,W/Hz$ for the discrete signal $x[n] =$ [3 7 9 −10 2 0 8] at $f = 500\,Hz$ with $f_s = 4\,KHz$ which we wish to get.

The clear-cut functional calling is as follows:

>>x=[3 7 9 -10 2 0 8]; ↵   ← x holds the $x[n]$ as a row matrix, x is
user-chosen

>>f=500; fs=4e3; ↵ ← Assigning frequency values, f⇔ f , fs⇔ f_s

>>P=pwsd(x,f,fs) ↵ ← Calling the function, P⇔ $P[f]$

P =
170.7990

✦ ✦ Power spectral density at a range of frequencies

The **pwsd** also keeps provision for returning several or range of $P[f]$ values. For example for the single value mentioned $x[n]$ and f_s, the $P[f]$ is $361\,W/Hz$, $170.799\,W/Hz$, and $433\,W/Hz$ at $f =0\,Hz$, $500\,Hz$, and $1000\,Hz$ respectively which we obtain by the following:

>>f=[0 500 1000]; ↵ ← Three frequencies assigned as a row matrix to f

>>P=pwsd(x,f,fs) ↵ ← Calling the function

P =
361.0000 170.7990 433.0000   ← The $P[f]$ values as a row
matrix respectively

Incidentally if we are interested about the $P[f]$ at a range of frequencies for example over $0 \le f \le f_s$ with frequency step $50\,Hz$, we need to execute **f=0:50:fs**; first and **P=pwsd(x,f,fs)**; afterwards.

✦ ✦ Power spectral density in dB

If we perform the $10\log_{10} P[f]$ operation, the PSD becomes in dBW/Hz that is we need to execute further **Pd=10*log10(P)**; after finding the **P** where the **Pd** holds the dBW/Hz value(s) and the **Pd** is a user-chosen variable.

✦ ✦ Power spectral density of a digital audio by built-in function

The **pwsd** function is designed just to provide the basics. When we have digital audio with million samples, execution of the function takes longer which is impractical. In that case we may apply MATLAB built-in object formation technique to compute or graph the PSD.

The built-in command **psd** computes and graphs the PSD samples with a syntax **psd**(spectrum type as an object, audio samples as a column matrix, reserved word **Fs** under quote, sampling frequency). The spectrum type is written by another reserve word **spectrum** and method with the syntax **spectrum**.method. There are many methods of power spectrum computing, each of which has a reserve word for example **burg** for Burg, **welch** for Welch, etc. Without any output argument, the **psd** graphs the PSD in dBW/Hz whereas with output argument it returns the PSD absolute values in W/Hz .

Suppose we wish to exercise the PSD on the digital audio **bird.wav** based on the Welch method.

We first execute the following:

>>[x,fs]=wavread('bird.wav'); ⏎   ← x holds $x[n]$ values as a column

                                               matrix and fs⇔ f_s

Then the power spectrum object type is formed by:

>>O=spectrum.welch; ⏎   ← O is user-chosen variable and

                                             holds the spectrum type

>>P=psd(O,x,'Fs',fs) ⏎   ← Calling the function, P is user-chosen

P =

                     Name: 'Power Spectral Density'
                     Data: [16385x1 double]
              SpectrumType: 'Onesided'
             Frequencies: [16385x1 double]
       NormalizedFrequency: false
                       Fs: 22050

The P return is rather an object than a variable whose explanation is displayed as seen above. The **Data** and **Frequencies** in above display hold the PSD values in W/Hz and frequencies in Hz respectively. You may extract those values by:

>>Pd=P.Data; fr=P.Frequencies; ⏎

The **Pd** and **fr** in above execution are user-chosen variables and retain the PSD values and frequencies respectively.

Finally power spectrum display is conducted by:

>>psd(O,x,'Fs',fs) ⏎   ← Figure 11.2(c) is the result

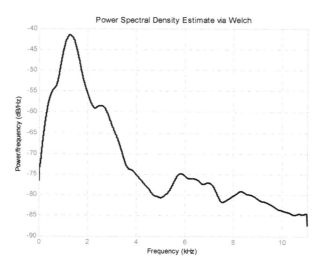

Figure 11.2(c) PSD of the audio **bird.wav** using
Welch method

-212-

The PSD so obtained might be of interest for instance the first two values of the PSD we call by:

>>Pd(1:2) ↵

ans =
    1.0e-007 *

        0.2150
        0.4301

Above return says that the first two PSD values are 0.215×10^{-7} W/Hz and 0.4301×10^{-7} W/Hz respectively. Similarly the first two frequency values we see by:

>>fr(1:2) ↵

ans =
        0
        0.6729

According to above execution the two frequencies are $0\,Hz$ and $0.6729\,Hz$ respectively.

11.4 Encoding a digital audio

A softcopy digital audio file size is dependent on bit number necessary to represent it. We know that the bit is binary digit meaning 0 or 1. Bit form representation is only possible if we can have the audio amplitude in integer form.

Let us say the modular digital audio $f[n]$ has the amplitude representation [1 2 255 2 1 3 254 3 2 3 255 1] in [0,255]. In order to represent these 256 amplitude levels, we need 8 bits (because 2^8=256). Concerning the $f[n]$, there are only five levels out of the 256 levels (which are 1, 2, 3, 254, and 255). The next power of 2 for the number five is 3 hence 3 bits are enough to represent the modular $f[n]$. The rest five bits out of eight are redundant. If we can code these five amplitude levels with suitable scheme, the redundancy is removed without losing any audio information. This redundancy removal idea is termed as the audio encoding.

According to the information theory, audio encoding is termed as the source coding. Audio amplitude levels are assumed to be the source symbol. There are many audio encoding schemes available whose reference can be [6], [11], and [12]. As an exemplary approach we introduce some audio encoding schemes namely natural encoding, run length encoding, and Huffman encoding in the following.

♦ ♦ **Natural encoding**

Natural code means the binary code of the audio amplitude levels. If there are 256 amplitude levels in the audio $f[n]$ in [0,255], the number of binary digits required is $\log_2 256$ =8.

MATLAB function **dec2bin** (abbreviation for the decimal to (2) binary) converts an integer to its equivalent binary number as a string or character array. For example the number 5 has the binary representation 101 which we exercise as follows:

With 3 bits:	With 8 bits:
>>dec2bin(5) ↵	>>dec2bin(5,8) ↵
ans =	ans =
101	00000101

If we intend to convert the integer number out of 8 bits, five zeroes are placed before the number equivalent of 5 and for which the second input argument of the **dec2bin** should be 8. Its implementation is also shown above on the right side.

For a set of numbers for example R=[5 7], the **dec2bin** returns a rectangular character matrix in which the first and second rows are the binary equivalents **101** and **111** respectively.

The reverse conversion that is binary-to-decimal takes place using the **bin2dec** (abbreviation for the binary to (2) decimal) but the input argument must be under quote. Just to verify the binary 101, we perform the following:

>>bin2dec('101') ↵

ans =
    5

For a digital audio the reader first needs the audio amplitudes to be in integer level for which section 3.2 cited functions are useful. Let us consider the digital audio **b.wav** of section 2.7 and the machine read (section 2.3) the audio as follows:

>>[x,fs]=wavread('b.wav'); ↵   ← x holds $f[n]$ values as a column matrix
                                    and fs⇔ f_s

Exercising section 2.6 mentioned **datastats** on x (i.e. execute **datastats(x)**), we come to know that the samples are in [−0.9822,0.9939]. Let us consider 256 levels for coding the audio samples meaning the data should be mapped to [0,255] and do so by:

>>xn=round(255*mat2gray(x)); ↵   ← xn holds $f[n]$ values in [0,255], xn is
                                     user-chosen variable

We code the audio amplitude integers to binary then by:

>>xb=dec2bin(xn) ↵   ← xb holds the coded bits, xb is user-chosen

xb =

01110011	← 1st sample of the audio
01110011	← 2nd sample of the audio
01110011	← 3rd sample of the audio
⋮	

✦✦ Run length encoding

The basic procedure in the run length encoding is to scan a digital audio frame and map each amplitude in the frame to the j-th amplitude level g_j and the j-th run length l_j. Along the scan frame, amplitudes having same level is the g_j and the number of repetitions of g_j in the frame is the l_j.

For discrete signal $f[n]$:

As usual we proceed with the modular discrete signal $f[n]=[1 \quad 2 \quad 2 \quad 2 \quad 3 \quad 3 \quad 254 \quad 254 \quad 1 \quad 1 \quad 3 \quad 3]$ in $[0,255]$. Assume that the scanning is to be occurred with 4 samples at a time. In the first 4 samples of $f[n]$, the amplitude levels 1 and 2 occur 1 and 3 times respectively hence $\begin{Bmatrix} g_1 = 1 \\ l_1 = 1 \end{Bmatrix}$

and $\begin{Bmatrix} g_2 = 2 \\ l_2 = 3 \end{Bmatrix}$. Again in the next 4 samples, each of the amplitude levels 3 and 254 appears two times so the coding for the second frame is $\begin{Bmatrix} g_1 = 3 \\ l_1 = 2 \end{Bmatrix}$ and $\begin{Bmatrix} g_2 = 254 \\ l_2 = 2 \end{Bmatrix}$

similarly the third frame coding is $\begin{Bmatrix} g_1 = 1 \\ l_1 = 2 \end{Bmatrix}$ and $\begin{Bmatrix} g_2 = 3 \\ l_2 = 2 \end{Bmatrix}$.

```
function code=rle(row)
%Input must be a row matrix
row=[row NaN];
code=[ ];
n=1;
while n~=length(row)
    if row(n)~=row(n+1)
        code=[code;[row(n) 1]];
    else
        k=n;
        cc=0;
        while ~(row(n)~=row(k))
            k=k+1;
            cc=cc+1;
            if k==length(row)
                break;
            end
        end
        code=[code;[row(n) cc]];
        n=k-1;
    end
    n=n+1;
end
```

Figure 11.2(d) Function rle for the run length encoding of a row matrix

For the sake of programming, we place the g_j and l_j as the row of another matrix therefore for the whole $f[n]$ we are supposed to have $\begin{matrix} g_j & l_j \\ \begin{bmatrix} 1 & 1 \\ 2 & 3 \end{bmatrix} \end{matrix}$, $\begin{matrix} g_j & l_j \\ \begin{bmatrix} 3 & 2 \\ 254 & 2 \end{bmatrix} \end{matrix}$, and $\begin{matrix} g_j & l_j \\ \begin{bmatrix} 1 & 2 \\ 3 & 2 \end{bmatrix} \end{matrix}$ for the first, second, and third frames respectively. That is our expectation from MATLAB implementation.

To the context of MATLAB, the storing of run length codes best happen by using a cell array whose discussion is seen in appendix B.21. Our written function (as seen in figure 11.2(d)) rle returns the run length codes as a rectangular matrix by taking a row matrix as the input. Type the codes of the figure 11.2(d) in the M-file editor (section 1.3) and save the file by the name **rle** in your working path of MATLAB or get the softcopy file from the link of section 2.7. Let us verify the run length codes for the first frame of the $f[n]$ as follows:

>>r1=[1 2 2 2]; ↵ ← Assigning the first frame to r1, r1 is user-chosen
>>c1=rle(r1) ↵   ← Calling the **rle** and assigning the return to the
                         variable c1, c1 is user-chosen variable

c1 =

$$\begin{array}{cc} 1 & 1 \\ 2 & 3 \end{array}$$ ← Exactly the first frame code i.e. $\begin{array}{cc} g_j & l_j \\ \begin{bmatrix} 1 & 1 \\ 2 & 3 \end{bmatrix} \end{array}$

Now we get every frame of $f[n]$, find its code **c1**, and assign the **c1** to one cell (the order is maintained as the first frame of $f[n]$ in the first cell of **code**, the second frame is in the second cell of **code**, and so on) of the cell array **code** (**code** is user-chosen variable) as follows:

>>f=[1 2 2 2 3 3 254 254 1 1 3 3]; ↵ ← Assigning the $f[n]$ to f as a
                                              row matrix, f is user-chosen
>>N=length(f); ↵   ← Finding the number of elements N in $f[n]$ by
                         length, N is user-chosen variable

In section 3.3 we illustrated how to exercise frame by frame approach. Here we apply the similar approach by:
>>for k=1:4:N, r1=f(k:k+3); c1=rle(r1); l=(k+3)/4; code{l}=c1; end ↵

A for-loop (appendix B.4) is employed in above execution. The for-loop counter **k** changes from 1 to the number of elements in $f[n]$ and the increment of which is 4 and which is the chosen frame length. Also we assumed that the **N** is multiple of frame length. The command **f(k:k+3)** selects 4 samples at a time or the frame from **f** and which is held by **r1**.

Note that the **3** is appearing in **k:k+3** from frame element number−1. The **c1** holds the code of the **r1** upon calling the function **rle**. There should be 3 run lengths and likewise the cell array **code** holds them as a row (user-chosen, could be column) cell. The integer index of the cell array is determined by (k+(frame length−1))/frame length that is why **l=(k+3)/4**; is used where the **l** is user-chosen variable for the integer index. The command **code{l}=c1**; assigns the consecutive **c1** as the l-th element to the cell. However we may call the **code** to see its contents:
>>code ↵

code =

| [2x2 double] | [2x2 double] | [2x2 double] |

$$
\begin{array}{cc} g_j & l_j \\ \begin{bmatrix} 1 & 1 \\ 2 & 3 \end{bmatrix} \end{array}
\qquad
\begin{array}{cc} g_j & l_j \\ \begin{bmatrix} 3 & 2 \\ 254 & 2 \end{bmatrix} \end{array}
\qquad
\begin{array}{cc} g_j & l_j \\ \begin{bmatrix} 1 & 2 \\ 3 & 2 \end{bmatrix} \end{array}
$$

Just to verify that the coding actually happened, we call the third cell elements by:

```
>>code{3}(:,:) ⤶
```

ans =

```
    1   2
    3   2
```

Run length coding applied to a digital audio:

From the implemented discrete signal $f[n]$, we intend to extend the encoding for the audio **b.wav** of section 2.7 with a frame length 25.

Going back to the natural code discussion in this section, the integer level audio amplitudes are stored in the workspace **xn** as a column matrix in [0,255]. Our written **rle** works on row matrix input so the **xn** is transposed by **xn'** and then assigned to **f** by:

```
>>f=xn'; ⤶        ← Basically the f holds the f[n]
```

Noticing the frame length 25, similar executions to that of $f[n]$ immediately follow as:

```
>>N=length(f); ⤶
>>for k=1:25:N, r1=f(k:k+24); c1=rle(r1); l=(k+24)/25; code{l}=c1; end ⤶
```

The workspace variable **code** now holds the run length codes for the digital audio. Referring to the workspace browser (section 1.4), the length of the **code** is 112 obviously from 2800/25 where 2800 is the total sample number of the audio. Just to see the run length code for instance the second cell element in the cell array, we perform the following:

```
>>code{2}(:,:) ⤶
```

ans =

```
    121   2
    122   2
    123   9
    124   4
    125   7
    126   1
     ↑    ↑
    g_j   l_j        ← The theory-discussed symbology
```

❖ ❖ Huffman encoding

This scheme has the following steps in the encoding:

(a) The Huffman coding starts with the probability vector of the amplitude levels of a discrete signal.

(b) The sum of all probabilities of the amplitude levels in a discrete signal is 1.

(c) To start the coding, we sort the probabilities in descending order and assign 1 to the last element and 0 to the element before the last element or vice versa.

(d) The last two probabilities are added and the new probability set is descendingly sorted again. This operation is continued until we end up with two probabilities and the assignment takes place in all intermediate probability vectors.

(e) For any unassigned probability in the vector, we check where it is occurring in the next set and bring the code for it in the current position. That is from the last sorted two, we go back to the sorted three and from the last sorted three to the last four and so on until we reach to the original probability vector we started with.

(f) The appearance of equal probability or user-assigned bit for the last probability vector does not allow us to have a unique code but the codes are optimum.

Way to implement the Huffman coding:

    The built-in function **huffmandict** determines the Huffman coding with a syntax [user-supplied variable for code dictionary, user-supplied variable for the average bit length in the coding]= **huffmandict**(discrete signal amplitude levels as a row or column matrix, probability vector as a row or column matrix).

Huffman coding applied to a discrete signal:

Example 1:

    In Huffman coding the first thing we need is the probability of the amplitude levels of a discrete signal. Let us consider the discrete signal $f[n]=[1 \quad 2 \quad 255 \quad 2 \quad 1 \quad 3 \quad 254 \quad 3 \quad 2 \quad 3 \quad 255 \quad 1]$ which holds five amplitude levels and whose probabilities are

$$\begin{pmatrix} amplitude & its \\ level & probability \\ 1 & 0.25 \\ 2 & 0.25 \\ 3 & 0.25 \\ 254 & 0.0833 \\ 255 & 0.1667 \end{pmatrix}$$. We wish to obtain these probabilities.

    Despite sections 8.1 and 8.4 explained the probability of a discrete signal, extra points are needed to be addressed while these probabilities are used in Huffman coding. Drawing the functions and symbology of those sections we execute the following:

```
>>I=[1 2 255 2 1 3 254 3 2 3 255 1]; ⏎      ← I holds f[n]
>>T=tabulate(I); ⏎      ← T holds amplitude frequency table of f[n]
>>A=T(:,1); ⏎    ← A holds amplitude levels, A is a user-chosen variable
>>p=T(:,3); ⏎    ← p holds percentage probability, p is user-chosen
>>p=p/100; ⏎    ← This p holds probability in 0-1 scale
```

One problem is associated with the **tabulate** which is inbetween level probabilities are set as 0 for example between amplitude levels 3 and 254 there is no level but the **tabulate** turns the probability of amplitude level 4, 5, 6, etc as 0. That is unnecessary in the coding. What we do is conduct the **find** of appendix B.9 to determine the 0 probability integer index and remove those index elements from the p by assigning empty matrix [ ] as follows:

```
>>c=find(p==0); ⏎          ← c holds 0 probability integer index, c is
                                 user-chosen variable
>>A(c)=[ ]; p(c)=[ ]; ⏎    ← Removing those 0 probability amplitudes
                               and probabilities
>>[A p] ⏎                  ← Just to see the outcome, appendix B.3

ans =
            1.0000    0.2500
            2.0000    0.2500
            3.0000    0.2500
          254.0000    0.0833
          255.0000    0.1667
              ↑          ↑
          amplitudes  probabilities
```

Example 2:

In this example we show how Huffman code is obtained starting from amplitude level and probability vector. Example 1 mentioned $f[n]$ has the probability vector, code, and average bit length as

$$\begin{bmatrix} amplitude & its \\ level & probability \\ 1 & 0.25 \\ 2 & 0.25 \\ 3 & 0.25 \\ 254 & 0.0833 \\ 255 & 0.1667 \end{bmatrix}, \quad \begin{bmatrix} amplitude & Huffman \\ level & code \\ 1 & 00 \\ 2 & 11 \\ 3 & 10 \\ 254 & 011 \\ 255 & 010 \end{bmatrix}, \quad \text{and} \quad 2.25$$

respectively which we intend to implement.

First we enter the amplitude levels as a row matrix by:

```
>>A=[1 2 3 254 255]; ⏎    ← A holds amplitude levels, A is
                               user-chosen
```

Then the probability entering takes place as a row matrix by:

```
>>p=[0.25 0.25 0.25 0.0833 0.1667]; ⏎ ← p holds
                               probabilities, p is user-chosen
```

The reader could have used example 1 mentioned A and p instead of the last two command lines typing. However we call the earlier function for the coding as:

>>[D,L]=huffmandict(A,p) ⏎   ← D or L is user-chosen

D =

[1]	[1x2 double]
[2]	[1x2 double]
[3]	[1x2 double]
[254]	[1x3 double]
[255]	[1x3 double]

L =

2.2500

As you see the D and L hold the Huffman code as 5×2 cell array and average bit length respectively. The first column of the D contains the amplitude levels which are also obvious from the display. The first and second columns of the cell array D you access by D{:,1} and D{:,2} respectively. All elements in the D{:,2} are seen by calling:

>>D{:,2} ⏎
ans =
      0   0   ← Code for amplitude level 1
ans =
      1   1   ← Code for amplitude level 2
ans =
      1   0   ← Code for amplitude level 3
ans =
      0   1   1   ← Code for amplitude level 254
ans =
      0   1   0   ← Code for amplitude level 255

Example 3:

When the $f[n]$ samples are not in integer amplitude level, we can not apply ongoing functions. In such event we transform the given data range to integer amplitude level in accordance with the section 3.2 cited technique. That is important for audio data which is usually in $[-1,1]$ range and decimal.

Huffman coding applied to a digital audio:

Now we intend to find the Huffman codes for the digital audio **b.wav** of section 2.7 with 8-bit allocation.

The 8-bit allocation indicates 256 amplitude levels or the audio data should be in [0,255]. Referring to the natural code discussion, the workspace **xn** holds the $f[n]$ or the audio samples as a column matrix in [0,255].

Example 1 of Huffman coding on discrete signal illustrates the probability finding which we exercise with the same symbology by:

```
>>T=tabulate(xn); ↵        ← T holds amplitude frequency table of f[n]
>>A=T(:,1); ↵  ← A holds integer amplitude levels
>>p=T(:,3); ↵  ← p holds percentage probability
>>p=p/100; ↵   ← This p holds probability in 0-1 scale
>>c=find(p==0); ↵          ← c holds 0 probability integer index
>>A(c)=[ ]; p(c)=[ ]; ↵    ← Removing those 0 probability amplitudes
                                               and probabilities
>>[A p] ↵                  ← To see the amplitudes and probabilities

ans =
          0    0.0004
    42.0000    0.0004
    43.0000    0.0004
    47.0000    0.0004
    50.0000    0.0004
    53.0000    0.0007     ← Amplitude level 53 has probability 0.0007
          ⋮
       ↑           ↑
  amplitudes   probabilities
>>[D,L]=huffmandict(A,p) ↵

D =
    [  0]    [1x11 double]
    [ 42]    [1x11 double]
    [ 43]    [1x11 double]
          ⋮
    [247]    [1x11 double]
    [250]    [1x10 double]
    [252]    [1x11 double]
    [255]    [1x10 double]
L =
      6.3418
```

As seen in above for example amplitude 53 in the audio has the probability 0.0007. Average coding bit length is 6.3418 bits and the codes are available in the variable D. Let us see the first two codes by:

```
>>D{1:2,2} ↵

ans =
     1   1   0   1   1   0   1   1   1   1   1

ans =
     1   1   0   1   1   0   1   1   1   1   0
```

Above two returns indicate that the first two amplitude levels of the digital audio (which are 0 and 42, shown before) are coded by 11011011111 and 11011011110 respectively.

✦ ✦ Coded digital audio generation

Instead of amplitude level, alphanumeric character for each amplitude can be used to represent a discrete signal or digital audio. For espionage reason or audio encryption, the coded audio is transmitted. Also the coded audio can be a relief for a system which does not support audio file transmission directly. The alphanumeric characters are completely user defined.

Character coding applied to a discrete signal:

Let us say an 8 level discrete signal in [0,7] is to be coded in terms of the capital letters A through H respectively. Considering the discrete signal $f[n]$=[0   2   5   7   4   2   7   1   5] and applying the coding, we are supposed to have the discrete signal as [A   C   F   H   E   C   H   B   F] which we wish to obtain.

Our computer keyboard has specific ASCII coding for example the character A has the code 65. MATLAB function **char** displays the ASCII character coding. Let us perform the following:

for the character A,
>>char(65) ↵

ans =

A

for the characters A through H,
>>char(65:72) ↵

ans =

ABCDEFGH

for the discrete signal $f[n]$,
>>f=[0 2 5 7 4 2 7 1 5]; ↵ ← $f[n]$ is assigned to f as a row matrix, f is
                                                            user-chosen variable
>>fc=char(f+65) ↵   ← fc is user-chosen variable, holds the codes

fc =

ACFHECHBF

Amplitude levels 65 through 72 are written by **65:72** in above middle implementation. For the given signal, the amplitude level starts with 0 or 0 is the minimum in $f[n]$.

If we add 65 to every amplitude level of $f[n]$, we follow the keyboard ASCII character coding which is carried out and assigned to the **fc** in above implementation (conducted by the command **fc=f+65**). The **fc** is a row character array which contains 9 characters exactly the number of elements in $f[n]$.

Character coding applied to a digital audio:

A good resolution digital audio does not possess 8 amplitude levels frequently 256, 512, ⋯ etc levels are seen. But that many characters may not be available for the transmission of the audio so we map the audio to predecided levels. Considering the **b.wav** of section 2.7, the audio is to be coded in [0,31] by using the characters 0 through 9 and A through V.

The problem statement requires 32 amplitude levels to represent the audio. From natural code discussion, we know that the audio samples are available in the workspace **x** (which is $f[n]$ here) as a column matrix in [−0.9822,0.9939]. We map the audio data in [0,31] complying the technique of section 3.2 by the following:

>>f=round(31*mat2gray(x)); ↵   ← f is user-chosen variable

From above execution the workspace **f** holds the audio data in [0,31]. The integers 0 through 9 have the ASCII coding 48 though 57 and the capital letters A through V have the ASCII coding 65 through 86 respectively. The number allocation is not consecutive that is why two different checkings are required for the $f[n]$.

The programming is not straightforward understandably there is a procedure. First we generate a character array of the same size as that of original audio. The total sample number in the audio we find by **length(x)**. Any character we may choose say 1. The **1** and **char(1)** are different − the former is number 1 and the latter is character 1. The command **ones(length(x),1)** means a column matrix of 1s whose size is the same as that of **f** (appendix B.10). Let us call the new character array as **fc** (which will be our coded audio subsequently) so its generation takes place by:

>>fc=char(ones(length(x),1)); ↵   ← fc is user-chosen variable

Then applying comparative operators (appendix B.1) we determine a logical (meaning element in the array is either 0 or 1) column array for which $0 \le f[n] \le 9$ is satisfied. The interval $0 \le f[n] \le 9$ is split as $0 \le f[n]$ (i.e **0<=f**) and $f[n] \le 9$ (i.e. **f<=9**) and combined by the AND operator (i.e. **0<=f&f<=9**) as follows:

>>l1=0<=f&f<=9; ↵   ← l1 is user-chosen variable

As conducted, the **l1** is a column array holding just 1s and 0s. The 1s indicate satisfying $0 \le f[n] \le 9$ whereas 0s do not. Only will be the elements having position indexes corresponding to 1s of **l1** coded by 0 through 9 which we determine by **find** of appendix B.9 as follows:

>>c1=find(l1==1); ↵   ← c1 is user-chosen variable

In terms of character, 48 should be added to 0-9 coding and we do so on the **c1** indexed elements on **f** as well as on **fc** by:

>>fc(c1)=char(48+f(c1)); ↵

For the other interval $10 \le f[n] \le 31$ analogous execution is also required and following is the continuation:

```
>>I2=10<=f&f<=31;  ⌐         ← I2 is user-chosen variable
>>c2=find(I2==1);  ⌐         ← c2 is user-chosen variable
>>fc(c2)=char(65-10+f(c2));  ⌐
```

We wrote 65-10 in the last command line because 10 in $f[n]$ should start with the character of code 65. Finally the coded audio is available in fc as a column character array. If you call the fc at the command prompt, you see the array which is not convenient for viewing. We may transpose the array from column to row (section 1.3) which is easier to perceive as follows:

```
>>fc'  ⌐
```

ans =

EEEEEEEEEE... ... ... FGGGGH
         ↑
A long row character array which is the audio b.wav

Instead of transmitting the audio amplitude level, we can transmit the character array during audio communication. Looking at the codes of the audio in the MATLAB workspace, there is no way of knowing how the audio wave shape looks until we decode it. That is the technology of secret audio transmission and reception whether military or commercial purpose.

11.5 Audio data saving and writing to a softcopy file

Two important points relevant to audio data are addressed in this section – audio data saving and writing audio soft files in MATLAB.

❖ ❖ Data saving in MATLAB workspace

Suppose you finished some calculation or simulation on audio data, wish to close MATLAB, and work on workspace variables later. In section 1.4 we explained the workspace saving and loading, the same function you may apply.

In the last section we exercised three variables namely x, f, and fc and which retain the audio b.wav samples in decimal in [−0.9822,0.9939], integer [0,31], and character coded form respectively. We wish to save these three variables. Reexecute the commands of the last section until you get the three variables. We need to select a file name say test. Then execute the following:

```
>>save test x f fc  ⌐
```

The syntax we applied in above execution is each word or variable is separated by one space. Following the save, the chosen file name is written, then are all variables. Quit MATLAB, open it later, and execute the following:

```
>>load test ↵
```

You must see the variables **x**, **f**, and **fc** in the workspace browser or execute the following to see what variables are present in the workspace:

```
>>who ↵
```

Your variables are:

f   fc   x

Note that for bringing the saved variables back, calling the file name is enough, you do not have to remember the variable names.

✦ ✦ Writing a softcopy digital audio file

Once audio analysis has been performed, the reader may need to write the audio data to a supportable softcopy file such as **wav** which we have been exercising almost in every chapter. MATLAB keeps provision for writing audio matrix data and relevant information of a digital audio to a portable softcopy file with the aid of function **wavwrite** (abbreviation for wave write). It is important to know about supportable audio formats and data classes prior to softfile writing.

The **wavwrite** employs a syntax **wavwrite**(audio matrix, sampling frequency, user-supplied quantization bits, user-supplied file name under quote). Usually audio data exists in MATLAB workspace as a row or column matrix. The quantization bits must be 8, 16, 24, or 32. If the audio matrix data is within −1 and 1 and decimal, the 32 quantization bit is applied.

✦ ✦ Example 1

Let us say we have the digital audio modular matrix $f[n]=[1$ $0\ 0\ -1\ 1\ 1\ 1\ 0\ 0\ -1]$, wish to write the audio matrix as a **wav** file based on sampling frequency $10\,KHz$ with 32-bit quantization, and name the softfile as **Myaudio**. The execution is the following:

```
>>f=[1 0 0 -1 1 1 1 0 0 -1]; ↵   ← Assigning the audio matrix f[n] to f,
                                    f is user-chosen variable
>>wavwrite(f,10e3,32,'Myaudio') ↵ ← Writing the audio data to the file
```

Note that whether the **f** content is a row or column matrix, the audio data is always saved as a column matrix. To see all files present in the working directory, you may exercise the command **dir**:

```
>>dir ↵
.        f1.m       test.m    test1.wav
..       mmax.m     Myaudio.wav  test2.wav
                        ↑
```
This is the file which has been written just now

How to verify that the audio data is written? Well the audio reader of section 2.3 may be invoked to check:

```
>>[x,fs]=wavread('Myaudio.wav') ↵
```

x =

1
0
0
-1
1
1
1
0
0
-1

fs =

10000

❖ ❖ Example 2

In the syntax of **wavwrite**, the other quantization bits 8, 16, and 24 need extra attention. For 8 bits, there are $2^8=256$ amplitude levels. Half of that level is 128. The audio data must be between −127 and 127 divided by 128. As an example the $f[n]$ can be [10   5   3   −10   127   127   25   0   0   −127]/128. If the given audio data is not so, the range mapping of section 3.2 is applied. Similar explanation goes for the 16 and 24 bits. With the 16 bits quantization the audio data must be between −255 and 255 divided by 256. Also note that without the division by 8 or 16 the audio signal amplitude is all integers.

❖ ❖ Example 3

For stereo type audio data we have two sets of audio samples – one for the left speaker and the other for the right. Suppose the left and right speaker samples are $f[n]|_{left}$=[10   5   3   −10   127   25   0   −127]/128 and $f[n]|_{right}$=[11   5   3   −10   126   25   0   −127]/128 respectively with sampling frequency 10 *KHz* and 8-bit quantization. The softfile name should be **Myaudio1**.

From the stereo samples we form a two column (section 1.3 and appendix B.3) rectangular matrix as follows:

>>f1=[10 5 3 -10 127 25 0 -127]'/128; ↵   ← Assigning the left stereo
                                data as a column matrix to f1, f1 is user-chosen variable
>>f2=[11 5 3 -10 126 25 0 -127]'/128; ↵   ← Assigning the right stereo
                                data as a column matrix to f2, f2 is user-chosen variable
>>f=[f1 f2]; ↵ ← Forming a two column matrix f, f is user-chosen
>>wavwrite(f,10e3,8,'Myaudio1') ↵ ← Writing the audio data to the file

MATLAB as an open working platform keeps enormous programming resources for study and research of digital audio problems. Application dependent digital audio processing needs graphical user interface design which is also implementable in MATLAB. We hope pertinent minute details of the digital audio linked programming would inspire the reader to conduct his/her study and research in the platform. However we bring an end to the chapter with this.

Chapter 12

 ## Mini Problems on Discrete Signal-System and Digital Audio

In this chapter discrete signal-system or digital audio problems from all chapter headings are randomly accumulated. Instead of organizing the problems on chapter specific term, we treat similar signal or audio processing problem as a mini project because each one needs MATLAB computing or graphing. Attached answer and associated clue in every mini project would provide some level of confidence to the reader on audio study in MATLAB. Although the mini projects seem simple, they pave the way for insight or study tools of much more intricate audio related phenomena in the field. However the mini project collection is as follows:

❖❖ **Project 1 (on sampling fundamentals of a digital audio)**
Suppose the softcopy digital audio **test.wav** is given:
(a) obtain the discrete audio data $v[n]$ as a column matrix, sampling frequency f_s, and number of samples M in the **test.wav**.
(b) play the audio using MATLAB audio player.
(c) obtain the sampling period T_s of the audio.
(d) what is the duration of the audio assuming $t = 0$ start?

Answers: (a) $f_s=22.05\ KHz$ and $M=161002$, (c) $T_s=0.0454\ m\sec$, and (d) 7.3016 sec over $0\le t\le 7.3016$ sec

Hint: sections 2.1-2.5

✦✦ Project 2 (on forward Z transform)

Apply the MATLAB built-in function **ztrans** to determine the unilateral forward Z transform for each of the following discrete signals:

(a) $f[n]=\sin 2n-\cos 2n$ (b) $f[n]=4^{-n}\cos n$

(c) $f[n]=2^n-3^{-3n}$ (d) $f[n]=n^3 3^{-n}$

(e) $f[n]=4\delta[n-3]$ (f) $f[n]=2^{-n}(u[n]-u[n-3])$

Answers: (a) $F(z)=\dfrac{z\sin 2+z\cos 2-z^2}{1-2z\cos 2+z^2}$ (b) $F(z)=\dfrac{4z(4z-\cos 1)}{16z^2-8z\cos 1+1}$

(c) $F(z)=\dfrac{53z}{(z-2)(27z-1)}$ (d) $F(z)=\dfrac{3z(9z^2+12z+1)}{(3z-1)^4}$

(e) $F(z)=\dfrac{4}{z^3}$ (f) $F(z)=1+\dfrac{1}{2z}+\dfrac{1}{4z^2}$

Hint: section 6.1

✦✦ Project 3 (on discrete Fourier transform)

Compute the forward discrete Fourier transform $F[k]$ for each of the following discrete signals:

(a) $f[n]=[9\quad 3\quad 5\quad 0\quad 1\quad 7\quad -12]$

(b) $f[n]=3\,e^{-2n}$ over $0\le n\le 5$

(c) $f[n]=3\sin\dfrac{2\pi n}{5}+5\cos\dfrac{2\pi n}{5}$ over $0\le n\le 4$

(d) $f[n]$ of (b) in magnitude-angle form

(e) $f[n]$ of (c) only for $F[1]$ in magnitude-angle form

Answers:

(a) $F[k]=[13\qquad -0.1826-j\,9.3437\qquad 0.8146-j\,16.2735$
$24.3681-j\,7.0970\qquad 24.3681+j\,7.0970\qquad 0.8146+j\,16.2735$
$-0.1826+j\,9.3437]$

(b) $F[k]=[3.4695\quad 3.1677-j\,0.3982\quad 2.7764-j\,0.3048\quad 2.6424$
$2.7764+j\,0.3048\quad 3.1677+j\,0.3982]$

(c) $F[k]=[0\quad 12.5-j\,7.5\quad 0\quad 0\quad 12.5+j\,7.5]$

(d) $F[k]=[3.4695\angle 0^\circ\quad 3.1926\angle-7.1651^\circ\quad 2.7931\angle-6.2646^\circ$
$2.6424\angle 0^\circ\quad 2.7931\angle 6.2646^\circ\quad 3.1926\angle 7.1651^\circ]$

(e) $14.5774\angle-30.9638^\circ$, use **abs(F(2))** and **180*angle(F(2))/pi** after **F=fft(f)**

Hint: section 4.1

✦✦ Project 4 (on forward Z transform of finite sequences)

Determine the forward Z transform for each of the following finite sequences by using MATLAB built-in functions:

(a) $f[n]=\begin{cases} 3^{-n} & for\ 0\le n<4 \\ 0 & elsewhere \end{cases}$

(b) $\begin{cases} f[n] \to & -2 & -4 & 3 & 0 & 12 & 16 \\ n \to & 2 & 3 & 4 & 5 & 6 & 7 \end{cases}$

(c) $f[n]=\begin{cases} e^{-n}\sin\dfrac{\pi(n-1)}{5} & for\ 1\le n<5 \\ 0 & elsewhere \end{cases}$

(d) The finite sequence envelope of the figure 12.1(a) follows $f[n]= 100-n^2$ over $0\le n\le7$

Answers:

(a) $F(z)=1+\dfrac{1}{3z}+\dfrac{1}{9z^2}+\dfrac{1}{27z^3}$

(b) $F(z)=\dfrac{-2z^5-4z^4+3z^3+12z+16}{z^7}$

(c) $F(z)=\dfrac{e^{-2}\left(z^2\sin\dfrac{\pi}{5}+e^{-1}z\sin\dfrac{2\pi}{5}+e^{-2}\sin\dfrac{2\pi}{5}\right)}{z^4}$

(d) $F(z)=100+\dfrac{99}{z}+\dfrac{96}{z^2}+\dfrac{91}{z^3}+\dfrac{84}{z^4}+\dfrac{75}{z^5}+\dfrac{64}{z^6}+\dfrac{51}{z^7}$

Hint: section 6.1

✦✦ **Project 5 (on digital audio amplitudes)**

Following questions are concerning the digital audio **test.wav** (section 2.7):

(a) obtain the first amplitude of the audio assuming $t=0$ start.

(b) obtain the first five amplitudes of the audio assuming $t=0$ start.

(c) obtain the 100-th through 105-th amplitudes of the audio assuming $t=0$ start.

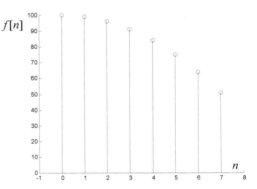

Figure 12.1(a) A finite sequence follows parabolic envelope

(d) what is the maximum audio amplitude?

(e) what is the minimum audio amplitude?

(f) what does continuous time interval of the audio cover if amplitudes 100000-th through 120000-th are selected assuming $t=0$ start?

(g) plot discrete amplitude versus sample number of the digital audio signal.

(h) plot amplitude versus time of the digital audio signal assuming $t = 0$ start.

Answers: (a) 0.0313, (b) [0.0313 0.0469 0.0547 0.0625 0.0625],
(c) [0.0469 0.0391 0.0391 0.0313 0.0156 0.0078]
(d) $v(t)|_{max} = 0.1484$ (e) $v(t)|_{min} = -0.1172$ (f) $4.5351 \sec \leq t \leq 5.4421 \sec$
(g) figure 12.1(b) (h) figure 12.1(c)
Hint: section 2.5

Figure 12.1(b) Plot of the amplitudes of audio test.wav versus sample number – right side figure

Figure 12.1(c) Plot of the amplitudes of audio test.wav versus continuous time – right side figure

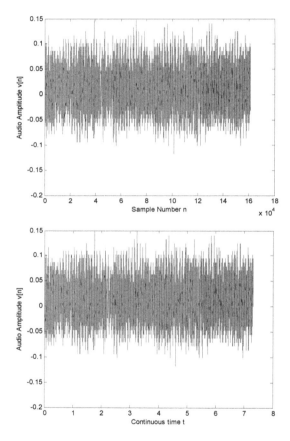

✦ ✦ Project 6 (on discrete Fourier transform)

Following questions are related to the digital audio test.wav (section 2.7):

(a) Graph the magnitude spectrum $|F[k]|$ versus integer frequency index k of the digital audio

(b) Graph the magnitude spectrum $|F[k]|$ versus discrete frequency of the digital audio

(c) Graph the magnitude, phase, real, and imaginary spectra versus discrete frequency of the digital audio in the same plot

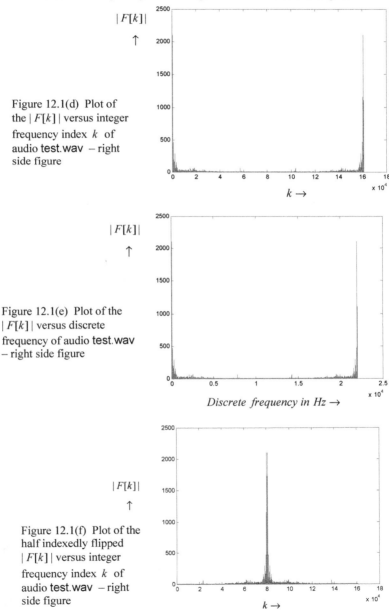

Figure 12.1(d) Plot of the $|F[k]|$ versus integer frequency index k of audio test.wav — right side figure

Figure 12.1(e) Plot of the $|F[k]|$ versus discrete frequency of audio test.wav — right side figure

Figure 12.1(f) Plot of the half indexedly flipped $|F[k]|$ versus integer frequency index k of audio test.wav — right side figure

(d) Graph the magnitude spectrum of the audio following the half index flipping

Answers: (a) Figure 12.1(d), (b) Figure 12.1(e), (c) Figure 12.1(g), and (d) Figure 12.1(f)

Hint: sections 4.4 and 4.5

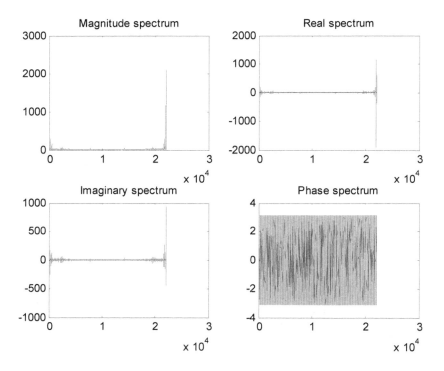

Figure 12.1(g) Four spectra of digital audio test.wav

✦✦ Project 7 (on discrete Fourier transform)

Following questions are related to the digital audio test.wav (section 2.7):

(a) determine the pitch or fundamental frequency of the audio
(b) determine the $|F[k]|_{\max}$ value from the magnitude spectrum of the audio
(c) how many frequencies constitute the digital audio?
(d) find the first five frequencies that constitute the digital audio
(e) find the 101-th through 105-th frequencies that constitute the digital audio
(f) find the first five amplitudes of $|F[k]|$ from the magnitude spectrum of the audio

Answers: (a) 59.9866 Hz (b) $|F[k]|_{max}$=2.0999×10^3 (c) 161002 (d)
[0 Hz 0.1370 Hz 0.2739 Hz 0.4109 Hz 0.5478 Hz] (e)
[13.6956 Hz 13.8325 Hz 13.9695 Hz 14.1064 Hz 14.2434 Hz]
(f) [1.7298×10^3 0.0319×10^3 0.0381×10^3 0.0117×10^3
0.0462×10^3]
Hint: sections 4.6 and 4.9

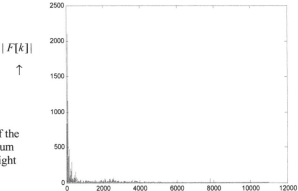

Figure 12.2(a) Plot of the
half magnitude spectrum
of audio test.wav – right
side figure

Discrete frequencies in Hz →

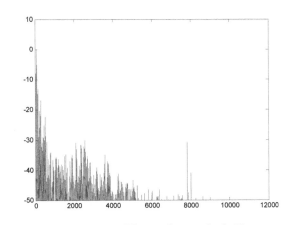

Figure 12.2(b) Plot of
the half magnitude
spectrum of audio
test.wav in decibel scale
clipped to −50dB – right
side figure

Discrete frequencies in Hz→

✦ ✦ Project 8 (on discrete Fourier transform)

Following questions are concerning the digital audio **test.wav**
(section 2.7):

(a) obtain the half magnitude spectrum of the audio and plot the
spectrum against discrete frequencies.

(b) plot the spectrum of the part (a) in decibel scale but a −50 dB clipping must be conducted before plotting.

(c) obtain the following basic statistics on the whole magnitude spectrum of the audio: number of samples, mean, median, maximum, minimum, and standard deviation.

(d) what inference do you make from the graphs of the parts (a) and (b) regarding discrete frequency?

Answers: (a) Figure 12.2(a), (b) Figure 12.2(b), (c) 161002, 4.1055, 2.0389, 2.0999×10^3, 0.0019, and 13.5002 respectively (d) insignificant frequency components in the spectrum are pronounced due to the logarithmic operation

Hint: sections 4.7, 4.8, and 4.10

✦✦ **Project 9 (on discrete STFT)**

(a) Suppose the window function $w[m] = \begin{cases} 1 & 0 \le m \le 4 \\ 0 & elsewhere \end{cases}$ is used to compute the STFT. Write a function such that the softfile returns the discrete window functional value for one m at a time and verify the written function file by calling for different value of m at the command prompt.

(b) By applying the window function of the part (a) and the direct formula of the transform, compute the discrete STFT coefficient $F[3,2]$ for the discrete signal $f[n]=[9\ -32\ -7\ -5\ -19\ 3]$.

(c) Obtain the whole $F[n,k]$ matrix for the part (b).

(d) Verify that $f[n]=-7$ for $n=2$ starting from the $F[n,k]$ matrix of part (c).

(e) Verify that you are able to recover the signal $f[n]$ from the $F[n,k]$ matrix of the part (c).

Answers: (a) Similar to the window function of section 5.1, only difference is the interval, (b) $F[3,2]=23.5+j\,21.6506$, and (c) $F[n,k]=$

$$\begin{bmatrix} 9 & 9 & 9 & 9 & 9 & 9 \\ -23 & -7+j27.7128 & 25+j27.7128 & 41 & 25-j27.7128 & -7-j27.7128 \\ -30 & -3.5+j33.775 & 28.5+j21.6506 & 34 & 28.5-j21.6506 & -3.5-j33.775 \\ -35 & 1.5+j33.775 & 23.5+j21.6506 & 39 & 23.5-j21.6506 & 1.5-j33.775 \\ -54 & 11+j17.3205 & 33+j38.1051 & 20 & 33-j38.1051 & 11-j17.3205 \\ -60 & 3.5+j19.9186 & 22.5+j40.7032 & 8 & 22.5-j40.7032 & 3.5-j19.9186 \end{bmatrix}$$

Hint: section 5.1

✦✦ **Project 10 (on windowing for discrete STFT)**

(a) Generate the functional values of a 6 point Hamming window as a column matrix.

(b) Generate the functional values of a 6 point Hanning window as a column matrix.

(c) Generate the functional values of a 7 point Blackman window as a column matrix.

(d) Apply the Hamming window of the part (a) to determine the discrete STFT coefficients $F[n,k]$ for the discrete signal $f[n]=[2\ -5\ -9\ 0\ 3]$.

(e) Verify that $f[n]=-9$ for $n=2$ starting from the $F[n,k]$ matrix of part (d).

(f) Verify that you are able to recover the signal $f[n]$ from the $F[n,k]$ matrix of the part (d).

Answers: (a) $w[m]=[0.08\quad 0.3979\quad 0.9121\quad 0.9121\quad 0.3979\quad 0.08]$, (b) $w[m]=[0\quad 0.3455\quad 0.9045\quad 0.9045\quad 0.3455\quad 0]$, (c) $w[m]=[0\quad 0.13\quad 0.63\quad 1\quad 0.63\quad 0.13\quad 0]$, (d) $F[n,k]=$

$$\begin{bmatrix} 0.16 & 0.16 & 0.16 & 0.16 & 0.16 \\ 0.3957 & 0.6721+j0.3804 & 1.1193+j0.2351 & 1.1193-j0.2351 & 0.6721-j0.3804 \\ -0.885 & 1.7921+j2.3151 & 3.2111+j0.4845 & 3.2111-j0.4845 & 1.7921-j2.3151 \\ -6.3171 & 3.3118+j6.4422 & 4.4075-j0.7247 & 4.4075+j0.7247 & 3.3118-j6.4422 \\ -11.7344 & 6.1020+j9.3911 & 1.7544-j4.9857 & 1.7544+j4.9857 & 6.1020-j9.3911 \end{bmatrix}$$

Hint: section 5.2

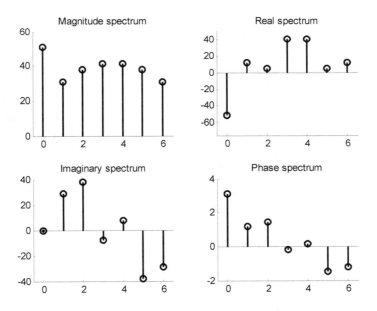

Figure 12.2(c) Plots of the four discrete spectra of the discrete signal in project 11(a) – each horizontal axis is k variation

✦✦ Project 11 (on discrete Fourier transform)

(a) Graph the $|F[k]|$, $\mathrm{Re}\{F[k]\}$, $\mathrm{Im}\{F[k]\}$, and $\angle F[k]$ spectra of the discrete signal $f[n]=[9\ -32\ -7\ -5\ -19\ 3\ 0]$.

(b) Consider the continuous sine function $x(t)$ of $\left\{\begin{array}{l}amplitude \pm 5 \\ frequency\ 20 Hz\end{array}\right\}$
over $0 \le t \le 0.2 \sec$. By choosing a sampling frequency $f_s = 200\ Hz$, verify that one strong peak exists in the half magnitude spectrum of $|F[k]|$ against discrete frequencies at exactly 20 Hz.

(c) Consider the identical phase and two frequency continuous sine function $x(t)$ of $\left\{\begin{array}{l}amplitude \pm 10 \\ frequency\ 30 Hz\end{array}\right\}$ and $\left\{\begin{array}{l}amplitude \pm 2 \\ frequency\ 60 Hz\end{array}\right\}$ over $0 \le t \le 0.1 \sec$. By choosing a sampling frequency $f_s = 300\ Hz$, verify that two strong peaks exist in the half magnitude spectrum of $|F[k]|$ against discrete frequencies at exactly 30 Hz and 60 Hz.

(d) Redo the part (c) with interchanged amplitudes i.e. 60 Hz and 30 Hz frequencies have amplitudes 10 and 2 respectively.

(e) What do you infer from the results of the parts (c) and (d)?

Answers: (a) Figure 12.2(c) and (e) larger amplitude of the two frequencies shows stronger peak in the spectrum.

Hint: sections 4.2 and 4.3

❖❖ **Project 12 (on forming a Toeplitz from discrete signals)**

(a) Consider the discrete signal $x[n] = [-1\ 0\ 3\ -70\ 4\ 3\ 9\ 8]$ for singular value decomposition. By choosing $L=3$, obtain the special Toeplitz matrix $f[m,n]$ for the decomposition.

(b) In part (a) now $L=4$

(c) In part (a) now $L=5$

Answers: (a) $f[m,n] = \begin{bmatrix} 3 & 0 & -1 \\ -70 & 3 & 0 \\ 4 & -70 & 3 \\ 3 & 4 & -70 \\ 9 & 3 & 4 \\ 8 & 9 & 3 \end{bmatrix}$, (b) $f[m,n] = \begin{bmatrix} -70 & 3 & 0 & -1 \\ 4 & -70 & 3 & 0 \\ 3 & 4 & -70 & 3 \\ 9 & 3 & 4 & -70 \\ 8 & 9 & 3 & 4 \end{bmatrix}$,

and (c) $f[m,n] = \begin{bmatrix} 4 & -70 & 3 & 0 & -1 \\ 3 & 4 & -70 & 3 & 0 \\ 9 & 3 & 4 & -70 & 3 \\ 8 & 9 & 3 & 4 & -70 \end{bmatrix}$

Hint: section 7.7

❖❖ **Project 13 (on range mapping of discrete signals)**

(a) A discrete signal $v[n]$ is given as $[-8\ \ 3\ \ \ 61\ \ -29\ \ -10\ \ 9\ \ 13]$. Map the signal data in $[-1,1]$ range by using the mathematical expression approach and by applying the built-in function mat2gray.

(b) Map the signal in part (a) to $[0,255]$ integer range for unsigned 8-bit integer storage.

(c) Map the signal in part (a) to [0,511] integer range.
Answers: (a) [−0.5333 −0.2889 1 −1 −0.5778 −0.1556 −0.0667],
(b) [60 91 255 0 54 108 119], and (c) [119 182 511 0
108 216 238]
Hint: section 3.2

❖❖ Project 14 (on discrete signal mixing)

(a) Two discrete signals of identical sampling frequency are given
as $v_1[n]$ =[0.9 −0.6 0.3 0.69] and $v_2[n]$=[−0.79 −0.6 0.46
0.49 1] each in the range [−1,1]. The dominance factors for the
two signals are c_1 =0.52 and c_2=0.32 respectively. Compute the
resultant discrete signal in the [−1,1] range.
(b) Compute the resultant discrete signal in part (a) in [0,255]
integer range.
(c) Compute the resultant discrete signal in part (a) in [0,511]
integer range.
Answers: (a) [0.4107 −1 0.5834 1 0.6163], (b) [180 0 202
255 206], and (c) [360 0 405 511 413]
Hint: section 3.6

❖❖ Project 15 (on STFT of mathematical functions)

(a) Compute the STFT of the discrete mathematical function
$f[n] = 2n+2$ over $0 \leq n \leq 5$ based on the rectangular window
$$w[m] = \begin{cases} 1 & -2 \leq m \leq 2 \\ 0 & elsewhere \end{cases}.$$
(b) Compute the STFT of the discrete mathematical function
$f[n] = \sin\frac{n}{5}$ over $0 \leq n \leq 5$ based on the rectangular window
$$w[m] = \begin{cases} 1 & -2 \leq m \leq 2 \\ 0 & elsewhere \end{cases}.$$

Answers:
(a)
$$F[n,k] = \begin{bmatrix} 12 & 1-j8.6603 & -3+j1.7321 & 4 & -3-j1.7321 & 1+j8.6603 \\ 20 & -7-j8.6603 & 5+j1.7321 & -4 & 5-j1.7321 & -7+j8.6603 \\ 30 & -12 & -j6.9282 & 6 & j6.9282 & -12 \\ 40 & -8+j10.3923 & -8+j3.4641 & -8 & -8-j3.4641 & -8-j10.3923 \\ 36 & -10+j13.8564 & -6+j6.9282 & -4 & -6-j6.9282 & -10-j13.8564 \\ 30 & -7+j19.0526 & -3+j1.7321 & -10 & -3-j1.7321 & -7-j19.0526 \end{bmatrix}$$

(b)
$$F[n,k] = \begin{bmatrix} 0.5881 & -0.0954-j0.5093 & -0.294+j0.1652 & 0.1907 \\ 1.1527 & -0.66-j0.5093 & 0.2706+j0.1652 & -0.3739 \\ 1.8701 & -1.0187+j0.1119 & -0.0881-j0.4561 & 0.3435 \\ 2.7116 & -0.598+j0.8407 & -0.5088+j0.2727 & -0.498 \\ 2.5129 & -0.6973+j1.0127 & -0.4095+j0.4447 & -0.2993 \\ 2.1235 & -0.5026+j1.35 & -0.2148+j0.1075 & -0.6888 \end{bmatrix}$$

Mohammad Nuruzzaman

$$\begin{bmatrix} -0.294 - j0.1652 & -0.0954 + j0.5093 \\ 0.2706 - j0.1652 & -0.66 + j0.5093 \\ -0.0881 + j0.4561 & -1.0187 - j0.1119 \\ -0.5088 - j0.2727 & -0.598 - j0.8407 \\ -0.4095 - j0.4447 & -0.6973 - j1.0127 \\ -0.2148 - j0.1075 & -0.5026 - j1.35 \end{bmatrix}$$

Hint: section 5.1

Figure 12.2(d)
Spectrogram of the
audio test.wav –
right side figure

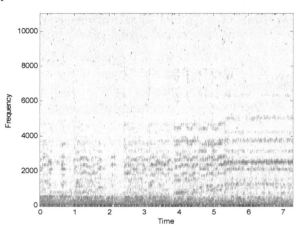

Figure 12.2(e)
Spectrogram of
the audio
hello.wav – right
side figure

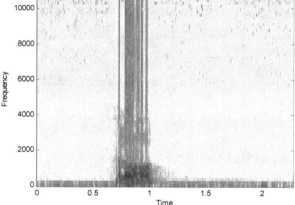

❖ ❖ Project 16 (on spectrographic analysis of an audio)

(a) Obtain the test.wav audio (section 2.7) and graph the spectrogram of the audio considering 256 as DFT point number.
(b) Obtain the hello.wav audio and graph the spectrogram of the audio considering 128 as DFT point number.
(c) What are the time and the frequency intervals of the spectrogram in part (a)?
(d) What are the time and the frequency intervals of the spectrogram in part (b)?

Answers: (a) Figure 12.2(d) (b) Figure 12.2(e)   (c) $0 \le t \le 7.2853\,\text{sec}$ and $0 \le f \le 11025Hz$ (d) $0 \le t \le 2.2785\,\text{sec}$ and $0 \le f \le 11025Hz$
Hint: section 5.5

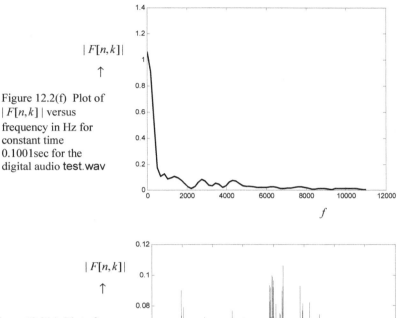

$|F[n,k]|$

↑

Figure 12.2(f) Plot of $|F[n,k]|$ versus frequency in Hz for constant time 0.1001sec for the digital audio **test.wav**

f

$|F[n,k]|$

↑

Figure 12.2(g) Plot of $|F[n,k]|$ versus time in sec for constant frequency 5.857 KHz for the digital audio **test.wav**

t

❖❖ Project 17 (on spectrographic analysis of an audio)

(a) Compute the discrete STFT $F[n,k]$ of the audio **test.wav** (section 2.7) by choosing the parameters N, M, and P as 128, 64, and 32 respectively and default Hanning window for computation where the symbols have section 5.3 cited meanings.

(b) From the $F[n,k]$ of the part (a), obtain the $F[n,k]$ versus time data for the constant frequency corresponding to $k = 34$ and graph the data. What is that constant frequency? What is the time interval of the graph?

(c) From the $F[n,k]$ of the part (a), obtain the $F[n,k]$ versus frequency data for the constant time corresponding to $n=69$ and graph the data. What is that constant time? What is the frequency interval of the graph?

(d) Do the same as in part (a) for the audio **hello.wav** by choosing the parameters N, M, and P as 256, 128, and 64 respectively.

(e) From the $F[n,k]$ of the part (d), obtain the $F[n,k]$ versus time data for the constant frequency corresponding to $k=124$ and graph that. What is that constant frequency? What is the time interval?

(f) From the $F[n,k]$ of the part (d), obtain the $F[n,k]$ versus frequency data for the constant time corresponding to $n=178$ and graph that. What is that constant time? What is the frequency interval of the graph?

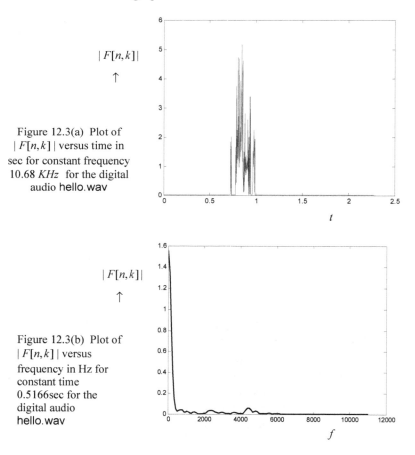

Figure 12.3(a) Plot of $|F[n,k]|$ versus time in sec for constant frequency 10.68 KHz for the digital audio hello.wav

Figure 12.3(b) Plot of $|F[n,k]|$ versus frequency in Hz for constant time 0.5166sec for the digital audio hello.wav

Answers: (a) Computation result is available in workspace variable (b) Figure 12.2(g), 5.857 KHz , and $0 \le t \le 7.2983$ sec (c) Figure 12.2(f), 0.1001sec, and $0 \le f \le 11025 Hz$ (d) Computation result is available in workspace variable (e) Figure 12.3(a), 10.68 KHz , and $0 \le t \le 2.2785$ sec (f) Figure 12.3(b), 0.5166sec, and $0 \le f \le 11025 Hz$

Hint: section 5.6

✦✦ Project 18 (on frame by frame selection of an audio)

(a) Write a program so that you can choose sequentially 4 samples at a time from the discrete signal $f[n]$=[1 4 5 1 −7 5 0 −1 3 9].

(b) If the frame duration of an audio signal is taken as 12 m sec at 22050 Hz sampling frequency, how many samples should be considered for the frame selection?

(c) Consider the softcopy audio **test.wav** of section 2.7 and choose a 12 m sec frame duration. Play the audio by the built-in function of MATLAB, select the audio frame duration as 12 m sec , select sequentially all frames in the audio, put the frames in order, and play the audio which is formed by placing the frames sequentially.

Answers: (b) 265

Hint: section 3.3

✦✦ Project 19 (on audio listening)

(a) Obtain the softcopy **test.wav** in your working path of MATLAB (section 2.7) and verify that the file exists in the working path.

(b) Let your computer read the softcopy audio of the part (a).

(c) Play the audio of the part (a) in MATLAB.

(d) Do the same as in parts (a), (b), and (c) for the audio **b.wav**.

(e) Do the same as in parts (a), (b), and (c) for the audio **hello.wav**.

Hint: section 2.3

✦✦ Project 20 (on audio amplitude characteristics)

(a) Following questions are concerning the digital audio **b.wav** (section 2.7), obtain the basic statistics of the audio amplitudes i.e. number of samples, mean, median, maximum, minimum, range, and standard deviation.

(b) Do the same as in part (a) for the **hello.wav**.

(c) Do the same as in part (a) for the **pitch.wav**.

Answers: (a) 2800, 9.7501×10^{-4}, 0.0013, 0.9939, −0.9822, 1.9761, and 0.1915 respectively (b) 50388, 0.0098, 0.0055, 0.9998, −1,

1.9998, and 0.1862 respectively (c) 6616, -8.3028×10^{-7}, 0, 1, -1, 2, and 0.7072 respectively

Hint: section 2.6

✦✦ Project 21 (on audio scaling)

(a) Obtain the softcopy test.wav (section 2.7) in your working path of MATLAB and scale up the amplitudes of the audio by a factor 3. Verify the scaling operation by playing both the original audio and the amplified one.

(b) Do the same as in part (a) for the audio b.wav but by a factor 1.5.

(c) Do the same as in part (a) for the audio hello.wav but scaling down operation of factor 0.5.

Hint: section 3.1

✦✦ Project 22 (on STFT of a discrete and an audio signals)

(a) By applying the built-in function of MATLAB, compute the forward STFT $F[n,k]$ of the discrete signal $f[n]=[-1\ \ 34\ \ 9\ \ -45\ \ 23\ \ 0\ \ 4\ \ -7]$ based on the Hanning window with $N=8$, $M=4$, and $P=2$ and a sampling frequency $f_s=2000\ Hz$ where the symbols have section 5.3 cited meanings.

(b) What are the discrete frequencies referring to the rows of the $F[n,k]$ in part (a)?

(c) What are the discrete times referring to the columns of the $F[n,k]$ in part (a)?

(d) Do the same as in part (a) for the discrete sine signal $f[n]=\sin\dfrac{2\pi n}{4}$ with $N=4$, $M=2$, and $P=1$ over the interval $0\le n\le 4$.

(e) What are the discrete frequencies referring to the rows of the $F[n,k]$ in part (d)?

(f) What are the discrete times referring to the columns of the $F[n,k]$ in part (d)?

(g) What do you infer from the values of $F[n,k]$ in part (d)?

(h) By applying the built-in function of MATLAB, determine the forward STFT $F[n,k]$ of the digital audio b.wav (section 2.7) based on the Hanning window with $N=128$, $M=64$, and $P=16$.

(i) What is the matrix dimension of $F[n,k]$ in part (h)?

(j) Do the same as in part (h) but for the audio hello.wav with $N=256$, $M=128$, and $P=32$.

(k) What is the matrix dimension of $F[n,k]$ in part (j)?

Answers: (a)

$$F[n,k] = \begin{bmatrix} 23.0013 & -16.7898 & 9.1459 \\ 32.3938 - j\,18.893 & -25.6719 + j\,7.9776 & 9.6564 - j\,1.9079 \\ -8.4861 - j\,46.3004 & -17.6943 + j\,40.7029 & 4.3283 - j\,2.4184 \\ -33.0848 - j\,2.6118 & 31.8907 + j\,49.5850 & 6.2362 + j\,5.3281 \\ -7.4111 & 64.616 & 13.9828 \end{bmatrix}$$

(b) $0\,Hz$, $250\,Hz$, $500\,Hz$, $750\,Hz$, and $1000\,Hz$ respectively
(c) 0sec, 0.001sec, and 0.002sec respectively
(d)

$$F[n,k] = \begin{bmatrix} 0.75 & 0.75 & -0.75 & -0.75 \\ -j\,0.75 & 0.75 & j\,0.75 & -0.75 \\ -0.75 & 0.75 & 0.75 & -0.75 \end{bmatrix}$$

(e) $0\,Hz$, $500\,Hz$, and $1000\,Hz$ respectively
(f) 0sec, 0.0005sec, 0.001sec, and 0.0015sec respectively
(g) The value of $F[n,k]$ is either real or imaginary for a sinusoidal signal.
(h) The computation result will be available in workspace.
(i) 65×58   (j) The computation result will be available in the workspace. (k) 129×524
Hint: sections 5.3 and 5.4

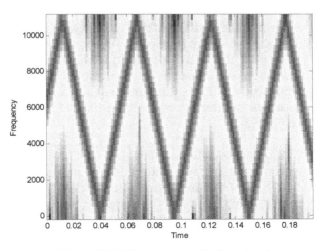

Figure 12.3(c) Spectrogram of a chirp signal

♦ ♦ Project 23 (on spectrogram of chirp signals)

A chirp signal is one whose frequency changes with time. The signal has different forms of representation for example sine, cosine, or exponential.

The variation of frequency may take place linearly, quadratically, or other with respect to time.

(a) Suppose a linear chirp signal is given by $v(t) = \sin[2\pi(f_0 + \beta t)t]$ where $f_0 = 6KHz$ and $\beta = 0.2KHz/m\sec$. Considering a sampling frequency $22050Hz$, generate the samples of the chirp signal over the interval $0 \le t \le 200m\sec$.

(b) Play the audio of the part (a) in MATLAB.

(c) Graph the spectrogram of the chirp signal in part (a) considering a 64 point DFT.

(d) What do you infer from the spectrogram of the chirp signal in part (c)?

Answers: (a) use scalar code to generate the samples which is fs= 22050; t=0:1/fs:200e-3; v=sin(2*pi*(6e3+0.2e3/1e-3*t).*t); (c) Figure 12.3(c) (d) Frequencies of the linear chirp signal show triangular variation over time.
Hint: appendix A, section 2.3, and section 5.5

◆◆ **Project 24 (on digital audio merging)**

(a) Obtain the softcopy audio test.wav, b.wav, nosound.wav, bird.wav, and hello.wav in your working path of MATLAB (section 2.7) and verify that the files exist in the working path.

(b) Let your computer read each of the softcopy audio of the part (a).

(c) First play the audio test.wav and b.wav separately and then play the two merged audio.

(d) First play the audio nosound.wav, hello.wav, and b.wav separately and then play the three merged audio.

(e) First play the audio nosound.wav, test.wav, hello.wav, and bird.wav separately and then play the four merged audio.

Hint: section 3.4

◆◆ **Project 25 (on digital audio splitting)**

(a) Obtain the softcopy audio hello.wav in your working path of MATLAB (section 2.7) and verify that the file exists in the working path.

(b) Let your computer read the softcopy audio of the part (a).

(c) Play the whole audio of the part (b).

(d) Play the first 35% duration of the audio in part (b).

(e) Play the last 35% duration of the audio in part (b).

(a) Play from the 25% to the 75% duration of the audio in the part (b).

Hint: section 3.5

✦✦ Project 26 (on eigenvalue and eigenvector)

(a) Employing the built-in function of MATLAB, determine the eigenvalues of the square matrix $A = \begin{bmatrix} 6 & 2 & 0 \\ 9 & 0 & 1 \\ -7 & 2 & -3 \end{bmatrix}$.

(b) Determine one set of eigenvectors for the matrix in part (a).

(c) Verify that for each eigenvalue-eigenvector set, the matrix relationship $A\,X = \lambda\,X$ holds true where the symbols have section 7.5 cited meanings.

(d) Do the same as in parts (a), (b), and (c) for the square matrix
$A = \begin{bmatrix} 6 & 2 & 0 & 1 \\ 9 & 0 & 1 & 2 \\ -7 & 2 & -3 & -1 \\ 8 & 0 & 8 & 1 \end{bmatrix}$.

(e) In parts (b) and (d), verify that each eigenvector has the magnitude 1.

Answers: (a) 8.1108, −0.8010, and −4.3097 (b) $\begin{bmatrix} -0.6583 \\ -0.6948 \\ 0.2897 \end{bmatrix}$, $\begin{bmatrix} -0.1387 \\ 0.4718 \\ 0.8707 \end{bmatrix}$,

and $\begin{bmatrix} 0.0703 \\ -0.3625 \\ 0.9293 \end{bmatrix}$ for 8.1108, −0.8010, and −4.3097 respectively (d)

eigenvalues: 8.7667, −2.8276+ j 1.9537, −2.8276− j 1.9537, and

0.8884 and eigenvectors: $\begin{bmatrix} -0.6043 \\ -0.6664 \\ 0.2751 \\ -0.3392 \end{bmatrix}$, $\begin{bmatrix} -0.0889 + j0.0285 \\ -0.0799 - j0.2127 \\ -0.3365 + j0.1887 \\ 0.8893 \end{bmatrix}$,

$\begin{bmatrix} -0.0889 - j0.0285 \\ -0.0799 + j0.2127 \\ -0.3365 - j0.1887 \\ 0.8893 \end{bmatrix}$, and $\begin{bmatrix} -0.2239 \\ 0.0991 \\ 0.2107 \\ 0.9464 \end{bmatrix}$ respectively.

Hint: section 7.5

✦✦ Project 27 (on singular value decomposition)

(a) By applying the built-in function of MATLAB for singular value decomposition, decompose the rectangular matrix $A = \begin{bmatrix} 6 & 2 & 0 & 1 \\ 9 & 0 & 1 & 2 \\ -7 & 2 & -3 & -1 \end{bmatrix}$ into three matrices − U, D, and V such that the matrix relationship $A = U \times D \times V^T$ is maintained.

(b) Do the same as in part (a) for the rectangular matrix $A = \begin{bmatrix} 0 & 2 & 1 \\ -4 & 0 & 1 \\ -7 & 2 & -3 \\ -1 & 7 & 9 \end{bmatrix}$.

(c) Determine the condition number of the matrix A in part (a). Is the matrix nearly a singular one?

(d) Do the same as in part (c) for the matrix in part (b).

Mohammad Nuruzzaman

Answers: (a) $U = \begin{bmatrix} -0.4465 & 0.6717 & -0.5911 \\ -0.6940 & 0.1571 & 0.7027 \\ 0.5648 & 0.7240 & 0.3960 \end{bmatrix}$,

$D = \begin{bmatrix} 13.3183 & 0 & 0 & 0 \\ 0 & 3.4729 & 0 & 0 \\ 0 & 0 & 0.7502 & 0 \end{bmatrix}$, and

$V = \begin{bmatrix} -0.9670 & 0.1083 & 0.0066 & -0.2306 \\ 0.0178 & 0.8037 & -0.5202 & 0.2882 \\ -0.1793 & -0.5802 & -0.6470 & 0.4611 \\ -0.1801 & 0.0754 & 0.5574 & 0.8069 \end{bmatrix}$

(b) $U = \begin{bmatrix} -0.1716 & -0.0181 & -0.4272 & 0.8875 \\ -0.0988 & -0.4062 & 0.8289 & 0.3716 \\ 0.0433 & -0.9136 & -0.3602 & -0.1836 \\ -0.9792 & 0.0037 & -0.0247 & -0.2011 \end{bmatrix}$,

$D = \begin{bmatrix} 11.6879 & 0 & 0 \\ 0 & 8.5533 & 0 \\ 0 & 0 & 2.2879 \\ 0 & 0 & 0 \end{bmatrix}$, and $V = \begin{bmatrix} 0.0916 & 0.9372 & -0.3365 \\ -0.6084 & -0.2148 & -0.7640 \\ -0.7883 & 0.2747 & 0.5505 \end{bmatrix}$

(c) 17.7533, No   (d) 5.1085, No

Hint: section 7.6

❖ ❖ Project 28 (on singular value decomposition of a discrete signal)

(a) By turning the discrete signal $v[n] = [-3\ 4\ 0\ -9\ 1]$ into semi Toeplitz $f[m,n]$ with $L=2$, singular value decompose the signal such that the matrix relationship $f[m,n] = U \times D \times V^T$ is maintained where U, D, and V are matrices and the T is matrix transposition operator.

(b) Do the same as in part (a) with $L=3$.

(c) Do the same as in part (a) but for the discrete signal $v[n] = [0\ 3\ 1\ -4\ 5\ 8\ 9]$ with $L=4$.

Answers: (a) $U = \begin{bmatrix} 0.4377 & 0.1302 & 0.8590 & -0.2315 \\ -0.2774 & 0.2840 & 0.3291 & 0.8568 \\ -0.5166 & -0.7722 & 0.3662 & -0.0520 \\ 0.6816 & -0.5532 & -0.1401 & 0.4579 \end{bmatrix}$,

$D = \begin{bmatrix} 11.1075 & 0 \\ 0 & 8.9790 \\ 0 & 0 \\ 0 & 0 \end{bmatrix}$, and $V = \begin{bmatrix} 0.6375 & 0.7704 \\ -0.7704 & 0.6375 \end{bmatrix}$

(b) $U = \begin{bmatrix} -0.2121 & -0.4108 & -0.8867 \\ -0.6952 & 0.7011 & -0.1585 \\ 0.6868 & 0.5828 & -0.4343 \end{bmatrix}$, $D = \begin{bmatrix} 10.1109 & 0 & 0 \\ 0 & 9.8260 & 0 \\ 0 & 0 & 2.2879 \end{bmatrix}$, and

$V = \begin{bmatrix} 0.6868 & -0.5828 & 0.4343 \\ -0.6952 & -0.7011 & 0.1585 \\ -0.2121 & 0.4108 & 0.8867 \end{bmatrix}$

(c) $U = \begin{bmatrix} -0.1530 & 0.4637 & 0.0952 & 0.8674 \\ 0.0940 & -0.5756 & 0.7764 & 0.2391 \\ 0.5540 & -0.4750 & -0.5459 & 0.4116 \\ 0.8129 & 0.4776 & 0.3002 & -0.1448 \end{bmatrix}$,

$$D = \begin{bmatrix} 15.7834 & 0 & 0 & 0 \\ 0 & 8.8669 & 0 & 0 \\ 0 & 0 & 6.1143 & 0 \\ 0 & 0 & 0 & 1.9692 \end{bmatrix}, \text{ and}$$

$$V = \begin{bmatrix} 0.8129 & -0.4776 & 0.3002 & -0.1448 \\ 0.5540 & 0.4750 & -0.5459 & 0.4116 \\ 0.0940 & 0.5756 & 0.7764 & 0.2391 \\ -0.1530 & -0.4637 & 0.0952 & 0.8674 \end{bmatrix}$$

Hint: section 7.7

✦✦ Project 29 (on digital audio mixing)

(a) Suppose three prototype digital audio signals of identical sampling frequency are given as $v_1[n] = [0.5 \quad -0.36 \quad 0.9 \quad 0.25]$, $v_2[n] = [-0.9 \quad -0.26 \quad 0.67 \quad 0.79 \quad 0.9]$, and $v_3[n] = [-0.27 \quad -0.7 \quad 0.9 \quad 0.13 \quad -1]$ each in the range $[-1,1]$. The dominance factors of mixing for the three signals are $c_1 = 0.2$, $c_2 = 0.9$, and $c_3 = 0.1$ respectively. Compute the resultant mixed digital signal in the $[-1,1]$ range assuming zero padding at the end for unequal length signal.

(b) Obtain the three digital audio files **b.wav**, **bird.wav**, and **test.wav** in your working path of MATLAB. Determine the number of samples in each audio.

(c) Mix the part (b) mentioned three audio by padding less sample number audio with appropriate zeroes where the dominance factors are 0.1, 0.5, and 1 respectively and play each audio as well as the mixed audio.

(d) Do the same as in part (c) for the dominance factors 1, 1, and 0 respectively.

(e) Do the same as in part (c) for the dominance factors 1, 0.1, and 1 respectively.

Answers: (a) $[-1 \quad -0.5516 \quad 1 \quad 0.877 \quad 0.7975]$ (b) 2800, 19456, and 161002 respectively

Hint: section 3.6

✦✦ Project 30 (on discrete signal sampling)

(a) The discrete signal $f[n] = [1 \quad -3 \quad 2 \quad 90 \quad 41 \quad 8 \quad 71 \quad 38]$ is to be downsampled by a factor of 2. Determine the sampled signal by employing built-in function of MATLAB.

$$25e^{-n}3^{-n} \longrightarrow \boxed{2\downarrow} \longrightarrow g[n]$$

Figure 12.3(d) A discrete signal is downsampled by a factor 2

(b) Do the same as in part (a) for a sampling factor 3.

(c) The discrete signal $f[n]=[1 \quad -3 \quad 2 \quad 90 \quad 41]$ is to be upsampled by a factor of 2. Determine the sampled signal by employing built-in function of MATLAB.

(d) Do the same as in part (c) for a sampling factor 3.

(e) A discrete signal over $0 \le n \le 6$ follows the function $25e^{-n}3^{-n}$ which is to be downsampled by a factor 2 as shown in figure 12.3(d). Obtain the resulting signal $g[n]$ by applying the built-in function of MATLAB.

(f) Do the same as in part (e) for an upsampling factor 2.

Answers: (a) $[1 \quad 2 \quad 41 \quad 71]$ (b) $[1 \quad 90 \quad 71]$ (c) $[1 \quad 0 \quad -3 \quad 0 \quad 2 \quad 0 \quad 90 \quad 0 \quad 41 \quad 0]$ (d) $[1 \quad 0 \quad 0 \quad -3 \quad 0 \quad 0 \quad 2 \quad 0 \quad 0 \quad 90 \quad 0 \quad 0 \quad 41 \quad 0 \quad 0]$ (e) $g[n]=[25 \quad 0.3759 \quad 0.0057 \quad 0.0001]$ (f) $g[n]=[25 \quad 0 \quad 3.0657 \quad 0 \quad 0.3759 \quad 0 \quad 0.0461 \quad 0 \quad 0.0057 \quad 0 \quad 0.0007 \quad 0 \quad 0.0001 \quad 0]$

Hint: section 3.7

Project 31 (on digital audio sampling)

(a) Obtain the digital audio hello.wav in your working path of MATLAB (section 2.7). Play the digital audio in MATLAB. Downsample the audio by a factor 2 and play the sampled audio.

(b) In part (a) now upsample the audio by a factor 2 and play both the original and the sampled audio.

(c) Resample the audio of the part (a) with $17\,KHz$ and play the original and the sampled audio.

(d) Resample the audio of the part (a) with $28\,KHz$ and play the original and the sampled audio.

(e) Do the same as in parts (a) through (d) but for the audio b.wav.

Hint: section 3.7

Project 32 (on histogram of integer type discrete signal)

(a) Apply the built-in function of MATLAB to obtain the frequencies of the following integer samples which are taken from a discrete signal: $-3, -2, -3, -3, -2, 1, 9, 1, 1$, and 5.

(b) Also determine the percentage of each sample integer occurrence in the signal of the part (a).

(c) Draw the continuous sense histogram of the integer samples in part (a).

(d) Do the same as in parts (a) through (c) for the integer samples which are given as: $-60, -72, 0, 56, 24, 32, 3, -60, 0, 24, 3, 32, -72, -72, -60, -72, 0, 0$, and -72.

Answers: (a) 3, 2, 3, 1, and 1 for −3, −2, 1, 5, and 9 respectively (b) 30%, 20%, 30%, 10%, and 10% for −3, −2, 1, 5, and 9 respectively (c) Figure 12.3(e) (d) Frequencies are 5, 3, 4, 2, 2, 2, and 1 and percentages are 26.32%, 15.79%, 21.05%, 10.53%, 10.53%, 10.53%, and 5.26% for −72, −60, 0, 3, 24, 32, and 56 respectively. Figure 12.3(f) for the continuous sense histogram
Hint: section 8.1

Figure 12.3(e) Continuous sense histogram for the data of the project 32(a)

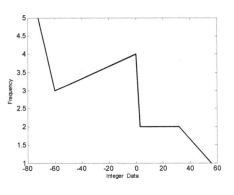

Figure 12.3(f) Continuous sense histogram for the data of the project 32(d)

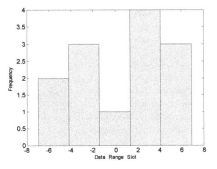

Figure 12.3(g) Histogram for the range data of the project 33(c) – right side figure

✦✦ Project 33 (on histogram of range type discrete signal)

(a) Apply built-in function of MATLAB to obtain the frequencies with 5 slots on the following range type data which is taken from a discrete signal: −3.91, −2.2, −3.12, −7, −4.7, 1.3, 6.9, 4.1, 2.1, 3.5, 2.56, 6.31, and 4.44.

(b) What are the range slots for the data in part (a)?

(c) Draw the histogram of the range data in part (a).

(d) What are the midpoints of the range slots for the data in part (a)?

Answers: (a) 2, 3, 1, 4, and 3 respectively (b) $-7 \leq X \leq -4.22$, $-4.22 < X \leq -1.44$, $-1.44 < X \leq 1.34$, $1.34 < X \leq 4.12$, and $4.12 < X \leq 6.9$ (c) Figure 12.3(g) (d) −5.61, −2.83, −0.05, 2.73, and 5.51 respectively

Hint: section 8.2

✦✦ Project 34 (on histogram of a digital audio)

(a) Obtain the softcopy audio b.wav in your working path of MATLAB (section 2.7) and determine the frequencies of the audio considering 8 range slots.

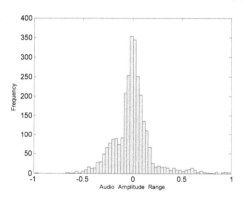

Figure 12.4(a) Histogram of the audio b.wav with 64 range slots

(b) Do the same as in part (a) but for 16 range slots.

(c) What are the midpoints of the slots in the part (a)?

(d) Graph the histogram of the b.wav considering 64 range slots.

(e) Graph the continuous sense histogram of the audio hello.wav

Figure 12.4(b) Continuous sense histogram of the audio hello.wav with 256 range slots – right side figure

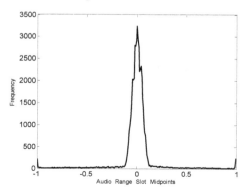

-250-

considering 256 range slots.

Answers: (a) 1, 15, 219, 1233, 1153, 103, 58, and 18 (b) 1, 0, 3, 12, 50, 169, 327, 906, 935, 218, 63, 40, 37, 21, 8, and 10 (c) −0.8587, −0.6117, −0.3646, −0.1176, 0.1294, 0.3764, 0.6234, and 0.8704 (d) Figure 12.4(a) (e) Figure 12.4(b)

Hint: section 8.3

❖❖ Project 35 (on quantization of a signal)

Apply built-in function of MATLAB to each of the following:

(a) quantize the single audio amplitude value $6.8\,mV$ on a $\Delta = 0.3\,mV$.

(b) quantize the single audio amplitude value $22.5\,mA$ on a $\Delta = 0.5\,mA$.

(c) quantize the discrete signal $f[n]=[-9.3\quad 7.3\quad 0\quad 6.8]$ in mV on a $\Delta =0.6\,mV$.

(d) discretize the continuous signal $f(t)=12e^{-2t}\sin 2t$ with a sampling period $T_s=0.05$sec over the interval $0\le t\le \dfrac{2}{\pi}$ sec and then quantize the discrete counterpart $f[n]$ on a $\Delta =0.5$.

(e) discretize the continuous signal $f(t)=3t-5$ with a sampling period $T_s=0.1$sec over the interval $0\le t\le 10$ sec and then quantize the discrete counterpart $f[n]$ on a $\Delta =0.25$.

Answers: (a) $6.9\,mV$ (b) $22.5\,mA$ (c) $q[n]=[-9.6\quad 7.2\quad 0\quad 6.6]$ in mV
(d) t=0:0.05:2/pi; f=12*exp(-2*t).*sin(2*t); Q=quant(f,0.5);
(e) t=0:0.1:10; f=3*t-5; Q=quant(f,0.25);

Hint: section 3.8

❖❖ Project 36 (on quantization of a signal)

(a) Suppose a continuous signal $f(t)=3t^2-7t+8$ exists over $0\le t\le 5$ secs. Digitize the signal by taking a sampling period 0.01sec and obtain it as a row matrix. The digitized signal is to be quantized in 16 levels. Obtain the signal as a row matrix after quantization. Graph the two digitized signals by using the **plot** of appendix D.

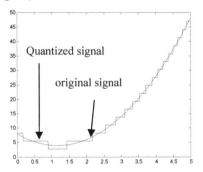

Figure 12.4(c) Plot of the original and quantized signals together

(b) Say another continuous signal $f(t)=e^{-0.2t}\sin t$ exists over $0\le t\le 2$ secs. Digitize the signal by taking a sampling frequency $100\,Hz$ and obtain it as a row matrix. The digitized signal is to be quantized by a 4-bit quantizer. Obtain the signal as a row matrix after quantization. From these two discrete signals, compute first the quantization error as a row matrix and then the signal to quantization noise power ratio in dB.

Answers: (a) The final output should be like the figure 12.4(c)

(b) SNR=33.301 dB

Hint: section 3.8

✦✦ Project 37 (on quantization of a digital audio)

(a) Obtain the digital audio **hello.wav** in your working path of MATLAB (section 2.7). Requantize the audio in 128 levels within the maximum and the minimum amplitudes.

(b) Play the original audio as well as the quantized one of part (a).

(c) Redo the part (a) if the operation is carried out by a 4-bit quantizer.

(d) Do part (b) for the quantized audio of the part (c). What do you infer from the execution of the part (c)?

(e) Calculate the SNR in dB due to the quantization of part (a) assuming that **hello.wav** audio data is the unquantized one.

(f) Do part (e) for the quantization of part (c).

(g) Do the parts (a) through (f) but for the audio **test.wav**.

Answers: (a) and (c) The audio data will be available in workspace variables (d) Background noise becomes pronounced due to the quantization (e) 32.287 dB (f) 14.4893 dB (g) Also background noise becomes pronounced, 35.3366 dB, and 16.9946 dB respectively

Hint: section 3.8

✦✦ Project 38 (on entropy and digital audio)

Apply built-in function or write code of MATLAB to determine the following:

(a) entropy of an audio signal which has 5 amplitude levels with probabilities 0.51, 0.02, 0.32, 0.12, and 0.03.

(b) entropy of a modular audio signal which shows integer type amplitudes as 22 34 22 43 50 50 34 77 98 77 77 8 77 255 22 255 43.

(c) entropy of the digital audio **b.wav** (section 2.7) considering 32 range slots.

(d) entropy of the digital audio **hello.wav** considering 32 range slots.

(e) entropies of the audio **bird.wav** considering 16, 32, 64, 128, and 256 range slots.

Answers: (a) 1.6532 bits   (b) 2.8666 bits, epsilon addition to probability vector might be necessary (c) 3.4221 bits   (d) 2.4903 bits (e) 2.518 bits, 3.4636 bits, 4.3027 bits, 5.1641 bits, and 6.0047 bits respectively

Hint: section 8.6

✦✦ **Project 39 (on probability and digital audio)**
 (a) Obtain the softcopy audio bird.wav (section 2.7) in your working
  path of MATLAB. Transform the audio amplitudes as integer
  level ranging from 0 to 7 and determine the probability of each
  integer level.
 (b) Do the same as in part (a) but for the audio b.wav.
 (c) Do the same as in part (a) but for the audio hello.wav.
 (d) Determine the probability of each range slot considering the
  audio of the part (a) with 8 range slots.
 (e) Do the same as in part (d) but for the audio b.wav.
 (f) Do the same as in part (d) but for the audio hello.wav.
 Answers: (a) 0.0014, 0.0128, 0.0506, 0.1750, 0.6534, 0.0883,
 0.0166, and 0.0020 respectively (b) 0.0004, 0.0021, 0.0546,
 0.4671, 0.4186, 0.0357, 0.0175, and 0.0039 respectively (c)
 0.0074, 0.0090, 0.0161, 0.4219, 0.5063, 0.0186, 0.0123, and 0.0083
 respectively (d) 0.0038, 0.0174, 0.0541, 0.1773, 0.6206, 0.0976,
 0.0232, and 0.0059 respectively (e) 0.0004, 0.0054, 0.0782,
 0.4404, 0.4118, 0.0368, 0.0207, and 0.0064 respectively (f) 0.0097,
 0.0099, 0.0158, 0.4193, 0.5024, 0.0187, 0.0124, and 0.0117
 respectively
 Hint: section 8.4
✦✦ **Project 40 (on moment and digital audio)**
 Apply built-in function of MATLAB to determine the following:
 (a) the 5th central moment of the discrete signal $X[n]=[-11 \quad 6 \quad 20 \quad 9 \quad 32]$.
 (b) the 4th central moment of the discrete signal $X[n]=[9 \quad 3 \quad 6 \quad 2 \quad 0]$.
 (c) the 1st central moment of the signal in part (b).
 (d) the 1st, 2nd, 3rd, and 5th central moments of the digital audio
  bird.wav (section 2.7) ignoring the sampling frequency effect.
 (e) part (d) considering the sampling frequency effect.
 (f) the 1st, 2nd, and 3rd central moments of the digital audio
  hello.wav ignoring the sampling frequency effect.
 (g) part (f) considering the sampling frequency effect.
 Answers: (a) -2.9×10^5 (b) 182.8 (c) 0 (d) 0, 0.0416, −0.003, and
 −0.0019 respectively (e) 0, 1.8863×10^{-6}, -6.1221×10^{-12}, and
 -8.2269×10^{-21} respectively (f) 0, 0.0347, and 0.0011 respectively
 (g) 0, 1.5718×10^{-6}, and 2.2074×10^{-12} respectively
 Hint: section 8.7
✦✦ **Project 41 (on probability density function and digital audio)**
 (a) Apply built-in function of MATLAB and obtain the discrete
  version pdf of the integer type audio amplitudes: 12 13 12 6
  7 7 9 9 13 20 20 21 13 13 9 10 7 12 21.

(b) Given the softcopy audio **hello**.wav (section 2.7) is in range form, apply built-in function of MATLAB and obtain the discrete version pdf of the audio considering 32 range slots.

(c) Do the same as in part (b) for the audio **b.wav**.

Answers: (a) Figure 12.4(d) (b) Figure 12.4(e) (c) Figure 12.4(f)

Hint: section 8.5

Figure 12.4(d)
Discretized pdf of the integer audio amplitudes in project 41(a) – right side figure

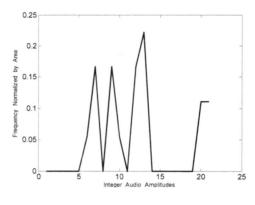

Figure 12.4(e) Discretized pdf of the audio **hello**.wav – right side figure

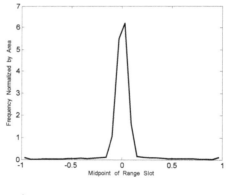

Figure 12.4(f) Discretized pdf of the audio **b.wav** – right side figure

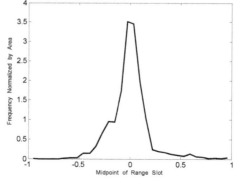

◆◆ Project 42 (on correlation and digital audio)

(a) Use MATLAB function to determine the correlation coefficient ρ of the discrete signals $X[n]=[9 \quad -5 \quad 8 \quad 12]$ and $Y[n]=[43 \quad 76 \quad 23 \quad 10]$.

(b) Do the same as in part (a) but for the unequal sample length signals $X[n]=[9 \quad 5 \quad 8 \quad 12]$ and $Y[n]=[43 \quad 76 \quad 23]$.

(c) Suppose one discrete signal follows linear variation $X[n]=n$ over the interval $0\le n\le 10$. Over the same interval another discrete signal also follows linear variation but with different slope which is $Y[n]=\dfrac{n}{3}$. Verify that these two signals are similar.

(d) Do the same as in part (c) but for $X[n]=\sin\dfrac{\pi n}{5}$ and $Y[n]=6\sin\dfrac{\pi n}{5}$.

(e) Find the correlation coefficient of the two digital audio **bird.wav** and **b.wav** (section 2.7).

(f) Do the same as in part (e) for the audio **test.wav** and **hello.wav**.

(g) Do the same as in part (e) for the audio **b.wav** and **hello.wav**.

Answers: (a) -0.926 (b) -0.9177 (c) Generate n over $0\le n\le 10$ by writing **n=0:10;**, generate the two discrete signals $X[n]$ and $Y[n]$ by n and **n/3** respectively, and prove that $\rho=1$ (d) Generate the two discrete signals $X[n]$ and $Y[n]$ by **sin(pi*n/5)** and **6*sin(pi*n/5)** respectively and prove that $\rho=1$ (e) 0.0044 (f) -0.0032 (g) -8.6147×10^{-4}

Hint: section 8.8

◆◆ Project 43 (on autocorrelation and digital audio)

Apply built-in function of MATLAB to determine the following:

(a) two sided autocorrelation $r[k]$ of the discrete signal $X[n]=[7 \quad 9 \quad 0 \quad 8 \quad 6 \quad -5]$ without scaling. What are the n and the k intervals?

(b) Do the same as in part (a) for unbiased and biased scalings.

(c) What are the one or right sided autocorrelation for the parts (a) and (b)? What are the n and the k intervals for the one sided autocorrelation?

(d) How can one obtain the one sided autocorrelation starting from the two sided one of part (a)?

Answers: (a) $r[k]=[-35 \quad -3 \quad 110 \quad 32 \quad 81 \quad 255 \quad 81 \quad 32 \quad 110 \quad -3 \quad -35]$, $0\le n\le 5$, and $-5\le k\le 5$ (b) $r[k]=[-35 \quad -1.5 \quad 36.6667 \quad 8 \quad 16.2 \quad 42.5 \quad 16.2 \quad 8 \quad 36.6667 \quad -1.5 \quad -35]$ and $r[k]=[-5.8333 \quad -0.5 \quad 18.3333 \quad 5.3333 \quad 13.5 \quad 42.5 \quad 13.5 \quad 5.3333 \quad 18.3333 \quad -0.5$

−5.8333] respectively   (c) $r[k]=[255$ 81   32   110   −3   −35],
$r[k]=[42.5$ 16.2   8   36.6667   −1.5   −35], and $r[k]=[42.5$ 13.5
5.3333   18.3333   −0.5   −5.8333] respectively and $0 \le n \le 5$ and
$0 \le k \le 5$ (d) r(6:11) assuming that the two sided autocorrelation is
stored in the r as a row or column matrix
Hint: section 8.9

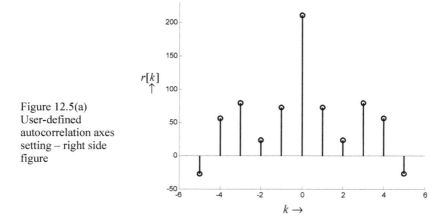

Figure 12.5(a)
User-defined
autocorrelation axes
setting – right side
figure

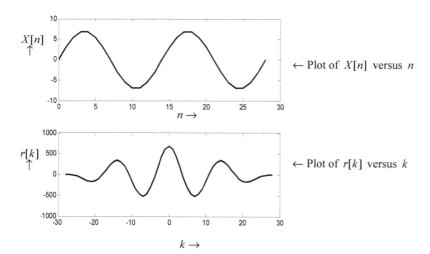

← Plot of $X[n]$ versus n

← Plot of $r[k]$ versus k

Figure 12.5(b) Plot of the discrete sine function and its autocorrelation in
continuous sense

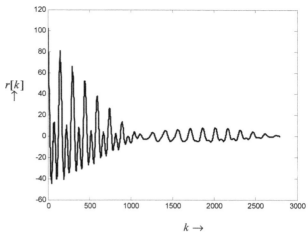

Figure 12.5(c) Plot
of one sided $r[k]$
versus k for the
audio b.wav – right
side figure

$k \rightarrow$

✦ ✦ Project 44 (on autocorrelation and digital audio)

(a) Graph the two sided autocorrelation $r[k]$ of the discrete random
process $X[n]=[9 \quad 2 \quad 2 \quad 8 \quad 7 \quad -3]$ without scaling by
using the function **stem** and set the axes of the graph
considering $-6 \le k \le 6$ and $-50 \le r[k] \le 230$ for better perception.

(b) Compute the two sided autocorrelation $r[k]$ of discrete sine
wave $X[n]=7\sin\dfrac{n\pi}{7}$ over $0 \le n \le 28$ without scaling and graph
both curves in continuous sense in the same window.

(c) Graph the one sided and unscaled autocorrelation $r[k]$ of digital
audio **b.wav** (section 2.7) in continuous sense.

(d) Do the same as in part (c) but for audio **hello.wav** with unbiased

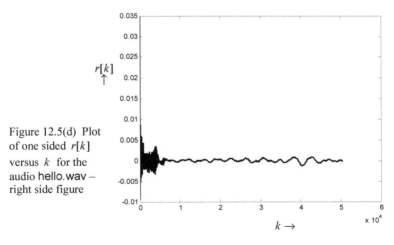

Figure 12.5(d) Plot
of one sided $r[k]$
versus k for the
audio hello.wav –
right side figure

$k \rightarrow$

scale.

Answers: (a) Figure 12.5(a)   (b) Figure 12.5(b)   (c) Figure 12.5(c)
(d) Figure 12.5(d)

Hint: section 8.9

✦✦ Project 45 (on inverse Z transform from system function)

Determine the unilateral inverse z transform for each of the following system functions:

(a) $F(z) = \dfrac{z}{z-1} + \dfrac{z}{z-4}$

(b) $F(z) = \dfrac{z^2}{(z-3)(z-8)}$

(c) $F(z) = \dfrac{4z^{-1} + 5z^{-2}}{\left(1 - \dfrac{1}{3}z^{-1}\right)\left(1 + \dfrac{1}{4}z^{-1}\right)}$

(d) $F(z) = \dfrac{2z^{-1} + 9z^{-2}}{(1 - \frac{9}{7}z^{-1})(1 - \frac{2}{5}z^{-1})(5 - \frac{1}{7}z^{-1})}$

(e) $F(z) = \dfrac{1}{(1-z^{-1})^3(1-z^{-2})}$

Answers (obviously for $n \geq 0$):

(a) $f[n] = 1 + 4^n$

(b) $f[n] = \dfrac{8^{n+1} - 3^{n+1}}{5}$

(c) $f[n] = -60\delta[n] + \dfrac{192}{7}\left(-\dfrac{1}{4}\right)^n + \dfrac{228}{7}\left(\dfrac{1}{3}\right)^n$

(d) $f[n] = \dfrac{2835}{1364}\left(\dfrac{9}{7}\right)^n - \dfrac{2401}{403}\left(\dfrac{2}{5}\right)^n + \dfrac{2219}{572}\left(\dfrac{1}{35}\right)^n$

(e) $f[n] = \dfrac{1}{16}(-1)^n + \dfrac{15}{16} + \dfrac{17n}{12} + \dfrac{5n^2}{8} + \dfrac{n^3}{12}$

Hint: section 6.2

✦✦ Project 46 (on sequence solution of a difference equation)

Determine the sequence solution (for k or $n \geq 0$) for each of the following difference equations by applying built-in MATLAB function:

(a) $y_{k+1} = -4y_{k-1} - 4y_k$, $y_0 = 4$, and $y_1 = 0$

(b) $7y[n+2] - 6y[n+1] - y[n] = 4 - n$, $y[0] = 2$, and $y[1] = 3$

(c) $y[n-3] - 2y[n-2] + y[n-1] = n+1$, $y[0] = 0$, and $y[1] = -1$

(d) $3x[n] - 2x[n-1] = 5\delta[n]$ and $x[0] = 0$

Answers:

(a) $y_k = 8(-2)^k + (-4k - 4)(-2)^k$

(b) $y[n] = \dfrac{1199}{512} - \dfrac{175}{512}\left(-\dfrac{1}{7}\right)^n + \dfrac{43n}{64} - \dfrac{n^2}{16}$

(c) $y[n] = \dfrac{n(n^2 + 9n - 16)}{6}$

(d) $x[n] = \dfrac{5}{3}\sum_{m=1}^{m=n}\left(\dfrac{2}{3}\right)^{n-m}\delta[m]$

Hint: section 6.3

◆◆ Project 47 (on Z transform system function from difference equations)

Determine the unilateral (before $k \geq 0$ or $n \geq 0$ the signal value is zero) Z transform system function for each of the following difference equations:

(a) the difference equation $y_{k+2} - 3y_{k+1} = 7y_k$ with initial values $y_0 = -1$ and $y_1 = -2$

(b) the difference equation $y[n+2] - 6y[n+1] + 11y[n] = 4 - n + n^2$ with initial values $y[0] = 3$ and $y[1] = -3$

(c) the difference equation $y[n-2] - 2y[n-1] + 3y[n] = 4\sin\dfrac{\pi n}{2}$

Answers:

(a) $Y(z) = -\dfrac{z(z-1)}{z^2 - 3z - 7}$

(b) $Y(z) = \dfrac{z(3z^4 - 30z^3 + 76z^2 - 74z + 27)}{(z-1)^3(z^2 - 6z + 11)}$

(c) $Y(z) = \dfrac{4z^3}{(z^2 + 1)(3z^2 - 2z + 1)}$

Hint: section 6.3

◆◆ Project 48 (on input-output on Z transform system)

(a) The discrete system of the figure 6.1(b) is characterized by $H(z) = \dfrac{3 - 2z^{-1}}{3 + 2z^{-1} - z^{-2}}$. Determine the output $y[n]$ when the input is $f[n] = $ [2 2 2 2].

(b) In part (a) now $H(z) = \dfrac{3z + 2z^2}{2 + 3z^4}$

(c) In part (a) now $f[n] = 3^{-n}$ over $0 \leq n \leq 5$

(d) In part (a) now $f[n] = e^{-n}\sin n$ over $0 \leq n \leq 7$

Answers:

(a) $y[n] = [2 \quad -0.6667 \quad 1.7778 \quad -0.7407]$

(b) $y[n] = [0 \quad 0 \quad 1.3333 \quad 3.3333]$

(c) $y[n] = [1 \quad -1 \quad 0.8889 \quad -0.9630 \quad 0.9259 \quad -0.9424]$

(d) $y[n] = [0 \quad 0.3096 \quad -0.2897 \quad 0.2213 \quad -0.2626 \quad 0.2516 \quad -0.2517 \quad 0.2527]$

Hint: section 6.4

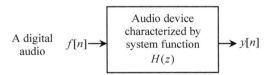

Figure 12.6(a) Audio signal passing through an audio device

✦✦ Project 49 (on input-output on Z transform system)

(a) The Z transform system of the figure 12.6(a) has $H(z) = \dfrac{3 - 2z^{-1}}{3 + 2z^{-1} - z^{-2}}$. When an input $f[n] = 2^{-n}$ is applied to the system in time index domain, what should be the output $y[n]$ in z domain as well as in index domain?

(b) Graph the output response $y[n]$ versus n over $0 \le n \le 7$ of the part (a).

Answers:

(a) $Y(z) = \dfrac{6z^3 - 4z^2}{6z^3 + z^2 - 4z + 1}$ and $y[n] = \dfrac{3^{-n}}{2} + \dfrac{5(-)^n}{6} - \dfrac{2^{-n}}{3}$

(b) suppose y holds the code of $y[n]$, then use the command stem(0:7, subs(y,0:7)) of appendix D, and the graph is not shown for the space reason

Hint: section 6.4

✦✦ Project 50 (on convolution of discrete signals)

Following questions are related to the system in figure 3.4(c):

(a) a discrete signal $f[n] = [3\ 3\ 0\ 5\ 2\ 4]$ is applied to the system with $h[n] = [2\ 2\ -3\ 8\ 9\ 2]$, what is the output signal $r[n]$ from the system?

(b) a discrete binary signal $f[n] = [1\ 0\ 1\ 0\ 1\ 1\ 0]$ is applied to the system with $h[n] = [1\ 1\ 1\ 1]$, what is the output signal $r[n]$ from the system?

(c) a discrete signal $f[n] = 23 \sin \dfrac{2n\pi}{7}$ over $0 \le n \le 4$ is applied to the system with $h[n] = 12\, e^{-\frac{n}{9}}$ over $0 \le n \le 4$, what is the output signal $r[n]$ from the system?

(d) a discrete signal $f[n] = 2n - 3$ over $1 \le n \le 4$ is applied to the system with $h[n] = 12 \ln n$ over $3 \le n \le 6$, what is the output signal $r[n]$ from the system?

(e) graph all three discrete signals $f[n]$, $h[n]$, and $r[n]$ of part (d) in a single plot.

Answers:

(a) $r[n] = [6\quad 12\quad -3\quad 25\quad 65\quad 30\quad 48\quad 49\quad 60\quad 40\quad 8]$

(b) $r[n] = [1\quad 1\quad 2\quad 2\quad 2\quad 3\quad 2\quad 2\quad 1\quad 0]$

(c) $r[n] = [0\quad 215.7855\quad 462.1734\quad 533.3229\quad 357.4864\quad 319.8929\quad 162.4450\quad -9.0234\quad -76.7826]$ where $0 \le n \le 8$ assuming the 0 starting

(d) $r[n] = [-13.1833\quad -3.4522\quad 36.8723\quad 113.6355\quad 162.6185\quad 161.0696\quad 107.5056]$ where $1 \le n \le 7$ assuming starting from $n = 1$

(e) figure 12.6(b)

Hint: section 3.10

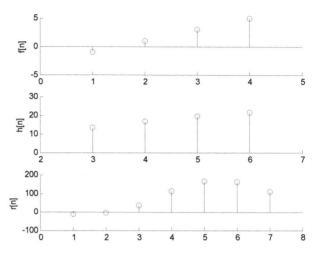

Figure 12.6(b) Plots of
the $f[n]$, $h[n]$, and $r[n]$
of the system in figure
3.4(c) (right side figure)

♦♦ Project 51 (on linear prediction of discrete signals)

(a) By applying built-in function of MATLAB, get full right sided autocorrelation vector $r[n]$ on biased scaling, Toeplitz matrix R, column vector V, and power E in the prediction error for the discrete signal $f[n]=[9\ \ 7\ \ -5\ \ 4\ \ -2\ \ 13\ \ \ 19]$ where the linear prediction order P is chosen to be 4.

(b) Compute the linear prediction coefficients from the quantities of the part (a) and verify the coefficients and the error power by employing the function lpc.

(c) In parts (a) and (b) now choose the prediction order $P=3$.

(d) In parts (a) and (b) now choose the prediction order $P=2$.

(e) What can be inferred about E from parts (a) through (d)?

Answers:

(a) $r[n]=[100.7143\ \ 31.5714\ \ 1\ \ 4.7143\ \ -3.1429\ \ 35.7143$

$24.4286]$, $R=\begin{bmatrix} 100.7143 & 31.5714 & 1 & 4.7143 \\ 31.5714 & 100.7143 & 31.5714 & 1 \\ 1 & 31.5714 & 100.7143 & 31.5714 \\ 4.7143 & 1 & 31.5714 & 100.7143 \end{bmatrix}$, $V=$

$\begin{bmatrix} 31.5714 \\ 1 \\ 4.7143 \\ -3.1429 \end{bmatrix}$, and $E=88.7306$

(b) $a_1=-0.3591$, $a_2=0.1369$, $a_3=-0.1118$, and $a_4=0.0817$

(c) $r[n]=[100.7143\ \ 31.5714\ \ 1\ \ 4.7143\ \ -3.1429\ \ 35.7143$

$$24.4286], \quad R = \begin{bmatrix} 100.7143 & 31.5714 & 1 \\ 31.5714 & 100.7143 & 31.5714 \\ 1 & 31.5714 & 100.7143 \end{bmatrix}, \quad V = \begin{bmatrix} 31.5714 \\ 1 \\ 4.7143 \end{bmatrix}, \quad E =$$

89.3266, and $a_1 = -0.3523$, $a_2 = 0.1265$, and $a_3 = -0.0830$

(d) $r[n] = [100.7143 \quad 31.5714 \quad 1 \quad 4.7143 \quad -3.1429 \quad 35.7143$

$24.4286], \quad R = \begin{bmatrix} 100.7143 & 31.5714 \\ 31.5714 & 100.7143 \end{bmatrix}, \quad V = \begin{bmatrix} 31.5714 \\ 1 \end{bmatrix}, \quad E = 89.9459,$ and

$a_1 = -0.3442$ and $a_2 = 0.098$

(e) Prediction error E increases with decreasing prediction order

Hint: sections 9.1-2

✦✦ Project 52 (on interpolation of discrete signals)

(a) Two signal points with the coordinates (7,–9) and (8,13) are given. Compute linearly interpolated values at 7.1 and 7.6.

(b) Three signal points with the coordinates (7,–9), (8,13), and (10,20) are given. Compute linearly interpolated values at 7.3, 8.4, and 9.2.

(c) In part (a) now over $7.2 \le n \le 7.7$ with n increment 0.1.

(d) In part (a) now the interpolation type is piecewise cubic spline.

(e) In part (b) now the interpolation type is piecewise cubic spline.

(f) In part (b) now the interpolation type is piecewise cubic spline and over the interval $7.2 \le n \le 7.7$ with n increment 0.1.

Answers:

(a) –6.8 and 4.2 respectively　　(b) –2.4, 14.4, and 17.2 respectively
(c) –4.6, –2.4, –0.2, 2, 4.2, and 6.4 respectively
(d) –6.8 and 4.2 respectively　　(e) –1.105, 18.3467, and 23.12
respectively　　(f) –3.6133, –1.105, 1.28, 3.5417, 5.68, and 7.695
respectively

Hint: section 3.9

✦✦ Project 53 (on interpolation of digital audio)

(a) Apply linear interpolation to the audio samples of the softcopy b.wav (section 2.7) on a step increment 0.5 in integer time index n. Play both the original and the interpolated audio.

(b) In part (a) now the step increment is 0.6. Can you name one science fiction movie where this sort of sound is heard?

(c) In part (a) now the audio is hello.wav and the time index increment is 1.7 over the whole interval.

Hint: section 3.9

✦✦ Project 54 (on frequency response of a Z transform system)

(a) For the system $F(z) = \dfrac{0.3z^{-1} + 0.4z^{-2}}{1.8 - 0.4z^{-1} - 0.2z^{-6}}$, compute the $F(e^{j\omega})$

values at angular frequency $\omega = 0$, 0.2, 0.4, 0.6, and π radians/sec.

(b) For the system function in part (a), compute the $F(e^{j\omega})$ for 6 uniformly spaced ω over $0 \le \omega < \pi$. What are the ω values for the 6 points?

(c) In part (b) if a sampling frequency $f_s = 10$ KHz is used, what are the Hertz frequencies corresponding to the ω values?

(d) For the system function in part (a), compute the $F(e^{j\omega})$ at frequencies $f = 0$ KHz, $f = 2$ KHz, $f = 3$ KHz, and $f = 7$ KHz where the sampling frequency is $f_s = 20$ KHz.

(e) In part (d) what are the related ω values?

(f) In part (a) obtain $\mathrm{Re}\{F(e^{j\omega})\}$, $\mathrm{Im}\{F(e^{j\omega})\}$, $|F(e^{j\omega})|$, and $\angle F(e^{j\omega})$ for different ω values.

(g) Graph the frequency spectra $20\log_{10}|F(e^{j\omega})|$ versus ω and $\angle F(e^{j\omega})$ versus ω in normalized ω frequency over $0 \le \omega \le \pi$ for the system in part (a).

(h) Graph the normalized magnitude spectrum in dB for the system in part (a) considering Hertz frequency interval $0 \le f \le 8KHz$ with $f_s = 16$ KHz and 0.05 KHz step.

Answers:

(a) $F(e^{j\omega}) = 0.5833$, $0.4462 - j\,0.2501$, $0.2944 - j\,0.3099$, $0.2092 - j\,0.3462$, and 0.05 respectively   (b) $F(e^{j\omega}) = 0.5833$, $0.2383-$

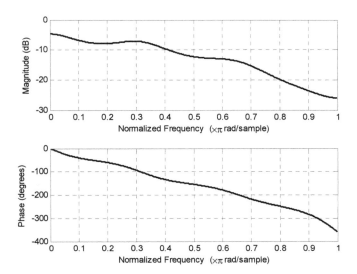

Figure 12.6(c) Magnitude and phase spectra of the system function $F(z)$ in project 54

$j\,0.3290$, $-0.1346-j\,0.3997$, $-0.2212-j\,0.1058$, $-0.1786+j\,0.0825$, and $-0.0182+j\,0.0853$ and $\omega=0$, 0.5236, 1.0472, 1.5708, 2.0944, and 2.6180 radians/sec respectively   (c) $0\;KHz$, $0.8333\;KHz$, $1.6667\;KHz$, $2.5\;KHz$, $3.3333\;KHz$, and $4.1667\;KHz$ respectively   (d) $F(e^{j\omega})=0.5833$, $0.1982-j\,0.3541$, $-0.0271-j\,0.4401$, and $-0.1353+j\,0.1054$ respectively   (e) $\omega=0$, 0.6283, 0.9425, and 2.1991 radians/sec respectively   (f) $\mathrm{Re}\{F(e^{j\omega})\}=0.5833$, 0.4462, 0.2944, 0.2092, and 0.05; $\mathrm{Im}\{F(e^{j\omega})\}=0$, -0.2501, -0.3099, -0.3462, and 0; $|F(e^{j\omega})|=0.5833$, 0.5115, 0.4275, 0.4045, and 0.05; and $\angle F(e^{j\omega})=0$, -0.5108, -0.8111, -1.0273, and 0 radians or 0^0, -29.2693^0, -46.473^0, -58.8573^0, and 0^0 respectively   (g) Figure 12.6(c)   (h) Figure 12.6(d)
Hint: section 6.7

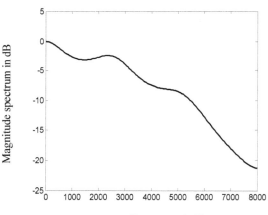

Figure 12.6(d)
Normalized magnitude
spectrum of the system
in project 54 – right side
figure

Frequency in Hertz

❖❖ Project 55 (on Butterworth filter design)

Apply built-in function to design the transform function $H(z)$ for each of the following Butterworth filters:

(a) lowpass, 2nd order, $120\;Hz$ crossover frequency, and $2\;KHz$ sampling frequency.

(b) in part (a) the type is now highpass.

(c) bandpass, 3rd order, $120\;Hz$ lower crossover frequency, $350\;Hz$ upper crossover frequency, and $2\;KHz$ sampling frequency.

(d) in part (c) now the filter type is bandstop.

Answers:

(a) $H(z)=\dfrac{0.0279+0.0557z^{-1}+0.0279z^{-2}}{1-1.4755z^{-1}+0.5869z^{-2}}$

(b) $H(z) = \dfrac{0.7656 - 1.5312z^{-1} + 0.7656z^{-2}}{1 - 1.4755z^{-1} + 0.5869z^{-2}}$

(c) $H(z) =$

$$\dfrac{0.0258 - 0.0773z^{-2} + 0.0773z^{-4} - 0.0258z^{-6}}{1 - 3.6205z^{-1} + 6.0614z^{-2} - 5.9846z^{-3} + 3.6906z^{-4} - 1.3390z^{-5} + 0.2271z^{-6}}$$

(d) $H(z) =$

$$\dfrac{0.4773 - 2.2642z^{-1} + 5.0123z^{-2} - 6.4157z^{-3} + 5.0123z^{-4} - 2.2642z^{-5} + 0.4773z^{-6}}{1 - 3.6205z^{-1} + 6.0614z^{-2} - 5.9846z^{-3} + 3.6906z^{-4} - 1.3390z^{-5} + 0.2271z^{-6}}$$

Hint: section 6.9

✦✦ Project 56 (on interpolation of digital audio)

(a) Apply linear interpolation to the audio samples of softcopy b.wav (section 2.7) on a step increment 0.5 in integer time index n. Play both the original and the interpolated audio.

(b) In part (a) now the step increment is 0.6. Can you name one science fiction movie where this kind of sound is heard?

(c) In part (a) now the audio is hello.wav and the time index increment is 1.7 over the whole interval.

Hint: section 3.9

✦✦ Project 57 (on numerical Z transform)

(a) A discrete signal is given as $\left\{ \begin{array}{cccccccc} f[n] & \to & 5 & -2 & 1 & 0 & 3 & 16 \\ n & \to & 4 & 5 & 6 & 7 & 8 & 9 \end{array} \right\}$.

Determine its numerical forward Z transform.

(b) A discrete signal is given as $f[n] = n^2 - 3$ over $0 \le n \le 4$. Determine its numerical forward Z transform.

(c) Determine the numerical inverse Z transform over $0 \le n \le 4$ for the system $F(z) = \dfrac{3z \sin 1}{z^2 - 3z \cos 1 + 3}$.

(d) Determine the numerical inverse Z transform over $0 \le n \le 6$ for the system $F(z) = \dfrac{0.7z^{-1}}{1 - 0.4z^{-2} - 0.7z^{-4}}$.

Answers:

(a) Numerator coefficients: [0 0 0 0 5 −2 1 0 3 16]
Denominator coefficient: 1

(b) Numerator coefficients: [−3 −2 1 6 13]
Denominator coefficient: 1

(c) $f[n]$=0, 2.5244, 4.0918, −0.9407, and −13.8004 for n=0, 1, 2, 3, and 4 respectively

(d) $f[n]$=0, 0.7, 0, 0.28, 0, 0.602, and 0 for n=0, 1, 2, 3, 4, 5, and 6 respectively

Hint: section 6.5

Mohammad Nuruzzaman

✦✦ Project 58 (on Chebyshev filter design)

Apply the built-in function **cheby1** or **cheby2** in order for determining the transform function of each of the following filters:

(a) a lowpass Chebyshev type I filter which has the following characteristics: order $N=2$, $5\,dB$ peak to peak ripple in the passband, crossover frequency $f_c=300\,Hz$, and sampling frequency $f_s=3\,KHz$.

(b) in part (a) now the filter type is highpass.

(c) a bandpass Chebyshev type I filter which has the following characteristics: order $N=3$, $5\,dB$ peak to peak ripple in the passband, lower and upper crossover frequencies are $300\,Hz$ and $600\,Hz$ respectively, and sampling frequency $f_s=3\,KHz$.

(d) in part (c) now the filter type is bandstop.

(e) in part (a) now the filter is type II with $30\,dB$ peak to peak stopband ripple.

(f) in part (b) now the filter is type II with $30\,dB$ peak to peak stopband ripple.

(g) in part (c) now the filter is type II with $30\,dB$ peak to peak stopband ripple.

(h) in part (d) now the filter is type II with $30\,dB$ peak to peak stopband ripple.

Answers:

(a) $H(z) = \dfrac{0.0296 + 0.0592z^{-1} + 0.0296z^{-2}}{1 - 1.5442z^{-1} + 0.7548z^{-2}}$

(b) $H(z) = \dfrac{0.3959 - 0.7918z^{-1} + 0.3959z^{-2}}{1 - 1.1622z^{-1} + 0.6538z^{-2}}$

(c) $H(z) =$
$$\dfrac{0.0047 - 0.0142z^{-2} + 0.0142z^{-4} - 0.0047z^{-6}}{1 - 3.3735z^{-1} + 6.3464z^{-2} - 7.3655z^{-3} + 5.8056z^{-4} - 2.8169z^{-5} + 0.7654z^{-6}}$$

(d) $H(z) =$
$$\dfrac{0.3252 - 1.2059z^{-1} + 2.4662z^{-2} - 3.026z^{-3} + 2.4662z^{-4} - 1.2059z^{-5} + 0.3252z^{-6}}{1 - 2.6052z^{-1} + 3.5048z^{-2} - 2.8126z^{-3} + 1.255z^{-4} - 0.02z^{-5} - 0.177z^{-6}}$$

(e) $H(z) = \dfrac{0.0342 - 0.0445z^{-1} + 0.0342z^{-2}}{1 - 1.773z^{-1} + 0.797z^{-2}}$

(f) $H(z) = \dfrac{0.2357 - 0.4241z^{-1} + 0.2357z^{-2}}{1 + 0.2996z^{-1} + 0.195z^{-2}}$

(g) $H(z) =$
$$\dfrac{0.0257 - 0.0556z^{-1} + 0.0473z^{-2} - 0.0474z^{-4} + 0.0556z^{-5} - 0.0257z^{-6}}{1 - 3.3206z^{-1} + 6.0682z^{-2} - 6.6944z^{-3} + 4.9175z^{-4} - 2.1766z^{-5} + 0.5309z^{-6}}$$

(h) $H(z) =$

$$\frac{0.3272 - 1.1538z^{-1} + 2.275z^{-2} - 2.7615z^{-3} + 2.275z^{-4} - 1.1538z^{-5} + 0.3272z^{-6}}{1 - 2.3384z^{-1} + 2.735z^{-2} - 2.2102z^{-3} + 1.369z^{-4} - 0.5205z^{-5} + 0.1003z^{-6}}$$

Hint: section 6.9

Figure 12.6(e)
Pole-zero map of
the system in
project 59(a) –
right side figure

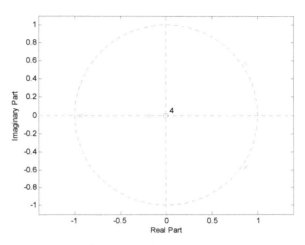

Figure 12.6(f)
Pole-zero map of
the system in
project 59(b) –
right side figure

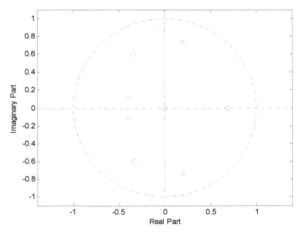

✦✦ Project 59 (on pole-zero map of a system)

(a) Draw the pole-zero map of a system which is characterized by

$$F(z) = \frac{1}{1 - 0.6z^{-1} - 0.7z^{-2} + 0.9z^{-3} + 0.2z^{-4}}.$$

(b) In part (a) now the $F(z)$ is $\dfrac{3 - z^{-3}}{2 + 0.8z^{-1} + 0.9z^{-2} + 0.8z^{-3} + 0.2z^{-4}}$.

(c) What are the poles and the zeroes of the system in part (a)?

(d) What are the poles and the zeroes of the system in part (b)?

Mohammad Nuruzzaman

(e) How many poles of the system in part (a) are outside the unit circle? Is the system in part (a) stable?

(f) Is the system in part (b) stable?

Answers:

(a) Figure 12.6(e) (b) Figure 12.6(f) (c) zeroes: 0 of order 4 and poles: $0.8665+ j\,0.5720$, $0.8665- j\,0.5720$, -0.9345, and -0.1985

(d) zeroes: 0, $-0.3467+ j\,0.6005$, $-0.3467- j\,0.6005$, and 0.6934 and poles: $0.1955+ j\,0.7425$, $0.1955- j\,0.7425$, $-0.3955+ j\,0.1150$, and $-0.3955- j\,0.1150$ (e) 2 and no (f) Yes

Hint: section 6.6

❖❖ Project 60 (on cascade and parallel filter equivalent)

(a) Define the filter system function $H_1(z) = \dfrac{7z^{-2} - 3z^{-1} + 5}{6z^{-3} - 3}$ as an object in MATLAB.

(b) Do the same as in part (a) for $H_2(z) = \dfrac{3.7(3z^{-1} - 5)}{(2z^{-1} - 5)(0.4z^{-1} - 5)}$.

(c) Do the same as in part (a) for $H_3(z) = \dfrac{2}{(z^{-1} - 1)(z^{-1} + 4)(2z^{-1} + 1)}$.

(d) If the systems in parts (a) and (b) are connected like figure 6.3(b), what is the equivalent system function $H_{eq}(z)$?

(e) If the systems in parts (a), (b), and (c) are connected like figure 6.3(c), what is the equivalent system function $H_{eq}(z)$?

(f) If the systems in parts (a) and (b) are connected like figure 6.3(d), what is the equivalent system function $H_{eq}(z)$?

(g) If the systems in parts (a), (b), and (c) are connected like figure 6.3(e), what is the equivalent system function $H_{eq}(z)$?

Answers:

(a) use filt function (b) same as (a) and also use conv function (c) same as (a) and (b) (d) $H_{eq}(z) = \dfrac{92.5 - 111z^{-1} + 162.8z^{-2} - 77.7z^{-3}}{75 - 36z^{-1} + 2.4z^{-2} - 150z^{-3} + 72z^{-4} - 4.8z^{-5}}$

(e) $H_{eq}(z) =$

$$\dfrac{-185 + 222z^{-1} - 325.6z^{-2} + 155.4z^{-3}}{300 + 231z^{-1} - 695.4z^{-2} - 486z^{-3} - 406.8z^{-4} + 1386z^{-5} - 228z^{-6} - 110.4z^{-7} + 9.6z^{-8}}$$

(f) $H_{eq}(z) = \dfrac{-180.5 + 168.3z^{-1} - 215z^{-2} + 197.4z^{-3} - 72.2z^{-4}}{75 - 36z^{-1} + 2.4z^{-2} - 150z^{-3} + 72z^{-4} - 4.8z^{-5}}$

(g) $H_{eq}(z) =$

$$\dfrac{-872 - 157.3z^{-1} + 1240z^{-2} - 802.5z^{-3} + 1723z^{-4} - 1303z^{-5} + 110.6z^{-6} + 144.4z^{-7}}{300 + 231z^{-1} - 695.4z^{-2} - 486z^{-3} - 406.8z^{-4} + 1386z^{-5} - 228z^{-6} - 110.4z^{-7} + 9.6z^{-8}}$$

Hint: section 6.9

✦✦ Project 61 (on parametric filter)

(a) Design a second order parametric filter with $f_c=250\,Hz$, $f_s=20\,KHz$, $G=7\,dB$, and $Q=0.9$ where the symbols have chapter 6 mentioned meanings.

(b) Do the same as in part (a) for $Q=0.6$.

(c) Do the same as in part (a) for $f_c=3\,KHz$.

Answers:

(a) $H(z) = \dfrac{1.052 - 1.911z^{-1} + 0.8648z^{-2}}{1 - 1.911z^{-1} + 0.9165z^{-2}}$

(b) $H(z) = \dfrac{1.076 - 1.872z^{-1} + 0.8013z^{-2}}{1 - 1.872z^{-1} + 0.8773z^{-2}}$

(c) $H(z) = \dfrac{1.372 - 0.8913z^{-1} + 0.02857z^{-2}}{1 - 0.8913z^{-1} + 0.4001z^{-2}}$

Hint: section 6.9

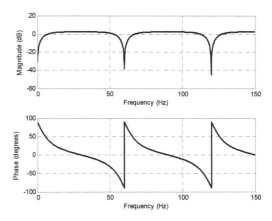

Figure 12.7(a) Magnitude and phase spectra of the filter in project 62(a)

✦✦ Project 62 (on digital filters)

(a) Design a periodic notch filter which will remove the $60\,Hz$ and its harmonics while the sampling frequency is $0.3\,KHz$. The z plane R of the filter should be 0.86.

(b) What is the system function of a 7-th order comb filter?

(c) What is the system function of a 7-point causal average filter?

(d) What is the system function of a 3rd order allpass filter which has the numerator coefficients forming an arithmetic progression with first term 0.2 and common difference 0.1?

(e) Magnitude and phase responses of the filters in parts (a) and (b) are to be graphed. The function **freqz** of section 6.7 is to be employed with 1024 points for each graphing. The sampling frequency of the filter in part (b) is $1\,KHz$.

Answers:

(a) $H(z) = \dfrac{1-z^{-5}}{1-\left(\dfrac{z}{0.86}\right)^{-5}}$ (b) $H(z) = 1 - z^{-7}$ (c) $H(z) = \dfrac{1-z^{-7}}{7(1-z^{-1})}$

(d) $H(z) = \dfrac{0.2 + 0.3z^{-1} + 0.4z^{-2} + 0.5z^{-3}}{0.5 + 0.4z^{-1} + 0.3z^{-2} + 0.2z^{-3}}$ (e) Figure 12.7(a) and

Figure 12.7(b) respectively

Hint: sections 6.7 and 6.9

Figure 12.7(b)
Magnitude and phase
spectra of the filter in
project 62(b)

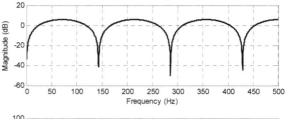

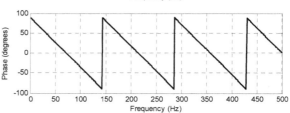

✦✦ Project 63 (on linear time invariant digital filters)

(a) Implement the IIR filter in MATLAB which is characterized by the system function $H(z) =$ $\dfrac{8 + 4z^{-2}}{5 - 3z^{-2} + 6z^{-5}}$.

(b) Implement the FIR filter in MATLAB which is characterized by the system function $H(z) =$ $4 - 3z^{-2} + 7z^{-6}$.

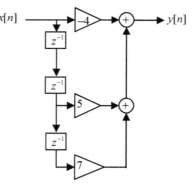

Figure 12.7(c) A digital filter

(c) Implement the digital filter which is depicted in figure 12.7(c).

(d) Implement the digital filter which is depicted in figure 12.7(d).

Answers (following symbols have section 6.8 mentioned meanings):

(a) B=[8 0 4]; A=[5 0 -3 0 0 6]; y=filter(B,A,f);

(b) B=[4 0 -3 0 0 0 7]; A=1; y=filter(B,A,f);

(c) B=[-4 0 5 7]; A=1; y=filter(B,A,f);

(d) B=[3 -5]; A=[1 -7 2]; y=filter(B,A,f);
Hint: section 6.8

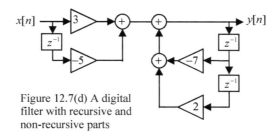

Figure 12.7(d) A digital
filter with recursive and
non-recursive parts

✦✦ Project 64 (on shelving filters)

(a) Design a second order low frequency shelving filter which has
the following specifications: boosting gain $10\,dB$, crossover
frequency 175 Hz, and sampling frequency 2 KHz.
(b) In part (a) now the gain is cutting.
(c) In part (a) now the filter is high frequency shelving.
(d) In part (c) now the gain is cutting.
Answers (following symbols have section 6.8 mentioned meanings):
(a) B=[1.3263   -1.0125   0.3668] and
    A=[1.0000   -1.2452   0.4604]
(b) B=[0.7540   -0.9389   0.3472] and
    A=[1.0000   -0.7634   0.2766]
(c) B=[2.6726   -4.1704   1.7130] and
    A=[1.0000   -1.2452   0.4604]
(d) B=[0.3742   -0.4659   0.1723] and
    A=[1.0000   -1.5605   0.6401]
Hint: section 6.9

✦✦ Project 65 (on digital filter application on audio)

(a) Apply the all pole filter $H(z)=\dfrac{1}{1+0.3z^{-1}+0.2z^{-2}+0.1z^{-3}+0.2z^{-4}}$ to
the audio **hello.wav** (section 2.7). Play both the original and the
filtered audio.
(b) In part (a) now the filter is second order bandpass Chebyshev
type I with lower and upper crossover frequencies 500 Hz and
2 KHz respectively and 4 dB peak to peak passband ripple.
(c) In part (a) now the filter is fifth order comb filter.
Hint: sections 6.9 and 6.10

✦✦ Project 66 (on basic matrix algebra)

(a) Apply the built-in function **rank** to find the rank of matrices
$$R=[7 \quad 88 \quad 0 \quad 88], \quad C=\begin{bmatrix}7\\8\\0\end{bmatrix}, \text{ and } A=\begin{bmatrix}7 & 3 & 2\\8 & 0 & 3\\16 & 0 & 6\end{bmatrix}.$$

(b) Determine the rank of audio **b.wav** (section 2.7).

(c) Determine matrix determinant of the square matrices $A =$

$$\begin{bmatrix} 1 & 2 & 7 \\ -2 & 5 & 2 \\ 4 & 8 & 7 \end{bmatrix} \text{ and } B = \begin{bmatrix} z & y^2 & y \\ 2y & x & 1 \\ 0 & 3 & 3 \end{bmatrix}.$$

(d) Determine the matrix inverse of $A = \begin{bmatrix} 3 & 2 & 5 \\ 0 & 3 & 6 \\ 1 & 0 & 5 \end{bmatrix}$ in rational and

decimal forms.

(e) Determine the inverse of symbolic element based matrix $A =$

$$\begin{bmatrix} 2y & 0 \\ z & x \end{bmatrix}.$$

(f) Determine the characteristic polynomial of the square matrices

$$A = \begin{bmatrix} 2 & 3 & -3 \\ -3 & -8 & 4 \\ 0 & 7 & 0 \end{bmatrix} \text{ and } B = \begin{bmatrix} y & x \\ -2 & -7 \end{bmatrix}.$$

Answers:

(a) 1, 1, and 2 respectively   (b) 1   (c) $|A| = -189$ and $|B| =$

$3zx - 3z - 6y^3 + 6y^2$

(d) $\begin{bmatrix} \frac{5}{14} & -\frac{5}{21} & -\frac{1}{14} \\ \frac{1}{7} & \frac{2}{21} & -\frac{3}{7} \\ -\frac{1}{14} & \frac{2}{21} & \frac{3}{14} \end{bmatrix} = \begin{bmatrix} 0.3571 & -0.2381 & -0.0714 \\ 0.1429 & 0.2381 & -0.4286 \\ -0.0714 & 0.0476 & 0.2143 \end{bmatrix}$

(e) $\begin{bmatrix} \frac{1}{2y} & 0 \\ -\frac{z}{2xy} & x \end{bmatrix}$ (f) $\lambda^3 + 6\lambda^2 - 35\lambda - 7$ and $\lambda^2 + 7\lambda - y\lambda - 7y + 2x$

respectively

Hint: sections 7.1-4

❖❖ **Project 67 (on time domain digital audio convolution)**

(a) Suppose time domain impulse response of a filter in relation to the figure 3.4(e) is given by $h[n] = u[n]$ over $0 \le n \le 35$. Apply the filter through convolution to the audio **bird.wav** (section 2.7) in time domain. Play both the original and the filtered audio.

(b) In part (a) now the audio is **hello.wav**.

(c) In part (b) now $h[n] = 2n - 100$ over $0 \le n \le 100$.

(d) In part (b) now $h[n] = e^{-\frac{n}{100}}$ over $0 \le n \le 100$.

Hint: section 3.10

❖❖ **Project 68 (on linear prediction and Z transform)**

(a) For the discrete signal $f[n] = [10 \quad 5 \quad -9 \quad 2 \quad 1 \quad 16]$ with prediction order 4, obtain the linear prediction system function $H(z)$.

(b) Obtain the zeroes and the poles of the system in part (a) in real-imaginary form.

(c) Do the same as in part (b) in magnitude-phase angle form.

(d) Graph the poles and the zeroes of the linear prediction system in part (a).

Answers:

(a) $\dfrac{\sqrt{69.0229}}{1-0.0297z^{-1}+0.0963z^{-2}+0.2522z^{-3}-0.1912z^{-4}}$ (b) zeroes: 0 of order 4 and poles: -0.7483, $0.1502+j\,0.7159$, $0.1502-j\,0.7159$, and 0.4776 (c) $0.7483\angle180°$, $0.7315\angle78.1506°$, $0.7315\angle-78.1506°$, and $0.4776\angle0°$ (d) Figure 12.8(a)

Hint: section 9.3

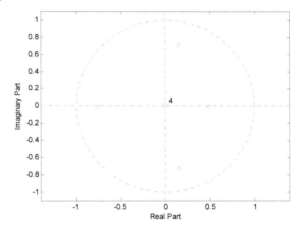

Figure 12.8(a) Pole-zero map of the linear prediction model for the project 68 – right side figure

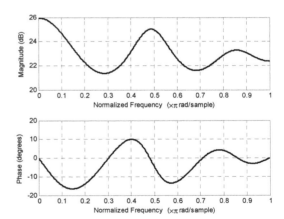

Figure 12.8(b) Frequency response of the linear prediction model $H(z)$ for project 69 – right side figure

♦♦ Project 69 (on frequency response of linear prediction model)

(a) Obtain the magnitude-phase frequency response of the linear prediction model $H(z)$ based on the prediction order 6 for the signal $f[n]=[1\quad 2\quad -5\quad 10\quad -1\quad 12\quad -13\quad 25\quad 29]$.

(b) Determine the resonator poles of the system in part (a).

(c) If a sampling frequency $f_s = 12$ *KHz* is used in part (a), what are the resonator frequencies?

(d) In 0-1 scale what are the resonator frequencies in part (c)?

(e) Do the normalized frequencies in part (d) seem to be verified from the figure 12.8(b)?

Answers:

(a) Figure 12.8(b)   (b) $0.6561\angle\pm152.2528°$, $0.7336\angle\pm87.6447°$, and $0.596\angle\pm11.5294°$ (c) 5.0751 *KHz*, 2.9215 *KHz*, and 0.3843 *KHz*   (d) 0.8458, 0.4869, and 0.0641 respectively   (e) Yes

Hint: section 9.4

❖❖ **Project 70 (on linear prediction on a digital audio)**

(a) Obtain the linear prediction model $H(z)$ for the digital audio b.wav (section 2.7) considering prediction order 8.

(b) Do the same as in part (a) for the first 5000 samples of the audio hello.wav.

(c) Do the same as in part (a) for the first 5000 samples of the audio test.wav.

Answers:

(a) $H(z) =$

$$\frac{\sqrt{0.0045}}{1-0.8968z^{-1}-0.0308z^{-2}+0.1891z^{-3}+0.0833z^{-4}-0.0641z^{-5}-0.1368z^{-6}-\overline{0.11z^{-7}+0.0031z^{-8}}}$$

(b) $H(z) =$

$$\frac{\sqrt{1.8992\times10^{-6}}}{1-1.5286z^{-1}+1.2102z^{-2}-0.7642z^{-3}+0.353z^{-4}-0.4879z^{-5}+0.4534z^{-6}-\overline{0.1872z^{-7}-0.0484z^{-8}}}$$

(c) $H(z) =$

$$\frac{\sqrt{5.5871\times10^{-5}}}{1-1.32z^{-1}+0.5999z^{-2}-0.162z^{-3}+0.1263z^{-4}-0.1852z^{-5}-0.0201z^{-6}+\overline{0.0433z^{-7}-0.0658z^{-8}}}$$

Hint: section 9.5

❖❖ **Project 71 (on spectrum envelope of digital audio by linear prediction)**

(a) Obtain the digital audio b.wav (section 2.7) and consider the first 500 samples of the audio.

(b) Determine the dB magnitude spectrum of the audio segment in part (a) by using discrete Fourier transform followed by normalization with respect to the maximum magnitude.

(c) Select the half magnitude spectrum of the part (b).

(d) Apply linear prediction with prediction order 20 to the audio segment in part (a) and obtain the system function $H(z)$.

(e) From the $H(z)$ of part (d), determine the half dB magnitude spectrum followed by normalization with respect to the maximum magnitude of $H(e^{j\omega})$.

(f) Graph the spectrum of the part (c), the spectrum of the part (e), and the spectra of the parts (c) and (e) together from up to down vertically in a single plot respectively where the vertical axis in each graph should be between −50 and 0 and horizontal axis of each should refer to integer frequency index.

Answers: Successful completion of the parts (a) through (f) should yield the graphical output like figure 12.8(c).

Hint: section 9.6

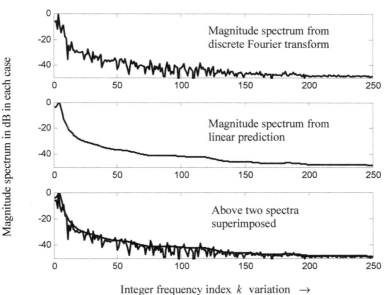

Figure 12.8(c) Magnitude spectrum envelop detection property of the linear prediction model in project 71

✦✦ Project 72 (on linear prediction error)

(a) Obtain the linear prediction error E for the samples of the digital audio **b.wav** (section 2.7) based on the prediction order 2, 200, 1000, and 2800.

(b) In part (a) now consider the prediction order P to be from 2 to 500 with increment 1. Graph the normalized E versus P.

(c) From the normalized plot of the part (b) what should be the approximate prediction order for 60% error?

Answers:

(a) E=0.0056, E=0.003, E=0.0027, and E=0.0026 respectively

(b) Figure 12.8(d)   (c) 50

Hint: section 9.7

Figure 12.8(d)
Normalized prediction
error versus order – right
side figure

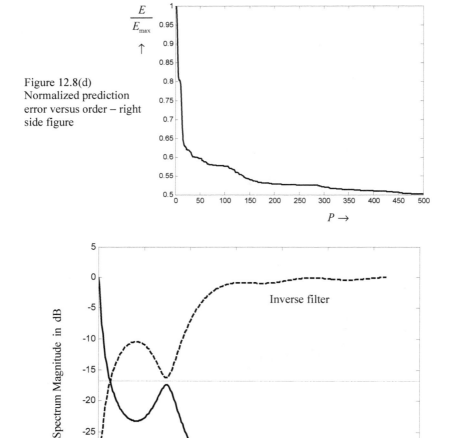

Figure 12.8(e) Image property of the inverse filter and linear prediction model

✦✦ Project 73 (on linear prediction and inverse filter)

(a) In projects 70(a) through 70(c) what are the expressions of the inverse filter $G(z)$?

(b) Verify that the dB normalized magnitude frequency responses of the 70(a) mentioned $H(z)$ and $G(z)$ are mutually reflective over the angular frequency interval $0 \le \omega \le \pi$ (choose ω step as 0.01 rad/sec).

(c) In part (b) what is the constant line dB value about which the reflection occurs?

Answers:

(a) $G(z) = 1 - 0.8968z^{-1} - 0.0308z^{-2} + 0.1891z^{-3} + 0.0833z^{-4} - 0.0641z^{-5} -$
$$0.1368z^{-6} - 0.11z^{-7} + 0.0031z^{-8}$$
$G(z) = 1 - 1.5286z^{-1} + 1.2102z^{-2} - 0.7642z^{-3} + 0.353z^{-4} - 0.4879z^{-5} +$
$$0.4534z^{-6} - 0.1872z^{-7} - 0.0484z^{-8}$$
$G(z) = 1 - 1.32z^{-1} + 0.5999z^{-2} - 0.162z^{-3} + 0.1263z^{-4} - 0.1852z^{-5} - 0.0201z^{-6}$
$$+ 0.0433z^{-7} - 0.0658z^{-8} \text{ respectively}$$

(b) You should be able to graph the figure 12.8(e) to prove the property.

(c) $-16.8629\,dB$

Hint: section 9.8

✦✦ Project 74 (on spectral subtraction algorithm)

(a) Suppose a noisy signal samples are given as $f[n]=[3\ \ 0\ \ 7\ 8\ -5\ \ 8]$. Apply the basic spectral subtraction algorithm with $D[k]=0.2$ to obtain the enhanced signal $\hat{f}[n]$.

(b) In part (a) now the $D[k]$ is 0.15.

(c) In part (a) now the $D[k]$ is 0.1 but use the author written function **basspec**.

(d) Consider the audio **test.wav** of section 2.7 which is a noisy one. Apply the basic spectral subtraction algorithm to the audio by choosing 200 samples at a time and by taking the $D[k]$ as 0.06. Play both the original and the enhanced audio. What do you infer from the playings?

Answers:

(a) $\hat{f}[n]=[3.0115\ \ -0.0345\ \ \ 6.8897\ \ \ 7.8782\ \ -4.9012\ \ \ 7.9564]$

(b) $\hat{f}[n]=[3.0086\ \ -0.0259\ \ \ 6.9173\ \ \ 7.9086\ \ -4.9259\ \ \ 7.9673]$

(c) $\hat{f}[n]=[3.0058\ \ -0.0173\ \ \ 6.9448\ \ \ 7.9391\ \ -4.9506\ \ \ 7.9782]$

(d) Background noise of the **test.wav** is slightly reduced.

Hint: sections 10.2 and 10.3

✦✦ Project 75 (on Wiener filtering in integer time index domain)

(a) Apply Wiener filtering to noisy signal $f[n]=[4 \quad 8 \quad -7 \quad 5 \quad 6 \quad 0 \quad 11]$ by choosing a frame length 4 and variance 2 in integer time index domain.

(b) In part (a) now the frame length is 3 with variance 2.5.

(c) In part (a) now the frame length is 3 with variance 2.

Answers:

(a) $r[n]=[3.8208 \quad 7.6589 \quad -6.4203 \quad 4.6981 \quad 5.9344 \quad 0.4012 \quad 10.2727]$

(b) $r[n]=[4 \quad 7.6064 \quad -6.4643 \quad 4.7373 \quad 5.1532 \quad 0.7005 \quad 10.3182]$

(c) $r[n]=[4 \quad 7.6851 \quad -6.5714 \quad 4.7898 \quad 5.3226 \quad 0.5604 \quad 10.4545]$

Hint: section 10.4

✦✦ Project 76 (on artificial noise generation)

(a) Generate a single random number between −1 and 1 which is uniformly distributed.

(b) Generate a five element column matrix of uniform random numbers in which each element is independently identically distributed (i.i.d) between −1 and 1.

(c) Do the same as in part (b) for a six element row matrix.

(d) Do the same as in part (b) for a 3×2 rectangular matrix.

(e) In parts (a) through (d) now the distribution is Gaussian with mean $\mu=3$ and standard deviation $\sigma=2$.

(f) Considering the softcopy bird.wav of section 2.7, add uniformly distributed noise between −0.07 and 0.07 to the audio and play the original audio, noise signal, and noisy audio.

(g) Do the same as in part (f) but the noise is Gaussian with mean $\mu=0.01$ and standard deviation $\sigma=0.2$.

Answers (Our executions returned the following certainly you will get different numbers due to randomness in the generation):

(a) 0.9003 (b) $\begin{bmatrix} -0.5377 \\ 0.2137 \\ -0.0280 \\ 0.7826 \\ 0.5242 \end{bmatrix}$ (c) $[-0.0871 \quad -0.9630 \quad 0.6428$

$-0.1106 \quad 0.2309 \quad 0.5839]$ (d) $\begin{bmatrix} -0.7222 & 0.2076 \\ -0.5945 & -0.4556 \\ -0.6026 & -0.6024 \end{bmatrix}$

(e) 3.2507, $\begin{bmatrix} 7.3664 \\ 2.7272 \\ 3.2279 \\ 5.1335 \\ 3.1186 \end{bmatrix}$, $[2.8087 \quad 1.3353 \quad 3.5888 \quad 0.3276 \quad 4.4286$

6.2471], and $\begin{bmatrix} 1.6164 & -0.1875 \\ 4.716 & 0.1181 \\ 5.5080 & 4.1423 \end{bmatrix}$ respectively

Hint: section 10.6

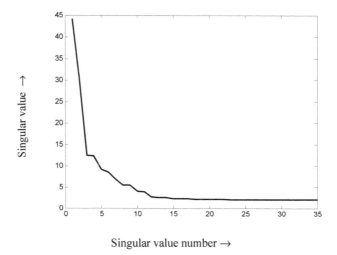

Singular value number →

Figure 12.8(f) Singular value characteristic of the
digital audio b.wav

◆ ◆ Project 77 (on singular value decomposition of digital audio)

(a) Considering the audio **b.wav** of section 2.7, form the special
Toeplitz matrix $f[m,n]$ with $L=35$ and factorize the $f[m,n]$ as
$U \times D \times V^T$ based on the singular value decomposition where the
symbols have section 7.6 cited meanings.

(b) What is the dimension of the matrix $f[m,n]$?

(c) What are the first 6 singular values of part (a)?

(d) What are the last 6 singular values of part (a)?

(e) Plot the singular values of the part (a) versus its integer position
number.

(f) Voice activity detection applies singular value decomposition.
Suppose **nosound.wav** of section 2.7 is the audio of a digital
recording ambient without speaker. Consider first 2800 samples
of the audio **test.wav**, choose Toeplitz matrix parameter L as
30, decompose the two Toeplitz matrices $f[m,n]$ of the two
digital audio based on singular values, and plot the singular
values of the both audio together. The singular value
characteristics then help in deciding the voice activity present in
the digital audio environment.

Answers:

(a) Factorization results will be available in U, D, and V in
accordance with section 7.8   (b) 2766×35   (c) 44.0927,

30.3022, 12.4415, 12.2768, 9.1214, and 8.5144, use the command S(1:6) where the S holds the singular values as a row or column matrix (d) 2.0313, 2.0269, 2.0235, 2.0233, 2.0181, and 2.0118, use the command S(30:end)   (e) Figure 12.8(f)   (f) Successful executions as mentioned in section 7.8 should provide you the graph 12.8(g)

Hint: section 7.8

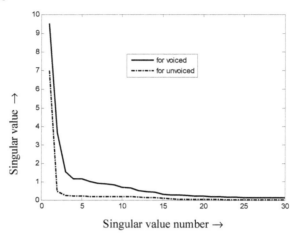

Figure 12.8(g)
Singular value
characteristics of
voiced and unvoiced
digital audio – right
side figure

◆◆ Project 78 (on covariance of random variables)

(a) Determine the covariance in rational and in decimal forms for the random variables which are placed in $A = \begin{bmatrix} 2 & 6 & 7 \\ 4 & -1 & -10 \end{bmatrix}$.

(b) Do the same as in part (a) for $A = \begin{bmatrix} -2 & 6 & 6 & 0 \\ 4 & 30 & 1 & 2 \\ 1 & -4 & -4 & 1 \end{bmatrix}$.

(c) In parts (a) and (b) the normalization now takes place by the number of rows in A.

(d) Determine the covariance between the random variables $X = [5 \ 6 \ -9 \ 0 \ 1]$ and $Y = [9 \ -7 \ 8 \ 2 \ 1]$.

(e) Do the same as in part (d) for $X = [5 \ 6 \ -9 \ 0 \ 1]$ and $Y = [10 \ 12 \ -18 \ 0 \ 2]$.

Answers:

(a) $V = \begin{bmatrix} 2 & -7 & -17 \\ -7 & 49/2 & 119/2 \\ -17 & 119/2 & 289/2 \end{bmatrix} = \begin{bmatrix} 2 & -7 & -17 \\ -7 & 24.5 & 59.5 \\ -17 & 59.5 & 144.5 \end{bmatrix}$

(b) $V = \begin{bmatrix} 9 & 36 & -15/2 & 3 \\ 36 & 916/3 & 25 & 12 \\ -15/2 & 25 & 25 & -5/2 \\ 3 & 12 & -5/2 & 1 \end{bmatrix} = \begin{bmatrix} 9 & 36 & -7.5 & 3 \\ 36 & 305.3333 & 25 & 12 \\ -7.5 & 25 & 25 & -2.5 \\ 3 & 12 & -2.5 & 1 \end{bmatrix}$

(c) $V = \begin{bmatrix} 1 & -7/2 & -17/2 \\ -7/2 & 49/4 & 119/4 \\ -17/2 & 119/4 & 289/4 \end{bmatrix} = \begin{bmatrix} 1 & -3.5 & -8.5 \\ -3.5 & 12.25 & 29.75 \\ -8.5 & 29.75 & 72.25 \end{bmatrix}$ and $V =$

$\begin{bmatrix} 6 & 24 & -5 & 2 \\ 24 & 1832/9 & 50/3 & 8 \\ -5 & 50/3 & 50/3 & -5/3 \\ 2 & 8 & -5/3 & 2/3 \end{bmatrix} = \begin{bmatrix} 6 & 24 & -5 & 2 \\ 24 & 203.5556 & 16.6667 & 8 \\ -5 & 16.6667 & 16.6667 & -1.6667 \\ 2 & 8 & -1.6667 & 0.6667 \end{bmatrix}$

respectively

(d) $c = -15.16$ (e) $c = 56.48$

Hint: section 7.9

✦✦ Project 79 (on covariance of digital audio)

(a) Determine the covariance between the audio **pitch.wav** and **b.wav** of section 2.7.

(b) Do the same as in part (a) for the audio **pitch.wav** and **nosound.wav**.

(c) Determine the variance-covariance matrix of audio **pitch.wav** by considering a Toeplitz with $L=3$ where L has section 7.7 cited meaning.

(d) Do the same as in part (c) for the audio **b.wav** but with $L=4$.

(e) Determine the variance-covariance matrix of the audio **b.wav**, **hello.wav**, and **pitch.wav**.

Answers:

(a) $c = -9.8673 \times 10^{-5}$ (b) $c = -2.0875 \times 10^{-5}$

(c) $V = \begin{bmatrix} 0.5002 & 0.48 & 0.4212 \\ 0.48 & 0.5002 & 0.48 \\ 0.4212 & 0.48 & 0.5002 \end{bmatrix}$ (d) $V = \begin{bmatrix} 0.0367 & 0.0337 & 0.0301 & 0.0264 \\ 0.0337 & 0.0367 & 0.0337 & 0.0301 \\ 0.0301 & 0.0337 & 0.0367 & 0.0337 \\ 0.0264 & 0.0301 & 0.0337 & 0.0367 \end{bmatrix}$

(e) $V = \begin{bmatrix} 0.002 & 0 & 0 \\ 0 & 0.0347 & 0 \\ 0 & 0 & 0.0657 \end{bmatrix}$ within 4 decimal accuracy

Hint: section 7.10

✦✦ Project 80 (on principal component analysis)

(a) Consider that the V is the variance-covariance matrix of $A = \begin{bmatrix} 21 & 5 \\ 18 & 7 \\ 25 & 9 \end{bmatrix}$. Decompose the V as $E \times D \times E^T$ where the symbols have section 7.11 cited meanings and obtain the E matrix and principal components (PCs).

(b) In part (a) now the A is $\begin{bmatrix} 21 & 5 & -3 & 4 \\ 18 & 7 & -5 & 9 \\ 25 & 9 & -7 & 0 \end{bmatrix}$.

(c) In part (a) now the A is $\begin{bmatrix} 21 & 5 & -3 & 4 & 8 \\ 18 & 7 & -5 & 9 & 2 \end{bmatrix}$.

(d) In parts (a) through (c) if the first component were held and the rest were discarded, how much percentage information would be retained?

Answers (note that the E is not unique):

(a) $E = \begin{bmatrix} 0.9277 & 0.3732 \\ 0.3732 & -0.9277 \end{bmatrix}$ and PCs: 13.9426 and 2.3908

(b) $E = \begin{bmatrix} 0.5946 & 0.0407 & 0.7882 & -0.1530 \\ 0.2006 & -0.6683 & -0.2474 & -0.6723 \\ -0.2006 & 0.6683 & -0.0222 & -0.7160 \\ -0.7522 & -0.3243 & 0.5630 & -0.1093 \end{bmatrix}$ and PCs: 34.8515,

5.8151, 0, and 0

(c) $E = \begin{bmatrix} -0.3397 & 0.2265 & -0.2265 & 0.5661 & -0.6794 \\ 0.2265 & 0.9617 & 0.0383 & -0.0957 & 0.1148 \\ -0.2265 & 0.0383 & 0.9617 & 0.0957 & -0.1148 \\ 0.5661 & -0.0957 & 0.0957 & 0.7608 & 0.2871 \\ -0.6794 & 0.1148 & -0.1148 & 0.2871 & 0.6555 \end{bmatrix}$ and PCs: 39,

0, 0, 0, and 0

(d) 85.36%, 85.7%, and 100% respectively

Hint: section 7.11

✦✦ **Project 81 (on principal component and discrete signal/audio)**

(a) Rearrange the discrete signal $f[n] = [10 \quad 8 \quad 20 \quad 9 \quad 2 \quad 15 \quad 17$ $19]$ to form a rectangular matrix A with 4 rows. Determine the principal components of the $f[n]$ based on the A.

(b) In part (a) now with 3 rows.

(c) Compute the principal components of digital audio **b.wav** (section 2.7) considering $M = 500$ where the M has section 7.11 cited meaning.

(d) Do the same as in part (c) for the audio **pitch.wav** and $M = 800$.

Answers:

(a) 61.7889 and 28.0444   (b) 185.94, 6.7266, and 0

(c) 0.0995, 0.0570, 0.0348, 0.0049, 0.0039, and 0.0005

(d) 2.2554, 1.788, 0.0965, 0, 0, 0, 0, 0, and 0 within 4 decimal accuracy

Hint: section 7.11

✦✦ **Project 82 (on Wiener filtering in integer time index domain)**

(a) Apply Wiener filtering to the audio **test.wav** (section 2.7) by choosing frame $f_b[n]$ length as 300 and variance v^2 as 0.0001 in integer time index domain. Play both the original and the Wiener filtered audio.

(b) Do the same as in part (a) for the audio **b.wav**.

(c) In part (a) now consider the frame by frame approach of the section 3.3. Rather than fixed variance, 5% frame variance is chosen for the Wiener filtering where the audio frame length is 331 and Wiener filtering frame $f_b[n]$ length is 40.

Answers:

(a) Background noise is reduced slightly

(b) No significant improvement due to the filtering

(c) Performance is better than in part (a). The L of section 3.3 should be 331. Between the lines starting with g and fn of figure 3.1(d), insert V=var(g); vn=wiener2(g,[40 1],0.05*V); but the fn is modified as fn=[fn;vn];. The V and vn are user-chosen variables. The V holds the consecutive frame variance stored in g.

Hint: section 10.5

✦✦ Project 83 (on noise reduction according to singular value)

(a) Determine the singular values of the signal $v[n]=[29 \quad 16 \quad -15 \quad 0 \quad 9 \quad -12 \quad 27 \quad 32 \quad -1 \quad 3]$ with semi Toeplitz parameter $L=6$.

(b) Suppose the $v[n]$ in part (a) represents a noise. What is the type of the noise?

(c) Suppose the $v[n]$ in part (a) now represents a noisy signal. There are five singular values labeled as λ_1, λ_2, λ_3, λ_4, and λ_5 in descending order. If you start discarding the singular values as follows: λ_5, $\lambda_5-\lambda_4$, $\lambda_5-\lambda_4-\lambda_3$, and so on, what is the SNR in absolute ratio as well as in dB?

(d) In part (c) suppose the SNR which is greater than $10\,dB$ is to be discarded. Obtain the approximated signal $\hat{v}[n]$.

(e) Verify that the author-written function svd_noise also returns the same signal as in part (d).

(f) Apply the singular value noise reduction algorithm through the svd_noise to the audio b.wav (section 2.7) with $L=50$ and $12\,dB$ SNR. Play the original and the approximated audio owing to singular value discarding.

Answers:

(a) 53.0132, 49.6334, 49.5464, 33.2178, and 17.8845

(b) Colored noise.

(c) Absolute ratio: 27.6131, 5.4302, 1.3599, and 0.4432

In dB: 14.4111, 7.3482, 1.3351, and −3.5343 respectively

(d) $\hat{v}[n]=[24.8019 \quad 16.1872 \quad -17.7646 \quad 2.5666 \quad 6.7733 \quad -10.1007 \quad 24.1448 \quad 33.9583 \quad -2.5012 \quad 4.0063]$

Hint: section 10.7

✦✦ Project 84 (on telephone dial pad signals)

(a) Display the telephone dial pad number 3 signal in time domain. Also display the power spectrum of the signal.

(b) Do the same as in part (a) for the number 5.

Hint: section 11.1

✦✦ Project 85 (on power-energy of discrete signal and digital audio)

(a) A discrete signal $x[n]=[14 \quad 9 \quad 21 \quad -10 \quad 0 \quad 2]$ is given, compute the power signal $p[n]$.

(b) Do the same as in part (a) for the signal $x[n]=[-16 \quad -8 \quad 0 \quad 3 \quad 7$ $9 \quad 10]$.

(c) If $T_s=1\sec$, what is the total energy E in Joule in the signal of the part (a).

(d) Do the same as in part (c) for the signal in part (b).

(e) Do the same as in part (c) at a sampling frequency $1.5\ KHz$.

(f) Do the same as in part (d) at a sampling frequency $1.5\ KHz$.

(g) After finding the power signal $p[n]$ of the audio **b.wav** (section 2.7) in integer time index domain, calculate the total energy in the audio.

(h) Do the same as in part (g) for the audio **test.wav**.

Answers:

(a) $p[n]=[196 \quad 81 \quad 441 \quad 100 \quad 0 \quad 4]$ in watt

(b) $p[n]=[256 \quad 64 \quad 0 \quad 9 \quad 49 \quad 81 \quad 100]$ in watt

(c) $E=822\ J$ (d) $E=559\ J$ (e) $E=0.548\ J$ (f) $E=0.3727\ J$

(g) $E=0.0047\ J$ (h) $E=0.009\ J$

Hint: section 11.3

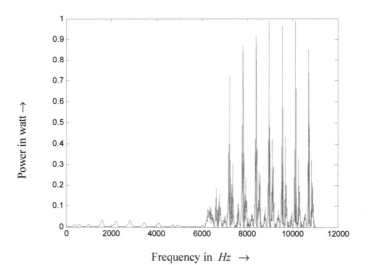

Figure 12.9(a) Power spectrum of the audio b.wav

✦✦ Project 86 (on power-energy of discrete signal and digital audio)

(a) From the time domain discrete signal $x[n]=[14 \quad 9 \quad 21 \quad -10 \quad 0$ $2]$, obtain the power signal $p[f]$ considering sampling frequency $f_s=400\ Hz$ and assuming that frequencies of $x[n]$ are distributed over $0 \le f < \frac{f_s}{2}$ where f is the frequency hidden in $x[n]$. Also determine the discrete frequencies corresponding to $p[f]$.

(b) Do the same as in part (a) for the signal $x[n] = [-16 \quad -8 \quad 0 \quad 3 \quad 7 \quad 9 \quad 10]$ and sampling frequency $f_s = 1.2\ KHz$.

(c) Subject to the assumption of the part (a), obtain the power signal $p[f]$ in watt from time domain audio signal **b.wav** (section 2.7) by linking it to its sampling frequency and plot the power $p[f]$ in watt versus frequency in Hertz in continuous sense.

(d) Do the same as in part (c) for the audio **nosound.wav**.

Answers:

(a) $p[f] = [196 \quad 81 \quad 441 \quad 100 \quad 0 \quad 4]$ in watt and for $f = 0\ Hz$, $40\ Hz$, $80\ Hz$, $120\ Hz$, $160\ Hz$, and $200\ Hz$ respectively

(b) $p[f] = [256 \quad 64 \quad 0 \quad 9 \quad 49 \quad 81 \quad 100]$ in watt and for $f = 0\ Hz$, $100\ Hz$, $200\ Hz$, $300\ Hz$, $400\ Hz$, $500\ Hz$, and $600\ Hz$ respectively

(c) Figure 12.9(a)   (d) Figure 12.9(b)

Hint: section 11.3

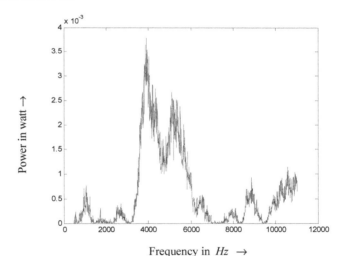

Figure 12.9(b) Power spectrum of the audio nosound.wav

✦✦ Project 87 (on power spectral density and digital audio)

(a) Based on the assumption and autocorrelation of basic power spectral density (PSD) definition as introduced in section 11.3, compute the power spectral density $P[f]$ of the discrete signal $x[n] = [14 \quad 9 \quad 21 \quad -10 \quad 0 \quad 2]$ at $f = 700\ Hz$ with sampling frequency $f_s = 3.5\ KHz$.

(b) Do the same as in part (a) for the signal $x[n]=[-16 \quad -8 \quad 0 \quad 3 \quad 7$ $9 \quad 10]$ at $f=900\,Hz$ with sampling frequency $f_s=3\,KHz$.

(c) In part (b) now at three frequencies $f=700\,Hz$, $f=800\,Hz$, and $f=900\,Hz$.

(d) In part (b) now over $0 \le f \le f_s$ with the frequency step $600\,Hz$.

(e) In part (d) now the PSD is required in dBW/Hz.

(f) Apply the built-in function **psd** to graph the power spectrum of the audio **b.wav** by Welch method.

(g) In part (f) what are the first four values of PSD and frequency?

Answers:

(a) $P[f]=814.8673\,W/Hz$ (b) $P[f]=369.8990\,W/Hz$

(c) $P[f]=223.7609\,W/Hz$, $454.0215\,W/Hz$, and

$369.8990\,W/Hz$ respectively

(d) $P[f]=[25 \quad 86.6606 \quad 178.3394 \quad 178.3394 \quad 86.6606 \quad 25]$ in W/Hz

(e) $[13.9794 \quad 19.3782 \quad 22.5125 \quad 22.5125 \quad 19.3782$
$\qquad 13.9794]$ in dBW/Hz

(f) Figure 12.9(c)

(g) $P[f]=0.3293\times10^{-4}\,W/Hz$, $0.6586\times10^{-4}\,W/Hz$, $0.6587\times10^{-4}\,W/Hz$, and $0.6587\times10^{-4}\,W/Hz$ and $f=0\,Hz$, $5.3833\,Hz$, $10.7666\,Hz$, and $16.1499\,Hz$ respectively

Hint: section 11.3

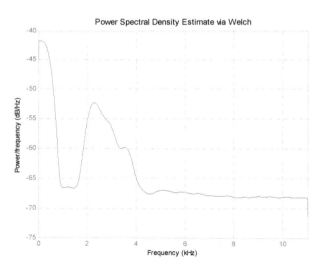

Figure 12.9(c) PSD of the audio **b.wav** using Welch method

✦✦ **Project 88 (on natural encoding)**

    (a) In a discrete signal if there are 64 amplitude levels, how many bits do you need to natural encode the signal?

    (b) Do the same as in part (a) for 128 and 200 levels.

    (c) A discrete signal $x[n]$ has the following amplitude levels: 7, 70, 289, and 29. What are the natural codes of these amplitude levels?

    (d) Each of the codes in part (c) is to be allocated in 10 bits. What is the representation of each code?

    (e) A naturally encoded discrete signal amplitude is given as 100101. What is the signal amplitude in integer decimal?

    (f) Do the same as in part (e) for each of the following strings: 1010, 1111111, 10101010, 10000001, and 100111111.

Answers:

(a) 6 bits, use $\log2(x)$ (b) 7 bits and 8 bits respectively

(c) 111, 1000110, 100100001, and 11101 respectively

(d) 0000000111, 0001000110, 0100100001,
    and 0000011101 respectively

(e) 37   (f) 10, 127, 170, 129, and 319 respectively

Hint: section 11.4

✦✦ **Project 89 (on natural encoding)**

    (a) Obtain the natural code for the digital audio **b.wav** of section 2.7 considering 128 amplitude levels in [0,127]. What are the first three amplitudes? What are the first three amplitude codes?

    (b) Do the same as in part (a) for 512 amplitude levels in [0,511].

    (c) Do the same as in part (a) for the audio **test.wav**.

    (d) Do the same as in part (a) for the audio **pitch.wav**.

Answers:

(a) Amplitudes: all three 57   Codes: all three 0111001

(b) Amplitudes: all three 230   Codes: all three 011100110

(c) Amplitudes: 71, 78, and 82
    Codes: 1000111, 1001110, and 1010010 respectively

(d) Amplitudes: 64, 81, and 98
    Codes: 1000000, 1010001, and 1100010 respectively

Hint: section 11.4

✦✦ **Project 90 (on run length encoding)**

    (a) Considering $f[n]$=[1 7 7 4 5 5 24 24 1 1 5 5] in [0,31], determine the run length codes of the discrete signal on a frame length 3.

    (b) Do the same as in part (a) on the frame length 4.

    (c) Apply the run length encoding on the digital audio **b.wav** of section 2.7 based on a frame length 10 and considering 128 amplitude levels. How many frames are in the coding? What are the codes in the first 3 frames?

(d) Do the same as in part (c) for the audio **pitch.wav** with the frame length 8.

Answers:

(a) $\begin{bmatrix} g_j & l_j \\ 1 & 1 \\ 7 & 2 \end{bmatrix}$, $\begin{bmatrix} g_j & l_j \\ 4 & 1 \\ 5 & 2 \end{bmatrix}$, $\begin{bmatrix} g_j & l_j \\ 24 & 2 \\ 1 & 1 \end{bmatrix}$, and $\begin{bmatrix} g_j & l_j \\ 1 & 1 \\ 5 & 2 \end{bmatrix}$ respectively

(b) $\begin{bmatrix} g_j & l_j \\ 1 & 1 \\ 7 & 2 \\ 4 & 1 \end{bmatrix}$, $\begin{bmatrix} g_j & l_j \\ 5 & 2 \\ 24 & 2 \end{bmatrix}$, and $\begin{bmatrix} g_j & l_j \\ 1 & 2 \\ 5 & 2 \end{bmatrix}$ respectively

(c) 280 frames and codes: $\begin{bmatrix} g_j & l_j \\ 57 & 7 \\ 58 & 3 \end{bmatrix}$, $\begin{bmatrix} g_j & l_j \\ 58 & 2 \\ 59 & 7 \\ 60 & 1 \end{bmatrix}$, and $\begin{bmatrix} g_j & l_j \\ 60 & 7 \\ 61 & 3 \end{bmatrix}$ respectively

(d) 827 frames and codes: $\begin{bmatrix} g_j & l_j \\ 64 & 1 \\ 81 & 1 \\ 98 & 1 \\ 112 & 1 \\ 121 & 1 \\ 127 & 2 \\ 122 & 1 \end{bmatrix}$, $\begin{bmatrix} g_j & l_j \\ 112 & 1 \\ 98 & 1 \\ 82 & 1 \\ 64 & 1 \\ 46 & 1 \\ 30 & 1 \\ 16 & 1 \\ 6 & 1 \end{bmatrix}$, and $\begin{bmatrix} g_j & l_j \\ 0 & 2 \\ 5 & 1 \\ 15 & 1 \\ 28 & 1 \\ 45 & 1 \\ 63 & 1 \\ 80 & 1 \end{bmatrix}$ respectively

Hint: section 11.4

◆◆ **Project 91 (on Huffman encoding)**

(a) For Huffman coding, determine the amplitude level probabilities of the discrete signal $f[n]=[2\ 3\ 25\ 2\ 1\ 3\ 25\ 3\ 2\ 3\ 25\ 1]$.

(b) Do the same as in part (a) for the signal $f[n]=[19\ 13\ 22\ 2\ 19\ 0\ 25\ 0\ 2\ 0\ 25\ 19]$.

(c) For the probability vector of the part (a) determine a Huffman code and average bit length in the coding.

(d) For the probability vector of the part (b) determine a Huffman code and average bit length in the coding.

Answers:

(a)
amplitude level	probability
1	0.1667
2	0.25
3	0.3333
25	0.25

(b)
amplitude level	probability
0	0.25
2	0.1667
13	0.0833
19	0.25
22	0.0833
25	0.1667

(c) $\begin{bmatrix} amplitude & Huffman \\ level & code \\ 1 & 11 \\ 2 & 10 \\ 3 & 00 \\ 25 & 01 \end{bmatrix}$ and 2 bits

(d) $\begin{bmatrix} amplitude & Huffman \\ level & code \\ 0 & 10 \\ 2 & 000 \\ 13 & 0011 \\ 19 & 01 \\ 22 & 0010 \\ 25 & 11 \end{bmatrix}$ and 2.5 bits

Hint: section 11.4

✦✦ **Project 92 (on Huffman encoding of a digital audio)**

(a) Find a Huffman code and average bit length for the digital audio **b.wav** of section 2.7 with 6-bit allocation. How many code words are in the scheme? What are the first three code words and what audio amplitudes do they correspond to?

(b) Do the same as in part (a) for the audio **test.wav** with 8-bit allocation.

(c) Do the same as in part (a) for the audio **pitch.wav**.

Answers:

(a) Codes will be available in the workspace variable, average bit length: 4.3957 bits, and number of code words: 55
first word: 11010011111   second word: 11010011110
third word: 101011000001   amplitudes: 0, 10, and 11 respectively

(b) Codes will be available in the workspace variable, average bit length: 4.1831 bits, and number of code words: 34
first word: 0011010111010011   second word: 001101011101000
third word: 001101011100   amplitudes: 0, 15, and 23 respectively

(c) Codes will be available in the workspace variable, average bit length: 5.8164 bits, and number of code words: 64
first word: 0110   second word: 00110
third word: 10110   amplitudes: 0, 1, and 2 respectively

Hint: section 11.4

✦✦ **Project 93 (on coded digital audio generation)**

(a) Code the 8-level discrete signal $f[n]=[1 \quad 3 \quad 4 \quad 7 \quad 4 \quad 3 \quad 7 \quad 1 \quad 5]$ in [0,7] by the characters A through H respectively.

(b) Do the same as in part (a) but with small characters, it is given that small a has the ASCII character code 97, and so on.

(c) Considering the **pitch.wav** of section 2.7, the digital audio is to be coded in [0,31] by the characters 0 through 9 and A through V respectively. Display the first and the last six characters of the coding.

(d) Do the same as in part (c) for the audio **nosound.wav**.

Answers:

(a) BDEHEDHBF   (b) bdehedhbf

(c) GKORUV... ... ... 0147BG

(d) AAAAAA... ... ... KKKJJJ

Hint: section 11.4

◆◆ Project 94 (on audio softcopy writing)

(a) Write the modular matrix $f[n]$=[0.9   0   0   −0.9   0.8   1   1   0   0   −0.9] as a **wav** file with 32-bit quantization based on the sampling frequency 16 *KHz* and name the softfile as **Audio2**. Verify that the written file **Audio2.wav** exists in your working folder. By machine reading of the **Audio2.wav**, check the saved data.

(b) Map the $f[n]$ data of part (a) in integer form in [−127,127] and then apply the 8 bit softcopy writing of the mapped data to a file by the name **Audio3.wav** with 20 *KHz* sampling frequency.

(c) Do the same as in part (b) in [−255,255] for 16 bit softcopy writing.

(d) Write the following modular stereo audio data in a softcopy by the name **Audio4.wav**: $f[n]|_{left}$=[11   4   1   −15   127   28   0   −127]/128 and $f[n]|_{right}$=[12   5   2   −14   126   25   0   −127]/128 with 20 *KHz* sampling frequency and 8-bit quantization.

Answers:

(b) **round(254*mat2gray(f)-127)**, [114   −7   −7   −127   100   127   127   −7   −7   −127]

(c) **round(510*mat2gray(f)-255)**, [228   −13   −13   −255   201   255   255   −13   −13   −255]

Hint: sections 3.2 and 11.5

———————— End of the chapter ————————

Appendices

Appendix A

Coding in MATLAB

     MATLAB executes the code of an expression in terms of string which is the set of keyboard characters placed consecutively. One distinguishing feature of MATLAB is that the workspace variable itself is a matrix. The strings adopted for computation are divided into two classes – scalar and vector. The scalar computation results the order of the output matrix same as that of the variable matrix. On the contrary, the order for the vector computation is determined in accordance with the matrix algebra rules. Some symbolic functions and their MATLAB counterparts are presented in table A.1. The operators for arithmetic computations are as follows:

addition	+
subtraction	−
multiplication	*
division	/
power	^

The operation sequence of different operators in a scalar or vector string observes the following order:

enclosing braces	( )	first,
power operator	^	then,
division operator	/	next,
multiplication operator	*	after that,
addition operator	+	then, and
subtraction operator	−	finally.

The syntax of the scalar computation urges us to use .*, ./, and .^ in lieu of *, /, and ^ respectively. The operators *, /, and ^ are never preceded by . for the vector computation. The vector string is the MATLAB code of any symbolic expression or function often found in mathematics. Starting from the simplest one, we present some examples on writing the long expressions in MATLAB for the scalar form and the vector form as well.

✦ ✦ **Write the MATLAB codes in scalar and vector forms for the following functions**

$A. \ \sin^3 x \cos^5 x$ $B. \ 2 + \ln x$ $C. \ x^4 + 3x - 5$ $D. \ \dfrac{x^3 - 5}{x^2 - 7x - 7}$

$E. \ \sqrt{|x^3| + \sec^{-1} x}$ $F. \ (1 + e^{\sin x})^{x^2 + 3}$ $G. \ \dfrac{\cosh x + 3}{\sqrt{\dfrac{x + 4}{\log_{10}(x^3 - 6)}}}$

$H. \ \dfrac{1}{(x - 3)(x + 4)(x - 2)}$ $I. \ \dfrac{1}{1 + \dfrac{1}{1 + \dfrac{1}{x}}}$ $J. \ \dfrac{a}{x + a} + \dfrac{b}{y + b} + \dfrac{c}{z + c}$

$K. \ \dfrac{u^2 v^3 w^9}{x^4 y^7 z^6}$

In tabular form, they are coded as follows:

Example	String for scalar computation	String for vector computation
A	sin(x).^3.*cos(x).^5	sin(x)^3*cos(x)^5
B	2+log(x)	2+log(x)
C	x.^4+3*x-5	x^4+3*x-5
D	(x.^3-5)./(x.^2-7*x-7)	(x^3-5)/(x^2-7*x-7)
E	sqrt(abs(x.^3)+asec(x))	sqrt(abs(x^3)+asec(x))
F	(1+exp(sin(x))).^(x.^2+3)	(1+exp(sin(x)))^(x^2+3)
G	(cosh(x)+3)./sqrt((x+4)./log10(x.^3-6))	(cosh(x)+3)/sqrt((x+4)/log10(x^3-6))
H	1./(x-3)./(x+4)./(x-2)	1/(x-3)/(x+4)/(x-2)
I	1./(1+1./(1+1./x))	1/(1+1/(1+1/x))
J	a./(x+a)+b./(y+b)+c./(z+c)	a/(x+a)+b/(y+b)+c/(z+c)
K	u.^2.*v.^3.*w.^9./x.^4./y.^7./z.^6	u^2*v^3*w^9/x^4/y^7/z^6

Digital audio programming circumstance dictates the type of code – whether scalar or vector should be employed.

Table A.1 Some mathematical functions and their MATLAB counterparts

Mathematical notation	MATLAB notation	Mathematical notation	MATLAB notation	Mathematical notation	MATLAB notation
$\sin x$	sin(x)	$\sin^{-1} x$	asin(x)	π	pi
$\cos x$	cos(x)	$\cos^{-1} x$	acos(x)	A+B	A+B
$\tan x$	tan(x)	$\tan^{-1} x$	atan(x)	A−B	A−B
$\cot x$	cot(x)	$\cot^{-1} x$	acot(x)	A×B	A*B
$\csc x$	csc(x)	$\sec^{-1} x$	asec(x)	e^x	exp(x)
$\sec x$	sec(x)	$\csc^{-1} x$	acsc(x)	A^B	A^B
$\sinh x$	sinh(x)	$\sinh^{-1} x$	asinh(x)	$\ln x$	log(x)
$\cosh x$	cosh(x)	$\cosh^{-1} x$	acosh(x)	$\log_{10} x$	log10(x)
$\sec hx$	sech(x)	$\sec h^{-1}x$	asech(x)	$\log_2 x$	log2(x)
$\csc hx$	csch(x)	$\csc ech^{-1}x$	acsch(x)	Σ	sum
$\tanh x$	tanh(x)	$\tanh^{-1} x$	atanh(x)	Π	prod
$\coth x$	coth(x)	$\coth^{-1} x$	acoth(x)	$\mid x \mid$	abs(x)
10^A	1e A e.g. 1e3	10^{-A}	1e- A e.g. 1e-3	$\sqrt{x}$	sqrt(x)

* In the six trigonometric functions for example sin(x), the x is in radian. If the x is in degree, we use sind(x). The other five functions also have the syntax cosd(x), tand(x), cotd(x), cscd(x), and secd(x) when the x is in degree. The default return from asin(x) is in radian, if you need the return to be in degree, use the command asind(x). Similar degree return is also possible from acosd(x), atand(x), acotd(x), asecd(x), and acscd(x).

We present some numerical examples to quote the difference between the scalar and vector computations in the following.

Let us say we have the matrices $A = \begin{bmatrix} 3 & 5 \\ 7 & 8 \end{bmatrix}$, $B = \begin{bmatrix} 5 & 2 & 1 \\ 0 & 1 & 7 \end{bmatrix}$, and $C = \begin{bmatrix} 3 & 2 & 9 \\ 4 & 0 & 2 \end{bmatrix}$. The scalar computation is not possible between the matrices A and B because of their unequal order, nor is between the matrices A and C for the same reason. On the contrary the scalar multiplication can be conducted between the B and C for having the same order and which is $B.*C = \begin{bmatrix} 15 & 4 & 9 \\ 0 & 0 & 14 \end{bmatrix}$ (element by element multiplication).

Matrix algebra rule says that any matrix A of order $M \times N$ can only be multiplied with another matrix B of order $N \times P$ so that the resulting matrix has the order $M \times P$. For the last paragraph cited A and B, we have $M = 2$, $N = 2$, and $P = 3$ and obtain the vector-multiplied matrix as $A \times B = \begin{bmatrix} 3 \times 5 + 5 \times 0 & 3 \times 2 + 5 \times 1 & 3 \times 1 + 5 \times 7 \\ 7 \times 5 + 8 \times 0 & 7 \times 2 + 1 \times 8 & 7 \times 1 + 8 \times 7 \end{bmatrix} = \begin{bmatrix} 15 & 11 & 38 \\ 35 & 22 & 63 \end{bmatrix}$, and which has the MATLAB code A*B not A.*B. Similar interpretation follows for the operators * and /.

Whenever we write the scalar codes A.*B, A./B, and A.^B, we make it certain that both the A and B are identical matrix in size. The 3*A means all elements of matrix A are multiplied by 3 and we do not use 3.*A. Also do we not use A./3 but do A/3. The signs + and - are never preceded by the operator . in the scalar codes. The command 4./A means 4 is divided by all elements in A. The A.^4 means power on all elements of A is raised by 4 and so on.

♣ ♣ Scale factor in resistor, inductor, or capacitor units

The standard units of resistance, inductance, and capacitance are Ohm(Ω), Henry(H), and Farad (F) respectively. In audio processing problems quantity of interest is often given in units which are in the power of 10. Table A.2 presents the engineering scale factor units and their MATLAB equivalences. For example the resistor $10.7\,K\Omega$ is coded as **10.7e3**. Again a capacitor of value $4.7\,\mu F$ is entered by writing **4.7e-6** in standard unit.

Table A.2 Engineering unit scale factors and their MATLAB counterparts

Scale factor	Symbol	As power of 10	MATLAB code
giga	G	10^9	e9
mega	M	10^6	e6
kilo	K	10^3	e3
milli	m	10^{-3}	e-3
micro	μ	10^{-6}	e-6
nano	n	10^{-9}	e-9
pico	p	10^{-12}	e-12

♣ ♣ Scale factor in voltage, current, frequency, or power units

Just quoted scale factor is also practiced in voltage, current, frequency, or power units. For instance the voltage $2.3\,mV$, the power

$1.11\,\mu W$, and the frequency $900\,MHz$ are coded by **2.3e-3, 1.11e-6,** and **900e6** respectively.

✦✦ Rational form coding

In symbolic or expression form computation it is recommended that the reader use the rational form instead of fractional form otherwise machine finds best digital representation of decimal data. For example the inductance $5.7\,mH$ is coded by **57/10000** instead of **5.7e-3**.

Appendix B

MATLAB functions/statements for digital audio study

While working on digital audio problems in MATLAB, we come across lots of built-in MATLAB functions or programming statements. In order to employ these elements for the audio analysis, we need to understand their input and output argument types and purpose of the elements. Functions or program elements exercised in the text with brief descriptions are in the following.

B.1 Comparative and logical operators

Comparative operators are used for the comparison of two scalar elements, one scalar and one matrix elements, or two identical size matrix elements. There are six comparative operators as presented in table B.1.

Table B.1   Equivalence of comparative operators

Comparative operation	Mathematical notation	MATLAB notation
equal to	$=$	==
not equal to	$\neq$	~=
greater than	$>$	>
greater than or equal to	$\geq$	>=
less than	$<$	<
less than or equal to	$\leq$	<=

The output of the expression pertaining to the comparative operators is logical – either true (indicated by 1) or false (indicated by 0). For example when A=3 and B=4, the comparisons A=B, $A \neq B$, A>B, $A \geq B$, A<B, and $A \leq B$ should be false(0), true(1), false(0), false(0), true(1), and true(1) respectively. We implement these comparative operations as presented in table B.2.

Table B.2   Scalar comparative operation

>>A=3; B=4; ⏎ >>A==B ⏎ ans = 0 >>A~=B ⏎ ans = 1	>>A>B ⏎ ans = 0 >>A>=B ⏎ ans = 0	>>A<B ⏎ ans = 1 >>A<=B ⏎ ans = 1

There are two operands A and B in table B.2, each of which is a single scalar. Each of the operands can be a matrix in general. In that case the logical decision takes place element by element on all elements in the matrix. For instance if $A=\begin{bmatrix} 5 & 8 \\ 5 & 7 \end{bmatrix}$ and $B=\begin{bmatrix} 2 & 1 \\ -2 & 9 \end{bmatrix}$, A>B should be $\begin{bmatrix} 5>2 & 8>1 \\ 5>-2 & 7>9 \end{bmatrix} = \begin{bmatrix} 1 & 1 \\ 1 & 0 \end{bmatrix}$. Again if the A happens to be a scalar (say A=4), the single scalar is compared to all elements in the B therefore $A \leq B$ should be $\begin{bmatrix} 4 \leq 2 & 4 \leq 1 \\ 4 \leq -2 & 4 \leq 9 \end{bmatrix} =$

Mohammad Nuruzzaman

$\begin{bmatrix} 0 & 0 \\ 0 & 1 \end{bmatrix}$. In a similar fashion the B also operates on A however the scalar and matrix related comparative implementation is presented in the table B.3.

Some basic logical operations are NOT, OR, and AND. The characters ~, |, and & of the keyboard are adopted for the logical NOT, OR, and AND respectively. In all logical outputs the 1 and 0 stand for true and false respectively. All logical operators apply to the matrices in general. For the matrix $A=\begin{bmatrix} 0 & 0 \\ 0 & 1 \end{bmatrix}$, NOT(A) operation

Table B.3 Scalar and matrix comparative operations

when A and B are matrices,	when A is scalar and B is matrix,
>>A=[5 8;5 7]; ↵	>>A=4; ↵
>>B=[2 1;-2 9]; ↵	>>B=[2 1;-2 9]; ↵
>>A>B ↵	>>A<=B ↵
ans = 1 1 1 0	ans = 0 0 0 1

should provide $\begin{bmatrix} 1 & 1 \\ 1 & 0 \end{bmatrix}$ (see table B.4). The logical OR and AND operations on the like positional elements of the two matrices $A=\begin{bmatrix} 1 & 1 \\ 0 & 1 \end{bmatrix}$ and $B=\begin{bmatrix} 0 & 1 \\ 1 & 1 \end{bmatrix}$ must return $\begin{bmatrix} 1 & 1 \\ 1 & 1 \end{bmatrix}$ and $\begin{bmatrix} 0 & 1 \\ 0 & 1 \end{bmatrix}$ respectively. Table B.4 shows both implementations.

Table B.4 Basic logical operations on matrix elements

for NOT(A) operation,	for A OR B,	for A AND B,	for A XOR B,
>>A=[0 0;0 1]; ↵	>>A=[1 1;0 1]; ↵	>>A&B ↵	>>xor(A,B) ↵
>>~A ↵	>>B=[0 1;1 1]; ↵		
	>>A|B ↵		ans =
ans =		ans =	1 0
1 1	ans =	0 1	1 0
1 0	1 1	0 1	
	1 1		

If the A or the B is a single 1 or 0, it operates on all elements of the other.

Sometimes we need to check the interval of the independent variable of mathematical functions for instance $-6 \leq x \leq 8$. The interval is split in two parts $-6 \leq x$ and $x \leq 8$. In terms of the logical statement one expresses the $-6 \leq x \leq 8$ as (-6<=x)&(x<=8).

There is no operator for the XOR logical operation instead the MATLAB function xor syntaxed by xor(A,B) implements the operation as presented in the table B.4.

B.2 Simple if/if-else/nested if syntax

Conditional commands are exercised by the **if-else** statements (reserve words). Also comparisons and checkings may need **if-else** statements. We can have different **if-else** structures namely simple-if, **if-else**, or nested-if depending on the programming circumstances, some of which we discuss in the following.

⊟ Simple if

The program syntax of the simple-**if** is as follows:

if *logical expression*
    Executable MATLAB command(s)
end

Logical expression usually requires the use of comparative operators whose reference is found in appendix B.1. If the logical expression beside the **if** is true, the command between the **if** and **end** is executed otherwise not. In tabular form the simple-**if** implementation is as follows:

Example: If $x \geq 1$, we compute $y = \sin x$. When $x = 2$, we should see $y = \sin 2 = 0.9093$.	Executable M-file: x=2; if x>=1 y=sin(x); end	Steps: Save the statements in a new M-file (section 1.3) by the name test and execute the following: >>test ⏎	Check from the command window after running the M-file: >>y ⏎ y = 0.9093

⊟ If-else

The general program syntax for the **if-else** structure is as follows:

if *logical expression*
    Executable MATLAB command(s)
else
    Executable MATLAB command(s)
end

If the logical expression beside the **if** is true, the command between the **if** and **else** is executed else the command between the **else** and **end** is executed. In tabular form, the **if-else-end** implementation is the following:

Example: When $x = 1$, we compute $y = \sin\dfrac{x\pi}{2} = 1$ otherwise $y = \cos\dfrac{x\pi}{2} = 0$.	Executable M-file: x=1; if x==1 y=sin(x*pi/2); else y=cos(x*pi/2); end	Steps: Save the statements in a new M-file by the name test and execute the following: >>test ⏎	Check from the command window after running the M-file: >>y ⏎ y = 1

If we had x=2; in the first line of the M-file in the last exercise, we would see y= $\cos \pi$ =−1.

🖻 Nested-if

The third type of the if structure is the nested-if whose general program syntax is attached in the right side text box. Clearly the syntax takes care of multiple logical expressions which we demonstrate by one example as shown in the following table.

> if *logical expression*
>   *Executable MATLAB command(s)*
> elseif *logical expression*
>   *Executable MATLAB command(s)*
>   ⋮
> elseif *logical expression*
>   *Executable MATLAB command(s)*
> else
>   *Executable MATLAB command(s)*
> end

| Example: The best example can be taking the decision of grades out of 100 based on the achieved number of a student. The grading policy is stated as if the achieved number of a student is greater than or equal 90, greater than or equal to 80 but less than 90, greater than or equal to 70 but less than 80, greater than or equal to 60 but less than 70, greater than or equal to 50 but less than 60, and less than 50, then the grade is decided as A, B, C, D, E, and F respectively. | Executable M-file:

 N=77;
 if N>=90
    g='A';
 elseif (N<90)&(N>=80)
    g='B';
 elseif (N<80)&(N>=70)
    g='C';
 elseif (N<70)&(N>=60)
    g='D';
 elseif (N<60)&(N>=50)
    g='E';
 else
    g='F';
 end | In the executable M-file, the N and g refer to the number achieved and the grade respectively. If the number N is 77, the grade g should be C. Any character is argumented under the single inverted comma.

 Steps: Save the left statements in a new M-file by the name test and execute the following:
 >>test ↵ | Check from the command window after running the M-file:
 >>g ↵

 g =

 C |

B.3 Data accumulation

Sometimes it is necessary that we perform appending operation on an existing matrix in MATLAB workspace.

✦ Appending rows

Assume that the $A = \begin{bmatrix} 1 & 3 & 5 \\ 2 & 6 & 8 \\ 9 & 5 & 0 \\ 4 & 7 & 8 \end{bmatrix}$ is formed by appending two row matrices [9 5 0] and [4 7 8] with the matrix $B = \begin{bmatrix} 1 & 3 & 5 \\ 2 & 6 & 8 \end{bmatrix}$.

We first enter the matrix B (section 1.3) into MATLAB and append one row after another by using the command as presented below:

for entering B, >>B=[1 3 5;2 6 8] ↵	for appending the first row, >>B=[B;[9 5 0]] ↵	for appending the second row, >>A=[B;[4 7 8]] ↵
B = 1 3 5 2 6 8	B = 1 3 5 2 6 8 9 5 0	A = 1 3 5 2 6 8 9 5 0 4 7 8

The command B=[B;[9 5 0]] in above execution says that the row [9 5 0] is to be appended with the existing B (inside the third bracket) and that the result is again assigned to B. You can append as many rows as you want. The important point is the number of elements in each row that is to be appended must be equal to the number of columns in the matrix B.

✦ Appending columns

Suppose $C = \begin{bmatrix} 1 & 3 & 5 & 9 & 3 \\ 2 & 6 & 8 & 0 & 1 \\ 9 & 5 & 0 & 1 & 9 \end{bmatrix}$ is formed by appending two column matrices $\begin{bmatrix} 9 \\ 0 \\ 1 \end{bmatrix}$ and $\begin{bmatrix} 3 \\ 1 \\ 9 \end{bmatrix}$ with matrix $D = \begin{bmatrix} 1 & 3 & 5 \\ 2 & 6 & 8 \\ 9 & 5 & 0 \end{bmatrix}$. We get the matrix D into MATLAB and append one column after another as follows:

for entering D, >>D=[1 3 5;2 6 8;9 5 0] ↵	for appending the first column, >>D=[D [9 0 1]'] ↵	for appending the second column, >>C=[D [3 1 9]'] ↵
D = 1 3 5 2 6 8 9 5 0	D = 1 3 5 9 2 6 8 0 9 5 0 1	C = 1 3 5 9 3 2 6 8 0 1 9 5 0 1 9

The column matrix [9 0 1]' and D in above execution has one space gap within the third bracket. In the second of above implementation, the resultant matrix is again assigned to D. Append as many columns as you want just remember that the number of elements in each column that is to be appended must be equal to the number of rows in the matrix D.

✦ Data accumulation by using the two appending techniques

Suppose initially there is nothing in the f matrix, which in MATLAB we write by the statement f=[ ]; (an empty matrix is assigned to f). An empty matrix does not have any size and completely empty, it follows the null symbol ∅ of the matrix algebra. Let us say k=2 and perform the assignment as follows:

>>f=[ ]; k=2; ↵

Now if we execute f=[f k] time and again first f=[f k] returns 2, second f=[f k] returns [2 2], third f=[f k] returns [2 2 2], and so on. This is called row directed data accumulation. Column directed data accumulation occurs by executing f=[f;k] each time.

The demonstrated k is just a scalar but it can be a return from some function, row matrix, or column matrix.

B.4 For-loop syntax

A for-loop performs similar operations for a specific number of times and must be started with the **for** and terminated by an **end** statements. Following the **for** there must be a counter. The counter of the for-loop can be any variable that counts integer or fractional values depending on the increment or decrement. If the MATLAB command statements between the **for** and **end** of a for-loop are few words lengthy, one can even write the whole for-loop in one line. The programming syntax and some examples on the for-loop are as follows:

✦ **Program syntax**

for *counter*=starting value:increment or decrement of the counter value:final value

Executable MATLAB command(s)

end

✦ **Example 1**

Our problem statement is to compute $y = \cos x$ for $x = 10^0$ to 70^0 with the increment 10^0. Let us assign the computed values to some variable y where y should be [$\cos 10^0$ $\cos 20^0$ $\cos 30^0$ $\cos 40^0$ $\cos 50^0$ $\cos 60^0$ $\cos 70^0$]=[0.9848   0.9397   0.866   0.766   0.6428   0.5   0.342].

In the programming context, y(1) means the first element in the row matrix y, y(2) means the second element in the row matrix y, and so on. MATLAB code for the $\cos x$ is cosd(x) where x is in degree. The for-loop counter expression should be k=1:1:7 or k=1:7 to have the control on the position index in the row matrix y (because there are 7 elements or indexes in y). Since the computation needs 10 to 70, one generates that by writing k*10. Following is the implementation:

Executable M-file:	*Or, as a one line:*
for k=1:1:7 y(k)=cosd(k*10); end	for k=1:1:7 y(k)=cosd(k*10); end

Steps we need:
Open a new M-file (section 1.3), type the executable M-file statements in the M-file editor, save the editor contents by the name **test** in your working path, and call the **test** as shown below.

```
>>test ↵
>>y ↵

y =
     0.9848 0.9397  0.8660  0.7660   0.6428  0.5000   0.3420
```

✦ Example 2

A for-loop helps us accumulate data (appendix B.3) consecutively controlled by the loop index. In this example we accumulate some data row directionally according to the for-loop counter index.

For $k =1$, 2, and 3, we intend to accumulate the k^2 side by side. At the end we should be having [1   4   9] assigned to some variable **f** – this is our problem statement.

for the right shifting,	for the left shifting,
>>f=[ ]; for k=1:3 f=[f k^2]; end ↵	>>f=[ ]; for k=1:3 f=[k^2 f]; end ↵
>>f ↵	>>f ↵
f =	f =
1 4 9	9 4 1

The for-loop for the accumulation is presented above. The accumulation may occur as right or left shifting. Corresponding to the right shifting, the vector code (appendix A) for k^2 is k^2. The statement f=[ ]; means that an empty matrix is assigned to f outside the for-loop but at the beginning. The k variation in our problem is put as the for-loop counter. How the for-loop accumulates is shown below:

```
When k=1,  f=[f k^2]; returns  f=[[ ] 1^2];  ⇒ f=1;
When k=2,  f=[f k^2]; returns  f=[1 2^2];    ⇒ f=[1  4];
When k=3,  f=[f k^2]; returns  f=[1 4 3^2];  ⇒ f=[1  4  9];
```

The accumulation is happening from the left to the right. A single change provides the shifting from the right to the left which is f=[k^2 f];. The complete code and its execution result are also shown above by the heading 'for the left shifting'.

✦ Example 3

Another accumulation can be column directed that is we wish to see the output like $\begin{bmatrix} 1 \\ 4 \\ 9 \end{bmatrix}$ in example 2.

We just insert the row separator of a rectangular matrix (done by the operator ;) in the command f=[f k^2];. Again the shifting can happen either from the up to down or from the down to up. Both implementations are shown below:

for the down shifting,	for the up shifting,
>>f=[ ]; for k=1:3 f=[f;k^2]; end ⏎	>>f=[ ]; for k=1:3 f=[k^2;f]; end ⏎
>>f ⏎	>>f ⏎
f =	f =
1	9
4	4
9	1

◆ Example 4

Many digital audio problems need writing multiple for-loops. Usually one loop is for one dimensional function, two loops are for two dimensional function, and so on. One dimensional functional data takes the form of a row or column matrix.

Suppose we have the one dimensional data as $y=[9\ 6\ 7\ 4\ 6]$. We wish to access to every data in y. A single for-loop helps us do that as shown below:

>>y=[9 6 7 4 6]; for k=1:length(y) v=y(k); end ⏎

First we assign the data to workspace **y** as a row matrix. The command **length** finds the number of elements in the row matrix **y**. The **y(k)** means the **k-th** element in the **y** which we assign to workspace **v** (any user-chosen variable). Every single data of the y is sequentially available in the **v**. The contents of **y** can be a column matrix too.

B.5 Finding the maximum/minimum numerically

Given a matrix, one finds the maximum element from the matrix by using the command **max** (min for the minimum). Let us say we have three matrices $R=[1\ -2\ 3\ 9]$, $C=\begin{bmatrix} 23 \\ -20 \\ 30 \\ 8 \end{bmatrix}$, and $A=\begin{bmatrix} 2 & 4 & 7 \\ -2 & 7 & 9 \\ 3 & 8 & -8 \end{bmatrix}$ whose maxima are 9, 30, and 9 (from all elements in the matrix) and minima are −2, −20, and −8 respectively. We find the maxima first entering (section 1.3) the respective matrices as follows:

for the row matrix,	for the column matrix,	for the rectangular matrix,
>>R=[1 -2 3 9]; ⏎	>>C=[23;-20;30;8]; ⏎	>>A=[2 4 7;-2 7 9;3 8 -8]; ⏎
>>max(R) ⏎	>>max(C) ⏎	>>max(max(A)) ⏎
ans =	ans =	ans =
9	30	9
>>min(R) ⏎	>>min(C) ⏎	>>min(min(A)) ⏎
ans =	ans =	ans =
-2	-20	-8

Font equivalence is maintained by using the same letter for example $A \Leftrightarrow A$ in the last implementation. If the matrix is a row or column one, we apply one **max** or **min**. For a rectangular matrix, the **max** or **min** separately operates on each column that is why two **max** or **min** functions are required. The functions are equally applicable for decimal number elements.

In the row matrix R, the maximum 9 is occurring as the fourth element in the matrix. Suppose we also intend to find the position index (that is 4) of the maximum element in the R. Now we need two output arguments – one for the maximum and the other for its index. Its implementation is shown in the right side attached text box of this paragraph in which the two output arguments **M** and **I** correspond to the maximum and its integer index respectively.

The function **min** also keeps this type of integer index returning option in a similar fashion.

```
for index finding in R,
>>[M,I]=max(R) ↵

M =
         9
I =
         4
```

B.6 Complex number basics

Symbolically the imaginary unit of a complex number is denoted by i, j, or $\sqrt{-1}$ whose MATLAB representation is i, j, or **sqrt(-1)**. As an example the complex number $4 + j5$ is entered into MATLAB by any of the following expressions **4+5i, 4+5*i, 4+i*5, 4+5*j**, or **4+5*sqrt(-1)**.

Matrix of complex numbers follows similar entering style to that of the integer or real number with little difference in conjugateness (section 1.3). Let us enter the complex number matrices $R = [\,3 - j \quad 4j \quad -4\,]$, $C = \begin{bmatrix} 7j \\ -4+5j \\ 8j \end{bmatrix}$, and $A = \begin{bmatrix} 2 & 5-j & 9j \\ 7j & 2+j & 11j \end{bmatrix}$ into MATLAB as conducted in the following:

```
for R,
>>R=[3-i   4i   -4] ↵

R =
      3.0000 - 1.0000i    0 + 4.0000i   -4.0000
```

```
for C,
>>C=[7i   -4+5i   8i].' ↵

C =
       0 + 7.0000i
      -4.0000 + 5.0000i
       0 + 8.0000i
```

```
for A,
>>A=[2 5-i 9i;7i 2+i 11i] ↵

A =
      2.0000              5.0000 - 1.0000i    0 + 9.0000i
      0 + 7.0000i         2.0000 + 1.0000i    0 +11.0000i
```

The operators .' and ' mean transpose without and with conjugates respectively. In the column matrix case if we use the operator ' at the end, we would assign $\begin{bmatrix} -7j \\ -4-5j \\ -8j \end{bmatrix}$ to C.

Modulus or absolute value of a complex number $A+jB$ is given by $\sqrt{A^2+B^2}$. To take the modulus of a complex number, we call the command **abs** (abbreviation for absolute value) with the syntax **abs**(complex scalar or matrix). For example the modulus of $4+j3$ and elements in $R=[12+j5 \quad -4-j3 \quad -8+j6]$ are 5 and [13 5 10] respectively which we compute by using the right side attached command. In both cases we assigned the return to workspace A which can be any user-supplied variable.

> modulus for the single complex number,
> >>A=abs(4+3i) ↵
>
> A =
>   5
> modulus for the complex row matrix elements in R,
> >>R=[12+5i -4-3i -8+6i]; ↵
> >>A=abs(R) ↵
>
> A =
>   13   5   10

Argument of a complex number $A+jB$ is given by $\tan^{-1}\frac{B}{A}$. To find the argument, we call the function **angle** with the syntax **angle**(complex scalar or matrix name). The function returns any value from $-\pi$ to π. For instance the arguments of $4+j3$ and each element in $R=[12+j5 \quad -4-j3 \quad -8+j6]$ are $\tan^{-1}\frac{3}{4}=0.6435^c$ and $[0.3948^c \quad -2.4981^c \quad 2.4981^c]$ respectively which we implement by right side attached text box commands. In both cases we assigned the return to the workspace P which can be any user-given variable. For degree to radian and radian to degree conversions we call the commands **deg2rad** and **rad2deg** respectively for example on P as **rad2deg(P)** for the degree.

> argument for the single complex number:
> >>angle(4+3i) ↵
>
> P =
>   0.6435
> argument for the complex row matrix elements in R,
> >>P=angle(R) ↵
>
> P =
>   0.3948 -2.4981 2.4981

The conjugate of a complex number $A+jB$ is given by $A-jB$. To find the conjugate of a complex number, we apply the function **conj** with the syntax **conj**(complex scalar or matrix name). As an example the conjugate of $4+j3$ and all elements in $R=[12+j5 \quad -4-j3 \quad -8+j6]$ are $4-j3$ and $[12-j5 \quad -4+j3 \quad -8-j6]$ respectively. Both implementations are shown below and assigned to the workspace C (user-given variable):

> conjugate for the single complex number,
> >>C=conj(4+3i) ↵
>
> C =
>   4.0000 - 3.0000i

conjugate of the elements in the row matrix R,
```
>>R=[12+5i -4-3i -8+6i]; ↵
>>C=conj(R) ↵
```

C =

    12.0000-5.0000i -4.0000+3.0000i -8.0000- 6.0000i

A complex number $A + jB$ has the real part A and the imaginary part B. To find the real and imaginary parts from complex number(s), we apply the functions **real** and **imag** with the syntax real(complex scalar or

```
>>real(R) ↵        ← for the real elements in R

ans =
        12  -4  -8
>>imag(R) ↵        ← for the imaginary elements in R

ans =
        5  -3  6
```

matrix name) and **imag**(complex scalar or matrix name) respectively. The real and imaginary parts for the elements in $R = [12 + j5 \quad -4 - j3 \quad -8 + j6]$ are $[12 \quad -4 \quad -8]$ and $[5 \quad -3 \quad 6]$ respectively whose findings are attached in the upper right side text box of this paragraph. The returns could have been assigned to some user-supplied variables.

Rectangular to polar conversion is widely used in audio study in Fourier domain.

Given a complex number in rectangular form $A + jB$, the polar or exponential form of the number is $re^{j\theta}$ where $r = \sqrt{A^2 + B^2}$ and $\theta = \tan^{-1}\frac{B}{A}$. Its MATLAB counterpart is **cart2pol** (abbreviation for the Cartesian to (2) polar) and we use the syntax $[\theta, r] =$ cart2pol(A, B) where θ is in radian.

```
from rectangular to polar conversion,
>>[t r]=cart2pol(5,4) ↵

t =
      0.6747
r =
      6.4031
from polar to rectangular conversion,
>>[A B]=pol2cart(0.6747,6.4031) ↵

A =
      5.0001
B =
      3.9998
```

Again given the polar form of a complex number $re^{j\theta}$, the reverse conversion is $A = r\cos\theta$ and $B = r\sin\theta$ and the resembling MATLAB function is **pol2cart** (abbreviation for the polar to (2) Cartesian) with the syntax $[A, B]$=pol2cart(θ in radian, r). Unavailability of the θ makes us write **t** instead of θ where the **t** is a user-supplied variable.

The rectangular form number $5 + j4$ has the polar form $(r, \theta) = (6.4031, 0.6747^c)$ which we intend to obtain and vice versa.

We see both implementations above in the lower right side text box of this page. Slight discrepancy is seen in the implementation, instead of 5 we are getting 5.0001. When we write 0.6747 as θ, we ignore the fifth digit and whatsoafter. The usage of [A B]=pol2cart(t,r) returns the perfect result because the t holds the complete data from the computation. Anyhow the input arguments of both functions can be matrix as well and the output arguments can be so.

Now we address how to turn some real data to complex number. Suppose we have two identical size row matrices $x=[5\ \ 6\ \ 7]$ and $y=[8\ \ 9\ \ -9]$ and wish to form the complex matrix $A=[5+j8\quad 6+j9\quad 7-j9]$ which needs us to exercise the commands x=[5 6 7]; y=[8 9 -9]; A=x+i*y;.

Again say we have some polar or exponential form data like $r=[6\ \ 4\ \ 3]$ and $\theta=[\frac{\pi}{3}\ \ -\frac{\pi}{7}\ \ \frac{\pi}{2}]$ and intend to form $A=[6e^{i\frac{\pi}{3}}\quad 4e^{-i\frac{\pi}{7}}\quad 3e^{i\frac{\pi}{2}}]$ which requires us to execute the following: r=[6 4 3]; t=[pi/3 -pi/7 pi/2]; A=r.*exp(i*t); where the t is for the θ and the scalar code .* is used for the multiplication (appendix A). If the t were in degrees, the command would be A=r.*exp(i*t*pi/180);.

B.7 Matrix data flipping

Usually two kinds of matrix data flipping are practiced in audio study as presented in the following (font equivalence is maintained using the same letter for example A⇔A and the flipped data is assigned to workspace F, which is a user-chosen variable).

✦✦ **Flipping from the left to right**

Flipping from the left to right of a row or rectangular matrix is performed by the command **fliplr** (abbreviation for flipping from left to right). Suppose we have the row matrix $R=[2\ \ 4\ \ 3\ \ -4\ \ 6\ \ 9\ \ 3\ \ 7\ \ 10]$. If you flip the elements of R from the left to right, the resulting matrix should be [10 7 3 9 6 −4 3 4 2]. For a rectangular matrix, the flipping operation from left to right takes place over each column that is turning the first column to the last, the second column to the second from the last, and so on. Considering A

$=\begin{bmatrix}4&23&85&34\\5&43&41&87\\8&65&76&71\end{bmatrix}$, just mentioned flipping should return

$\begin{bmatrix}34&85&23&4\\87&41&43&5\\71&76&65&8\end{bmatrix}$. Both implementations are exercised as follows:

Left to right flipping of the row matrix,	Left to right flipping of rectangular matrix,
>>R=[2 4 3 -4 6 9 3 7 10]; ↵	>>A=[4 23 85 34;5 43 41 87;8 65 76 71]; ↵
>>F=fliplr(R) ↵	>>F=fliplr(A) ↵
F =	F =
10 7 3 9 6 -4 3 4 2	34 85 23 4
	87 41 43 5
	71 76 65 8

Since a column matrix has only one column, no change to the column matrix is brought about by the **fliplr**.

◆ ◆ **Flipping from up to down**

The function **flipud** (abbreviation for <u>fli</u>pping from <u>u</u>p to <u>d</u>own) flips the elements of a column or rectangular matrix from up to down. Flipping the column matrix $C = \begin{bmatrix} 4 \\ 7 \\ 8 \\ 3 \\ 1 \end{bmatrix}$ from up to down results

the matrix $\begin{bmatrix} 1 \\ 3 \\ 8 \\ 7 \\ 4 \end{bmatrix}$. Flipping from up to down of a rectangular matrix

happens over each row for example $D = \begin{bmatrix} 4 & 23 & 85 \\ 5 & 43 & 41 \\ 8 & 65 & -1 \\ 3 & 12 & 13 \end{bmatrix}$ becomes

$\begin{bmatrix} 3 & 12 & 13 \\ 8 & 65 & -1 \\ 5 & 43 & 41 \\ 4 & 23 & 85 \end{bmatrix}$ due to the flipping. No change occurs to a row matrix

when the **flipud** is applied on it. Both examples are implemented as follows:

Up to down flipping of the column matrix,	Up to down flipping of the rectangular matrix,
>>C=[4 7 8 3 1]'; ↵	>>D=[4 23 85;5 43 41;8 65 -1;3 12 13]; ↵
>>F=flipud(C) ↵	>>F=flipud(D) ↵
F =	F =
1	3 12 13
3	8 65 -1
8	5 43 41
7	4 23 85
4	

B.8 Matrix data rounding and remainder after integer division

As the title articulates we introduce the truncation or rounding and remainder after integer division of matrix elements in this subappendix (font equivalence is maintained by using the same letter for example $R \Leftrightarrow R$).

◆ ◆ **Truncating matrix elements**

Built-in function **fix** discards the fractional parts of a matrix elements regardless of the magnitude and returns the integer parts on all elements in the matrix when the matrix is its input argument. Let us say we have the row matrix $R = [1.2578 \quad -9.3445 \quad -8.9999]$ which should return $[1 \quad -9 \quad -8]$ following the removal of the fractional parts and we carry out the implementation as follows:

>>R=[1.2578 -9.3445 -8.9999]; ↵   ← We assigned the row matrix to
                                        the workspace R

>>V=fix(R) ↵   ← Truncated elements are
                                        assigned to V as a row matrix

V =
    1 -9 -8

If the R were a rectangular matrix, the command would be applied equally to all elements in the matrix.

✦✦ **Rounding matrix elements**

Any fractional number can be rounded to its nearest integer by using the command **round** which means if the fractional part of the number is greater than or equal to 0.5, it is taken as 1 and if it is less

for rounding the C elements:
>>C=[1.5001 -9.5000 -8.4999]; ↵
>>V=round(C) ↵
V =
2 -10 -8

than 0.5, it is taken as 0. Referring to $C=[1.5001 \quad -9.5000 \quad -8.4999]$, the rounding operation on all elements on the C should provide us $[2 \quad -10 \quad -8]$ whose implementation is presented in the attached text box of this paragraph and the rounded elements are held in V. The C can be a rectangular matrix in general.

✦✦ **Remainder after integer division**

When an integer is divided by another integer, there is no fractional part in the integer division for example the integers 3 and 2 provide the quotients $\frac{2}{3}=0$ and $\frac{3}{2}=1$. Remainder after integer division is found by the command **rem** which basically computes integer–divider×quotient. When 2 is divided by 3, we should get $2-3\times0=2$ as the remainder after the integer division. Similarly 3 by 2 should provide us 1. Again the same operation on all elements of

$$A = \begin{bmatrix} 2 & 9 \\ -56 & -5 \\ 6 & 76 \\ 3 & 2 \end{bmatrix} \text{ by } -3 \text{ should return } \begin{bmatrix} 2 & 0 \\ -2 & -2 \\ 0 & 1 \\ 0 & 2 \end{bmatrix}. \text{ Also the same}$$

operation for the like positional elements of dividend $D = \begin{bmatrix} 2 & 3 & 4 \\ 7 & 9 & 2 \end{bmatrix}$

and divider $B = \begin{bmatrix} 3 & 4 & 2 \\ -3 & 2 & 3 \end{bmatrix}$ should return us $\begin{bmatrix} 2 & 3 & 0 \\ 1 & 1 & 2 \end{bmatrix}$. All these

are presented below (every result is assigned to V):

| when 2 divided by 3,
>>V=rem(2,3) ↵

V =
   2
when 3 divided by 2,
>>V=rem(3,2) ↵

V =
   1 | when A is divided by -3,
>>A=[2 9;-56 -5;6 76;3 2]; ↵
>>V=rem(A,-3) ↵

V =
  2   0
 -2 -2
  0   1
  0   2 | like positional elements
of D divided by B,
>>D=[2 3 4;7 9 2]; ↵
>>B=[3 4 2;-3 2 3]; ↵
>>V=rem(D,B) ↵

V =
  2 3 0
  1 1 2 |

B.9 Position indexes of matrix elements with conditions

MATLAB command **find** looks for the position indexes of matrix elements subject to some logical condition whose general format is [R C]= find(condition) where the indexes returned to the R and the C are meant to be for the row and the column directions respectively. The R and the C are user-chosen workspace variables. Let us consider $A = \begin{bmatrix} 11 & 10 & 11 & 10 \\ 12 & 10 & -2 & 0 \\ -7 & 17 & 1 & -1 \end{bmatrix}$

which we enter by the following:

>>A=[11 10 11 10;12 10 -2 0;-7 17 1 -1]; ↵   ← A is assigned to A

We would like to know what the position indexes of A where the elements are greater than 10 are. In matrix A the left-upper most element has the position index (1,1). The elements of A being greater than 10 have the position indexes (1,1), (2,1), (3,2), and (1,3). MATLAB finds the required index in accordance with columns. Placing the row and column indexes vertically, we have $\begin{bmatrix} 1 \\ 2 \\ 3 \\ 1 \end{bmatrix}$ and $\begin{bmatrix} 1 \\ 1 \\ 2 \\ 3 \end{bmatrix}$ respectively. The output arguments R and C of the **find** receive these two column matrices respectively. The input argument of the **find** must be a logical statement, any element in A greater than 10 is written as $A>10$ (appendix B.1). The position indexes are found as shown in the right side attached text box.

where elements of A are greater than 10,
>>[R C]=find(A>10) ↵

R =
1
2
3
1

C =
1
1
2
3

where elements of A =10,	where elements of A ≤0,	for the row matrix D,
>>[R C]=find(A==10) ↵	>>[R C]=find(A<=0) ↵	>>D=[-10 34 1 2 8 4]; ↵
		>>R=find(D>=8) ↵
R =	R =	
1	3	R =
2	2	2 5
1	2	for the column matrix E,
C =	3	>>E=[-2 8 -2 7]'; ↵
2	C =	>>C=find(E~=-2) ↵
2	1	
4	3	C =
	4	2
	4	4

To exercise more conditions, what are the position indexes in the matrix A where the elements are equal to 10? The answer is (1,2), (2,2), and (1,4). Again the position indexes where the elements are less than or equal to zero are (3,1), (2,3), (2,4), and (3,4).

The comparative operators $>$, $<$, $\geq$, $\leq$, and $\neq$ have the MATLAB counterparts $>$, $<$, $>=$, $<=$, and $\sim=$ respectively.

So far we considered a rectangular matrix for demonstration of the position index finding. Let us see how the **find** works for a row or column matrix. Let us take $D=[-10 \quad 34 \quad 1 \quad 2 \quad 8 \quad 4]$ from which we find the position indexes of the elements where they are greater than or equal to 8. Obviously they are the 2nd and 5th elements. Here we do not need to place two output arguments to the **find**.

Again let us find the position indexes of the elements of the column matrix $E = \begin{bmatrix} -2 \\ 8 \\ -2 \\ 7 \end{bmatrix}$ where the elements are not equal to -2. The 2nd and 4th elements are not equal to -2.

The output of the function **find** is a row one for the row matrix input and a column one for the column matrix input.

Presented on the lower half side of the last page are the executions for all these conditional findings.

B.10 Matrix of ones, zeroes, and constants

MATLAB built-in commands **ones** and **zeros** implement user-defined matrix of ones and zeroes respectively. Each function conceives two input arguments, the first and second of which are the required numbers of rows and columns respectively. Let us say we intend to form the matrices $A = \begin{bmatrix} 1 & 1 & 1 \\ 1 & 1 & 1 \\ 1 & 1 & 1 \\ 1 & 1 & 1 \end{bmatrix}$, $B = \begin{bmatrix} 1 & 1 & 1 \\ 1 & 1 & 1 \\ 1 & 1 & 1 \end{bmatrix}$, and $C = \begin{bmatrix} 1 & 1 & 1 & 1 \\ 1 & 1 & 1 & 1 \end{bmatrix}$. Their orders are 4×3, 3×3, and 2×4 respectively and the implementations are as follows:

for A,	for B,	for C,
>>A=ones(4,3) ↵	>>B=ones(3) ↵	>>C=ones(2,4) ↵

```
A =                  B =              C =
    1   1   1            1   1   1        1   1   1   1
    1   1   1            1   1   1        1   1   1   1
    1   1   1            1   1   1
    1   1   1
```

Either the number of rows or columns will do if the matrix is a square. For the row and column matrices of ones for example of length 6, the commands would be **ones(1,6)** and **ones(6,1)** respectively.

Formation of the matrix of zeroes is quite similar to that of the matrix of ones. Replacing the function **ones** by **zeros** does the formation. Matrix of zeroes like $A = \begin{bmatrix} 0 & 0 & 0 \\ 0 & 0 & 0 \\ 0 & 0 & 0 \\ 0 & 0 & 0 \end{bmatrix}$, $B = \begin{bmatrix} 0 & 0 & 0 \\ 0 & 0 & 0 \\ 0 & 0 & 0 \end{bmatrix}$, and $C = \begin{bmatrix} 0 & 0 & 0 & 0 \\ 0 & 0 & 0 & 0 \end{bmatrix}$ (whose orders are 4×3, 3×3, and 2×4) we form by the commands **A=zeros(4,3)**,

B=zeros(3), and C=zeros(2,4) respectively. A row and a column matrices of 6 zeroes are formed by the commands zeros(1,6) and zeros(6,1) respectively.

A matrix of constants is obtained by first creating a matrix of ones of the required size and then multiplying by the constant number. For example the matrix $\begin{bmatrix} 0.2 & 0.2 & 0.2 \\ 0.2 & 0.2 & 0.2 \\ 0.2 & 0.2 & 0.2 \\ 0.2 & 0.2 & 0.2 \end{bmatrix}$ is generated by the command 0.2*ones(4,3).

B.11 Summing matrix elements

MATLAB function sum adds all elements in a row, column, or rectangular matrix when the matrix is its input argument. Example matrices are $R = [1\ {-2}\ 3\ 9]$, $C = \begin{bmatrix} 23 \\ -20 \\ 30 \\ 8 \end{bmatrix}$, and $A = \begin{bmatrix} 2 & 4 & 7 \\ -2 & 7 & 9 \\ 3 & 8 & -8 \end{bmatrix}$ whose all element sums are 11, 41, and 30 for the R, C, and A respectively. We execute the summations as follows (font equivalence is maintained by using the same letter for example A$\Leftrightarrow$ A):

Sum for the row matrix,	Sum for the column matrix,	Sum for the rectangular matrix,
>>R=[1 -2 3 9]; ↲	>>C=[23 -20 30 8]'; ↲	>>A=[2 4 7;-2 7 9;3 8 -8]; ↲
>>sum(R) ↲	>>sum(C) ↲	>>sum(sum(A)) ↲
		ans =
ans =	ans =	30
11	41	

For a rectangular matrix, two functions are required because the inner sum performs the summing over each column and the result is a row matrix. The outer sum provides the sum over the resulting row matrix.

The function is operational for real, complex, even for symbolic variables like x or y.

B.12 Padding arrays by user-defined elements

If a matrix exists in MATLAB workspace, we pad the matrix by any element and by any number of the element with the help of the function padarray which applies a syntax padarray(matrix variable name, dimension to be padded as a two element row matrix, the value to be padded, the direction of the padding). The direction of padding is understood by the function by three reserve words pre, post, and both for padding at the beginning, after the ending, and both the beginning and the ending of the array respectively. Each of the three reserve words is put under a quote.

Even though the function handles row, column, or rectangular matrix in general, we wish to take a row or column matrix example because that best suits to digital audio data.

Let us take the example of the row matrix A=[4   6   7   0]. We wish to pad the row matrix by the number −2 and by 3 times of the number after the array A so that we will have the row matrix [4   6   7   0   −2   −2   −2] following the padding.

If we do not want extra row addition, we set the first dimension as 0 so as a two element row matrix basically we need [0   3], the second element 3 means three times requirement of the −2. In order to accomplish the padding, we execute the following:

```
>>A=[4 6 7 0];  ⏎        ← A holds the A as a row matrix
>>B=padarray(A,[0,3],-2,'post')  ⏎   ← B holds the padded A, B is user-chosen

B =
     4   6   7   0   -2   -2   -2
```

Similarly padding at the beginning should provide us [−2   −2   −2   4   6   7   0] which takes place by the following:

```
>>B=padarray(A,[0,3],-2,'pre')  ⏎

B =
     -2   -2   -2   4   6   7   0
```

Again padding before and after the array by the same number we get by:

```
>>B=padarray(A,[0,3],-2,'both')  ⏎

B =
     -2   -2   -2   4   6   7   0   -2   -2   -2
```

If the A were a column matrix, the padding would occur in a similar fashion. But the second input argument of the **padarray** would be [3   0] in each of the above three cases for the column matrix.

B.13 Mathematics readable form by pretty

There is a function called **pretty** which returns any functional expression as close as mathematical form provided that its input argument is in vector string or code (appendix A) form. For example in vector string form, the x^3 is coded as **x^3**. When **x^3** is the input argument of the **pretty**, the return of the **pretty** is x^3 provided that the independent variable x is defined before by using the command **syms**.

The **pretty** is applied only for the display reason, no computation is conducted on the expression. If a function has the code **2*x**, you will see the function as **2 x**. It does not work on numeric values but usage of the command **sym** on the numeric values shows the rational form. For example **pretty(sym(3.3))** displays $\frac{33}{10}$. Just to show by another example, **1/(x+1)** is $\frac{1}{x+1}$ which can be executed as:

```
>>syms x  ⏎       ← Declaring the related x of  1/(x+1)  by the syms
```

```
>>y=1/(x+1); ↵     ← Assigning the code of  1/(x+1)  to y where y is a user-chosen
```
variable
```
>>pretty(y) ↵     ← Applying the pretty on the codes stored in y
       1
     ------
     x + 1
```

B.14 Formation of a Toeplitz matrix

A Toeplitz matrix is one whose entries are constant along each diagonal. For example a 4×4 and a 3×5 Toeplitz matrices take the form as

$$T_{4\times4} = \begin{bmatrix} d & c & e & f \\ c & d & c & e \\ e & c & d & c \\ f & e & c & d \end{bmatrix} \text{ and } T_{3\times5} = \begin{bmatrix} d & c & e & f & b \\ c & d & c & e & f \\ e & c & d & c & e \end{bmatrix} \text{ where } a, b, c, d, e, \text{ and}$$

f are real numbers respectively. The Toeplitz is a symmetric matrix. To form a Toeplitz matrix, we apply the built-in function **toeplitz** with the syntax **toeplitz**(elements as a row or column matrix for square Toeplitz) or **toeplitz**(super diagonal elements as a row or column matrix, sub diagonal elements as a row or column matrix for rectangular Toeplitz).

Let us generate a square Toeplitz of order 5×5 from $C = \begin{bmatrix} 5 \\ 0 \\ -4 \\ 9 \\ 1 \end{bmatrix}$ so

$$T_{5\times5} = \begin{bmatrix} 5 & 0 & -4 & 9 & 1 \\ 0 & 5 & 0 & -4 & 9 \\ -4 & 0 & 5 & 0 & -4 \\ 9 & -4 & 0 & 5 & 0 \\ 1 & 9 & -4 & 0 & 5 \end{bmatrix}.$$ As a rectangular example, a Toeplitz of

order 3×5 is to be formed where the sub diagonal elements are $R = [10 \ -9 \ 7]$ and the super diagonal elements are $C = [10 \ 55 \ 5 \ 40 \ 6]$. The matrix

should look like $\begin{bmatrix} 10 & 55 & 5 & 40 & 6 \\ -9 & 10 & 55 & 5 & 40 \\ 7 & -9 & 10 & 55 & 5 \end{bmatrix}$. Both examples are presented

below:

for 5×5 Toeplitz matrix,
```
>>C=[5 0 -4 9 1]'; ↵
>>T=toeplitz(C) ↵

T =
    5    0   -4    9    1
    0    5    0   -4    9
   -4    0    5    0   -4
    9   -4    0    5    0
    1    9   -4    0    5
```

for 3×5 Toeplitz matrix,
```
>>R=[10 -9 7]; ↵
>>C=[10 55 5 40 6]; ↵
>>T=toeplitz(R,C) ↵

T =
   10   55    5   40    6
   -9   10   55    5   40
    7   -9   10   55    5
```

The first elements of R and C must be identical when the rectangular Toeplitz is formed otherwise error message is displayed by MATLAB. In the last executions the given matrix is assigned to like name variable for example R to R. Row or column matrix entering reference is seen in section 1.3. In either Toeplitz the return is assigned to T which is a user-chosen variable.

B.15 Roots of a polynomial from its equation

A polynomial equation of degree n has n roots. Depending on the coefficients of a polynomial, the roots can be real or complex. The polynomial equation $4x^2 + 4x + 1 = 0$ has the roots -0.5 and -0.5 and we intend to find those.

MATLAB function **roots** finds the roots of a polynomial equation when the polynomial coefficients in descending power of x as a row matrix is its input argument. If the equation were of higher degree than that of 2 for example $x^5 + 5.5x^4 + 7.5x^3 - 2.5x^2 - 8.5x - 3 = 0$ and whose roots are given by $x = -3, -2, -1, -0.5,$ and 1, it is solved by the function too.

Example of a polynomial equation bearing complex roots is $x^3 + x^2 + 4x + 4 = 0$ whose roots are -1, $2i$, and $-2i$. All root finding examples are shown below:

roots for $4x^2 + 4x + 1 = 0$: >>A=[4 4 1]; ↵ >>r=roots(A) ↵	roots for $x^5 + 5.5x^4 + 7.5x^3 - 2.5x^2 - 8.5x - 3 = 0$: >>A=[1 5.5 7.5 -2.5 -8.5 -3]; ↵ >>r=roots(A) ↵	roots for $x^3 + x^2 + 4x + 4 = 0$: >>A=[1 1 4 4]; ↵ >>r=roots(A) ↵
r = -0.5000 -0.5000	r = -3.0000 -2.0000 1.0000 -1.0000 -0.5000	r = - 0.0000 + 2.0000i - 0.0000 - 2.0000i - 1.0000

In above executions the workspace A (which is user-chosen variable) holds the polynomial coefficients as a row matrix and the roots are returned to workspace r as a column matrix for each polynomial where the r is a user-chosen variable. Consecutive roots (i.e. first, second, third, etc) from the r are picked up by the commands r(1), r(2), r(3), etc respectively.

B.16 Nonzero elements in a matrix

The number of nonzero elements in a matrix is found by the function nnz (which is the abbreviation for number non zero). To find the number of nonzero elements in a matrix, we apply the command nnz(matrix name). Like name variable is used for assigning in the following e.g. R to R.

Our chosen row, column, and rectangular matrices for the

implementation are $R = [2 \quad 3 \quad 0 \quad 8 \quad -8 \quad 0 \quad 7 \quad 0 \quad 13]$, $C = \begin{bmatrix} -27 \\ -9 \\ 0 \\ 67 \\ 0 \\ 0 \end{bmatrix}$,

and $A = \begin{bmatrix} 0 & 3 & 8 \\ 9 & 0 & 6 \\ 0 & 0 & 1 \end{bmatrix}$ respectively. There are six, three, and five elements which

are not equal to zero in R, C, and A respectively. See the executions as
follows:

for the row matrix,
```
>>R=[2 3 0 8 -8 0 7 0 13]; ↵
>>nnz(R) ↵
```

ans =
     6

for the column matrix,
```
>>C=[-27 -9 0 67 0 0]'; ↵
>>nnz(C) ↵
```

ans =
     3

for the rectangular matrix,
```
>>A=[0 3 8;9 0 6;0 0 1]; ↵
>>nnz(A) ↵
```

ans =
     5

In some circumstances we may have to pick up the nonzero elements from a
matrix. The built-in function **nonzeros** detects the nonzero elements in a
matrix. The search is carried out according to columns. The return from the
nonzeros is a column matrix regardless of the type of input matrix. To find
the nonzero elements in a matrix, we apply the command **nonzeros**(matrix
name).

For example the nonzero elements of row matrix $R = [3 \quad 0 \quad 18 \quad -8$

$0 \quad 7 \quad 0 \quad 11]$, column matrix $C = \begin{bmatrix} -3 \\ -7 \\ -1 \\ 0 \\ 7 \\ 0 \end{bmatrix}$, and rectangular matrix $A = \begin{bmatrix} 0 & 3 & 8 \\ 9 & 0 & 6 \end{bmatrix}$

are $[3 \quad 18 \quad -8 \quad 7 \quad 11]$, $[-3 \quad -7 \quad -1 \quad 7]$, and $[9 \quad 3 \quad 8 \quad 6]$ for R, C, and
A respectively. The findings are seen below:

for the row matrix,
```
>>R=[3 0 18 -8 0 7 0 11]; ↵
>>nonzeros(R) ↵
```

ans =
     3
     18

for the column matrix,
```
>>C=[-3 -7 -1 0 7 0]'; ↵
>>nonzeros(C) ↵
```

ans =
     -3
     -7

```
    -8                                    -1
     7                                     7
    11
```

for the rectangular matrix,
```
>>A=[0 3 8;9 0 6]; ↵
>>nonzeros(A) ↵
```

ans =
```
        9
        3
        8
        6
```

We could have assigned the return from the **nnz** or **nonzeros** to some other workspace variable.

B.17 Multiple maxima numerically

Appendix B.5 cited **max** has the limitation that it determines only one maximum. When multiple maxima are required, our written function **mmax** can be considered. Figure B.1(a) shows the code of the function **mmax**. Open a new M-file (appendix C), type the codes of the figure B.1(a) in the file, and save the file by the name **mmax**.

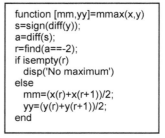

```
function [mm,yy]=mmax(x,y)
s=sign(diff(y));
a=diff(s);
r=find(a==-2);
if isempty(r)
    disp('No maximum')
else
    mm=(x(r)+x(r+1))/2;
    yy=(y(r)+y(r+1))/2;
end
```

Figure B.1(a) Function file for multiple maxima

As you see there are two input arguments in the **mmax** – **x** and **y**, each of which is identical size row or column matrix. Besides the two output arguments (indicated by **mm** and **yy**) correspond to the values stored in **x** for the maxima and the maximum functional values respectively.

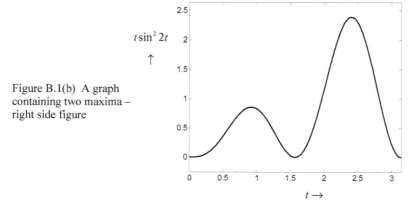

$t\sin^2 2t$

↑

Figure B.1(b) A graph containing two maxima – right side figure

$t \rightarrow$

As an example figure B.1(b) shows the plot of $t\sin^2 2t$ versus t. Given that the two maxima are occurring at $t=0.9183$ and $t=2.4078$ over $0 \le t \le \pi$ and the two local maximum functional values are 0.8549 and 2.3822 respectively which we intend to find.

Our function file works on step size (smaller – better, say 0.0001) selection. As a first step we generate a row vector (section 1.3) on the chosen step size over the given interval by writing **t=0:0.0001:pi;**. Then we calculate the functional values by using the scalar code (appendix A) which becomes **y=t.*sin(2*t).^2;**. After that we call the **mmax** with the **t** (same as **x**) and the **y** as the input arguments. You may execute the following at the command prompt:

```
>>t=0:0.0001:pi; ↵
>>y=t.*sin(2*t).^2; ↵
>>[tm,ym]=mmax(t,y) ↵  ← tm and ym hold the t  points and functional values
                          both as a row matrix respectively, where tm and ym
                          are user-chosen variables
tm =
        0.9183   2.4078
ym =
        0.8549   2.3822
```

There are two maxima that is why it is a two element row matrix. For three maxima, both returns would be a three element row matrix.

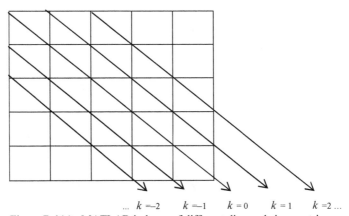

$\dots$ $k=-2$ $k=-1$ $k=0$ $k=1$ $k=2 \dots$

Figure B.1(c)   MATLAB indexes of different diagonals in a matrix

B.18 Picking up diagonal elements from a matrix

If a square/rectangular matrix is given, we pick up subdiagonal, diagonal, or superdiagonal elements by using the function **diag** which applies the syntax **diag**(matrix name, type of diagonal).

Referring to the figure B.1(c), the integers $k=0$, $k=1$, and $k=2$ correspond to the diagonal, the first superdiagonal, and the second superdiagonal respectively and so on. Again $k=-1$ and $k=-2$ correspond to the first and the second subdiagonals respectively and so on.

To work with diagonal elements, let us take $A = \begin{bmatrix} 1 & 2 & 3 & 7 \\ 7 & 8 & 9 & 6 \\ 13 & 9 & 2 & 1 \end{bmatrix}$ and

we wish to pick up the diagonal elements from A (shown by the dotted line, i.e. [1   8   2]) where $k=0$.

How if can the second superdiagonal elements $\Big($ shown by the

dotted line in $\begin{bmatrix} 1 & 2 & 3 & 7 \\ 7 & 8 & 9 & 6 \\ 13 & 9 & 2 & 1 \end{bmatrix} \Big)$ be? The answer is use the command

diag(A,2).

Again to pick up the first subdiagonal elements as shown by dotted

line in $\begin{bmatrix} 1 & 2 & 3 & 7 \\ 7 & 8 & 9 & 6 \\ 13 & 9 & 2 & 1 \end{bmatrix}$, we exercise diag(A,-1). Regardless of the diagonal

type, the return is a column matrix. All examples are shown below:

for the diagonal elements,
```
>>A=[1 2 3 7;7 8 9 6;13 9 2 1]; ↵
>>diag(A,0) ↵
```

ans =
```
   1
   8
   2
```
for the first subdiagonal elements,
```
>>diag(A,-1) ↵
```

ans =
```
   7
   9
```
We could have assigned the return to some user-chosen variable d e.g. d=diag(A,-1) for the last one.

for the 2nd superdiagonal elements,
```
>>diag(A,2) ↵
```

ans =
```
   3
   6
```

```
function y=top2m(x)
    [m,n]=size(x);
    y=[ ];
    for p=-(m-1):n-1
        y=[y mean(diag(x,p))];
    end
```

Figure B.1(d) Function file for averaging diagonal elements in a matrix

B.19 Averaging diagonal elements in a matrix

Suppose we have the rectangular matrix $A = \begin{bmatrix} 1 & 2 & 3 & 7 \\ 7 & 8 & 9 & 6 \\ 13 & 9 & 2 & 1 \end{bmatrix}$. We

wish to average each diagonal (whether sub or super) in the matrix. The main diagonal which is [1   8   2] has the average 3.6667. Again the first superdiagonal which is [2   9   1] has the average 4 and so on. Averaging all elements we should be having [13   8   3.6667   4   4.5   7].

Author written function file **top2m** as seen in figure B.1(d) performs the computation with the syntax **top2m**(rectangular matrix). The return from the function is the average of each diagonal as a row matrix. Type the codes of the figure B.1(d) in a new M-file (section 1.3 and appendix C) and save the file by the name **top2m** in your working path of MATLAB. You may get the softcopy from the link of section 2.7. After having the file, carry out the following:

```
>>A=[1 2 3 7;7 8 9 6;13 9 2 1]; ↵        ← Entering A to A, A is user-chosen
>>B=top2m(A) ↵           ← Calling the function and the return is assigned to B,
                                              B is user-chosen variable

B =
          13.0000   8.0000   3.6667   4.0000   4.5000   7.0000
```

B.20 Rearranging row or column matrix to a rectangular one

A long row or column matrix is converted to a rectangular matrix by using the function **reshape** with the syntax **reshape**(given matrix name, required number of rows, required number of columns) and the return is the rearranged matrix. The row-column number product of the new matrix must be identical with the total number of given elements.

Let us consider the row matrix $H =[3$ 14 -9 0   12   11   56   78   9   34   91   30] which has 12 elements. Whatever be the order of the rearranged matrix, the product of the order of the rearranged matrix must be 12. It is evident that we may have 3×4, 4×3, 6×2, or 2×6 matrices from H. When the elements of H are placed consecutively, they may be arranged either in column by column or in row by row.

Column by column:

From H, we wish to form a matrix N of order 3×4 in which the first column will be the first three elements of H, the second column will be the second three elements of H, and so will be the others i.e. H is rearranged as $N = \begin{bmatrix} 3 & 0 & 56 & 34 \\ 14 & 12 & 78 & 91 \\ -9 & 11 & 9 & 30 \end{bmatrix}$ and its implementation is shown as follows:

```
>>H=[3 14 -9 0 12 11 56 78 9 34 91 30]; ↵   ← Entering H to H
>>N=reshape(H,3,4) ↵

N =
          3    0   56   34
         14   12   78   91
         -9   11    9   30
```

Row by row:

Matrix M of order 3×4 is to be formed from H. In M, the first row will be the first four elements of H, the second row will be the second four elements of H, and so will be the other i.e. $M =$ $\begin{bmatrix} 3 & 14 & -9 & 0 \\ 12 & 11 & 56 & 78 \\ 9 & 34 & 91 & 30 \end{bmatrix}$ and the implementation is shown below:

>>M=reshape(H,4,3)' ↵

```
M =
        3   14   -9    0
       12   11   56   78
        9   34   91   30
```

Since the **reshape** always operates on column basis, the number of rows becomes the number of columns and vice versa and then transposition (') is conducted for the M formation.

Next we take the example of the column matrix $F = \begin{bmatrix} 5 \\ 7 \\ -9 \\ 7 \\ 23 \\ 11 \\ 9 \\ 10 \end{bmatrix}$ which has 8

elements therefore the product of the order of the rearranged matrix must be 8. Like the row one, there can be two possible rearranging.

Column by column:
We have two options to reshape F – either 2×4 or 4×2. Let us say the matrix O of order 2×4 is to be formed from F in which the columns are the consecutive elements of F that is $O = \begin{bmatrix} 5 & -9 & 23 & 9 \\ 7 & 7 & 11 & 10 \end{bmatrix}$ so we execute the following:

>>F=[5 7 -9 7 23 11 9 10]'; ↵
>>O=reshape(F,2,4) ↵

```
O =
        5   -9   23    9
        7    7   11   10
```

Row by row:
Again a matrix P of order 2×4 is to be formed from F in which the rows are consecutive elements of F and it should look like $P = \begin{bmatrix} 5 & 7 & -9 & 7 \\ 23 & 11 & 9 & 10 \end{bmatrix}$ for which we carry out the following:

>>P=reshape(F,4,2)' ↵

```
P =
        5    7   -9    7
       23   11    9   10
```

The numbers of columns and rows interchange as the input arguments of the **reshape** and transposition takes place afterwards in the last execution.

B.21 Cell arrays

A cell array is composed of cells where the cells may contain ordinary arrays (of real, integer, or complex numbers), structure arrays, multidimensional arrays, character arrays... etc. You may assume that the cell is an object rather than a variable. The cells of a cell array are indexed like a rectangular matrix but by using the second brace {...}. For example A{3,4} indicates that the A is a cell array and we are addressing the cell with the coordinates (3,4) – third row and fourth column. If we form a cell array A of order 2×3, the position indexes of different cells are A{1,1}, A{1,2}, A{1,3}, A{2,1}, A{2,2}, and A{2,3} in row direction respectively.

Cell(1,1): $\begin{bmatrix} 1+2i & 2-3i \\ 4+6i & 9+3i \end{bmatrix}$	Cell(1,2): $\begin{bmatrix} 7 & 2 \\ 4 & 9 \end{bmatrix}$
Cell(2,1): [2.34 34.5 4.6]	Cell(2,2): $\begin{bmatrix} x^2-4x+56 \\ 3x+7 \end{bmatrix}$

Figure B.1(e) Two dimensional cell array A of order 2×2

As an example we wish to assign the matrices $\begin{bmatrix} 1+2i & 2-3i \\ 4+6i & 9+3i \end{bmatrix}$, $\begin{bmatrix} 7 & 2 \\ 4 & 9 \end{bmatrix}$, [2.34 34.5 4.6], and $\begin{bmatrix} x^2-4x+56 \\ 3x+7 \end{bmatrix}$ row directionally to different cells of a 2×2 cell array A. Schematic representation for different cell contents is shown in the figure B.1(e) whose implementation is presented in the following:

>>A{1,1}=[1+2i 2-3i;4+6i 9+3i]; ↵ ← Assigning complex matrix $\begin{bmatrix} 1+2i & 2-3i \\ 4+6i & 9+3i \end{bmatrix}$ to cell(1,1), section 1.3 and appendix B.6

>>A{1,2}=[7 2;4 9]; ↵ ← Assigning the integer matrix $\begin{bmatrix} 7 & 2 \\ 4 & 9 \end{bmatrix}$ to cell(1,2)

>>A{2,1}=[2.34 34.5 4.6]; ↵ ← Assigning the decimal element matrix [2.34 34.5 4.6] to cell(2,1)

>>syms x, A{2,2}=[x^2-4*x+56;3*x+7]; ↵ ← Assigning the symbolic matrix $\begin{bmatrix} x^2-4x+56 \\ 3x+7 \end{bmatrix}$ to cell(2,2), appendix B.13

>>A ↵ ← Calling the A to see its contents

A =

    [2x2 double]   [2x2 double]
    [1x3 double]   [2x1 sym]

Instead of displaying the contents, calling of the A shows the type of the component cells.

Let us see some maneuverings of the cell array in the following. The cell(1,2) has the 2×2 integer matrix. Element having the position index (2,1) of this matrix is 4 and we access the element as follows:

>>A{1,2}(2,1) ↵

ans =

4

Choosing proper index similar to a rectangular array, access to the subset of a cell is also possible. For instance the cells of the array A taken from the intersection of the first and second rows and the second column which are shown in figure B.1(f) are invoked as follows:

>>A(1:2,2) ↵

ans =

[2x2 double]
[2x1 sym]

Cell(1,2): $\begin{bmatrix} 7 & 2 \\ 4 & 9 \end{bmatrix}$
Cell(2,2): $\begin{bmatrix} x^2 - 4x + 56 \\ 3x + 7 \end{bmatrix}$

Figure B.1(f) Subset of cell array A

Placing cell inside cell is also possible. Suppose another cell array B of order 1×2 is to be formed where the cell(1,1) and the cell(1,2) of B contain just mentioned 2×2 cell array A and a row matrix [47   31] respectively (figure B.1(g)). It is just the matter of assignment as follows:

Cell(1,1): A	Cell(1,2): [47 31]

Figure B.1(g) 1×2 cell array B showing cell inside cell

>>B{1,1}=A; ↵
>>B{1,2}=[47 31]; ↵
>>B ↵

B =

{2x2 cell}   [1x2 double]

Appendix C

Creating a function file

A function file is a special type of M-file (section 1.3) which has some user-defined input and output arguments. Both arguments can be single or multiple. The first line in a function file always starts with the reserve word **function**. A function file must be in your working path or its path must be defined in MATLAB. Depending on problem, a function file is written by the user and can be called from the MATLAB command prompt or from another M-file. For convenience, long and clumsy programs are split into smaller modules and these modules are written in a function file. The basic structure of a function file is as follows:

MATLAB Prompt   function file

>> g =call f $\Longrightarrow$ $g(\underbrace{y_1, y_2,....y_m})=f(\underbrace{x_1,x_2,x_3,...x_n})$
                                                                    output arguments   input arguments

We present following examples for illustration of function files keeping in mind that the arguments' order and type of the caller and the function file are identical.

🔲🔲 Example 1

Let us say the computation of $f(x)=x^2-x-8$ is to be implemented as a function file. When $x=-3$ and $x=5$, we should be having 4 and 12 respectively.

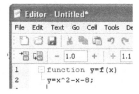

Figure C.1(a) Single input – single output function file

The vector code (appendix A) of the function is **x^2-x-8** assuming scalar **x** and obviously the **x** is for x. We have one input (which is x) and one output (which is $f(x)$). Open a new M-file editor, type the codes of the figure C.1(a) exactly as they appear in the M-file, and save the file by the name **f**. The assignee **y** and independent variable **x** can be any variable of your choice and which are the output and input arguments of the function respectively. Again the file and function name **f** can be any user-chosen name only the point is the chosen function or file name should not exist in MATLAB. Let us call the function **f(x)** to verify the programming as shown in the right side text box. You can write dozens of MATLAB executable statements in the file but whatever is assigned to the last **y** returns the function **f(x)** to **g**. Writing the = sign between the **y** and **f(x)** in the function file is compulsory.

```
Calling for example 1:
>>g=f(-3) ↵   ← call f(x) for x = -3

g =
        4
for x = 5,
>>g=f(5) ↵

g =
        12
```

⊟⊟ Example 2

Example 1 presents one input-one output function how if we handle multiple inputs and one output? The input argument variables are separated by commas in a function file. A three variable function $f(x_1, x_2, x_3) = x_1^2 - 2x_1 x_2 + x_3^2$ is to be computed by a function file. The input arguments (assuming all scalar) are x_1, x_2, and x_3 and

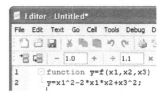

Figure C.1(b) Multiple inputs – single output function file

the output argument is the functional value of the function. The x_1 is written as **x1**, and so is the others. Follow the M-file procedure of the example 1 but the code should be as shown in the figure C.1(b). Let us inspect the function (with the specific $x_1 = 3$, $x_2 = 4$, and $x_3 = 5$, the output value of the three variable function must be $f(3,4,5) = 3^2 - 2 \times 3 \times 4 + 5^2 = 10$) as presented in the text box below.

```
Calling for example 2: when input arguments are all scalar:
>>g=f(3,4,5) ↵     ← calling f(x₁, x₂, x₃) for x₁=3, x₂=4, and x₃=5

g =
      10
Calling for the example 2: when input arguments are all column matrix:
>>x1=[2 3 4]'; ↵   ← x₁ values are assigned to x1 as a column matrix
>>x2=[-2 2 5]'; ↵  ← x₂ values are assigned to x2 as a column matrix
>>x3=[1 0 3]'; ↵   ← x₃ values are assigned to x3 as a column matrix
>>f(x1,x2,x3) ↵    ← calling f(x₁, x₂, x₃) using column matrix input arguments

ans =
      13
      -3
      -15
```

The **function** not only works for the scalar inputs but also does for matrices in general for example a set of input argument values are $x_1 = \begin{bmatrix} 2 \\ 3 \\ 4 \end{bmatrix}$, $x_2 = \begin{bmatrix} -2 \\ 2 \\ 5 \end{bmatrix}$, and $x_3 = \begin{bmatrix} 1 \\ 0 \\ 3 \end{bmatrix}$ for which the $f(x_1, x_2, x_3)$ values should be $\begin{bmatrix} 13 \\ -3 \\ -15 \end{bmatrix}$ respectively.

The computation needs the scalar code (appendix A) of $f(x_1, x_2, x_3)$ regarding x_1, x_2, and x_3. The modified second line statement of the figure C.1(b) now should be **y=x1.^2-2*x1.*x2+x3.^2;**. On making the modification and saving the file, let us carry out the commands which are placed in above text box of this page too. If it is necessary, the output can be assigned to user-supplied workspace variable **v** by writing **v=f(x1,x2,x3)** at the command prompt. The return from the function file also follows the same input matrix order. If the input arguments of **f(x1,x2,x3)** are rectangular matrix, so is the output. The input arguments of the function file do not have to be the mathematics symbol. Suppose x_1 =**ID**, x_2 =**Value**, and x_3 =**Data**, one could have written the first and second lines of the function file in the figure C.1(b) as function **y=f(ID,Value, Data)** and **y=ID.^2-2*ID.*Value+ Data.^2**; respectively.

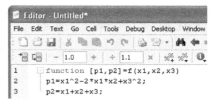

Figure C.1(c) Function file for three input and two output arguments

◱◱ **Example 3**

To illustrate a multi-input and multi-output function file, let us consider that p_1 and p_2 are to be found from three variables x_1, x_2, and x_3 (all are scalars) by employing the expressions $p_1 = x_1^2 - 2x_1 x_2 + x_3^2$ and $p_2 = x_1 + x_2 + x_3$ whose function file (type the codes in a new M-file editor and save the file by the name **f**) is presented in the figure C.1(c).

Choosing x_1 =4, x_2 =5, and x_3 =6, one should get p_1 =12 and p_2 =15 for which right side text box commands are conducted at the command prompt. More

```
Function file calling for the example 3:
>>[p1,p2]=f(4,5,6) ⏎ ← calling the function file f for p₁ and
                        p₂ using x₁=4, x₂=5, and x₃=6
p1 =
        12
p2 =
        15
```

than one output arguments (which are here p_1 are p_2 and represented by **p1** and **p2** respectively) are separated by commas and placed inside the third bracket following the word **function** of the figure C.1(c).

When we call the function from the command prompt, the output argument writing is similar to that of the function file (that is why we write **[p1,p2]** as output arguments at the command prompt). The output argument variable names do not have to be **p1** and **p2** and can be any name of user's choice. If there were three output arguments p_1, p_2, and p_3, the output arguments in the function file would be written as **[p1,p2,p3]** and their calling would happen in a like manner.

Note: We saved different function files by the same name **f** just for simplicity and maintaining unifying approach. By this action any previously saved file by the name **f** disappears. What we suggest is save the function file by other name like **f1** and call accordingly for instance the first line of figure C.1(c) would be function **[p1,p2]=f1(x1,x2,x3)** and calling would take place as **[p1,p2]=f1(4,5,6)** for the last illustration.

Appendix D

Some graphing functions of MATLAB

One of the nicest features of MATLAB is you can have your graphics drawn while you are programming digital audio related problems. There are so many easy accessible built-in graphics functions that one finds it very interesting when the input-output argumentation style of these functions is understood. Some graphing functions which we frequently applied in previous chapters are addressed here for the syntax details.

✦✦ y versus x data

The command plot graphs y versus x data. Let us say we have the attached (on the right side in this

Tabular data of y versus x type:							Command to graph the y vs x data:
x	-6	-4	0	4	5	7	>>x=[-6 -4 0 4 5 7]; ↵
y	9	3	-3	-5	2	0	>>y=[9 3 -3 -5 2 0]; ↵ >>plot(x,y) ↵

paragraph) tabular data. We intend to graph these data as y versus x graph.

Commands to graph the data are also presented beside the tabular data on the right side in the last paragraph. First we assign the x and the y data to the workspace variables x and y (some user-chosen variables) respectively and then call the command plot to see the figure D.1(a). The plot has two input arguments, the first and second of which are the x and y data both as a row or column matrix of identical size respectively.

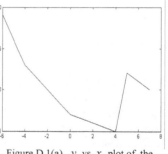

Figure D.1(a) y vs x plot of the tabular data

In order to graph the mathematical expression by using the plot, one first needs to calculate the functional values by using the scalar code (appendix A) and then applies the command. During the calculation, computational step selection is mandatory which is completely user-defined.

For instance we wish to graph the function $f(x) = x^2 - x + 2$ over $-2 \le x \le 3$.

Let us choose some x step say 0.1. The x vector as a row matrix is generated by x=-2:0.1:3; (section 1.3). At every element in x vector, the functional value is computed and assigned to workspace f by f=x.^2-x+2;. The f is any user-chosen variable. Now we call the command by writing plot(x,f) to see the graph (not shown for space reason).

The command **plot** just draws the graph, no graphical features such as x axis label or title are added to the graph. It is the user who is supposed to add these graphical features.

✦✦ Multiple y data versus common x data

The **plot** keeps many options, one of which is just discussed. We graph several y data versus common x data with the help of the same **plot** but with different number of input arguments. Let us choose the right side attached table

Tabular data for multiple y versus common x :						
x	-6	-4	0	4	5	7
y_1	9	3	-3	-5	2	0
y_2	0	-2	1	0	5	7.7
y_3	-1	2	8	1	0	-3

for the graphing. We intend to plot the y_1, the y_2, and the y_3 data on common x data. To do so,

```
>>x=[-6 -4 0 4 5 7]; ↵      ← Assigning the x data as a row matrix to x
>>y1=[9 3 -3 -5 2 0]; ↵     ← Assigning the y₁ data as a row matrix to y1
>>y2=[0 -2 1 0 5 7.7]; ↵    ← Assigning the y₂ data as a row matrix to y2
>>y3=[-1 2 8 1 0 -3]; ↵     ← Assigning the y₃ data as a row matrix to y3
>>plot(x,y1,x,y2,x,y3) ↵    ← Applying the command plot
```

The **plot** now has six input arguments – two for each graph, the first and second of which are the common x data and y data to be plotted respectively. If there were four y data, the command would be **plot(x,y1,x,y2,x,y3,x,y4)**. Once the data is plotted for several y, identifying the y traces is obvious and which is carried out by the command **legend**. The command **legend('y1', 'y2','y3')** puts identifying marks/colors

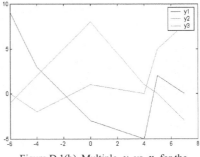

Figure D.1(b) Multiple y vs x for the tabular data

among various graphs. The input argument of the **legend** is any user-given word but under quote and separated by a comma. The number of y traces must be equal to the number of input arguments of the **legend**. We gave the names **y1**, **y2**, and **y3** for the three y traces respectively. In doing so, we end up with the figure D.1(b). You can even move the legend on the plot area by using the mouse. You see all graphics throughout the text as black and white because we did not include color graphics in the text (for expense reason). But MATLAB displays figures in color plots, which you can easily identify.

Another situation can be that we have several functions and intend to plot those on common x variation. For instance we wish to graph $y_1 = x^3 - x^2 + 4$ and $y_2 = x^2 - 7x - 5$ over the common $-1 \le x \le 3$.

Under these circumstances, the step selection of the x data is compulsory. Without calculating the functional values of the given y curves, we can not graph the functions and for which we use the scalar code. Let us choose the x step as 0.1. We first generate the common **x** vector as a row matrix by writing **x=-1:0.1:3;** and then calculate the y_1 and y_2 (**y1**$\Leftrightarrow y_1$ and **y2**$\Leftrightarrow y_2$) data by writing **y1=x.^3-x.^2+4; y2=x.^2-7*x-5;** and eventually the graph appears by executing **plot(x,y1,x,y2)**, graph is not shown for space reason. Thus you can graph three or more functions.

✦✦ Functions of the form $y = f(x)$

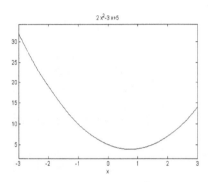

Figure D.1(c) Plot of $y = 2x^2 - 3x + 5$
versus x over $-3 \le x \le 3$

If any function is of the form $y = f(x)$ and the $f(x)$ versus x is to be graphed, the built-in function **ezplot** is the best option which uses a syntax **ezplot**(functional vector code under quote according to appendix A, interval bounds as a two element row matrix) where the first and second elements in the row matrix are the beginning and ending bounds of the interval respectively. The **ezplot** graphs $y = f(x)$ in the default interval $-2\pi \le x \le 2\pi$ when no interval description is argumented.

Let us say we intend to graph the function $y = 2x^2 - 3x + 5$ over the interval $-3 \le x \le 3$. We first give $2x^2 - 3x + 5$ MATLAB vector code and then assign that to **y** as follows:
```
>>y='2*x^2-3*x+5'; ↵
```
In above implementation the **y** is any user-chosen variable. The interval $-3 \le x \le 3$ is entered by **[-3 3]**. To obtain the plot of y in the given interval, we execute the following at the command prompt:
```
>>ezplot(y,[-3,3]) ↵
```
Above command results the figure D.1(c).

✦✦ Multiple graphs in the same window

The function **subplot** splits a figure window in subwindows based on the user definition. It accepts three positive integer numbers as the input arguments, the first and second of which indicate the number of subwindows in the horizontal and the number of subwindows in the vertical directions respectively. For example 22 means two subwindows horizontally and two subwindows vertically, 32 means three subwindows horizontally and two subwindows vertically, ... and so on. The third integer in the input argument numbered consecutively just offers the control on the subwindows so generated. If the first two digits are 32, there should be 6 subwindows

Commands for the figure D.1(d):	
>>subplot(121) ↵	← It handles the first graph
>>ezplot('x') ↵	← Plotting $y = x$
>>subplot(122) ↵	← It handles the second graph
>>ezplot('exp(-x)') ↵	← Plotting $y = e^{-x}$

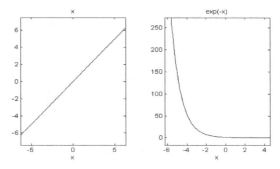

Figure D.1(d) Plots of $y = x$ and $y = e^{-x}$ side by side in the same window

and they are numbered and controlled by using 1 through 6. When you plot some graph in a subwindow, as if you are handling an independent figure window.

Let us say we intend to graph $y = x$ and $y = e^{-x}$ side by side as two different plots by using earlier mentioned **ezplot** but in the same window. If we imagine the subfigures as matrix elements, we have a figure matrix of size 1×2 (one row and two columns). That is why the first two integers of the input argument of the **subplot** should be 12. Attached commands in the upper right text box of the last paragraph show the figure D.1(d). The third integers 1 and 2 in the **subplot** give the control on the first and second subfigures respectively.

As another example we wish to plot $y = x$ and $y = e^{-x}$ in the upper row and only $y = (1 - e^{-x})$ in the lower row subfigures in the

Commands for the figure D.1(e):	
>>subplot(221) ↵	← Subfigure selection for $y = x$
>>ezplot('x') ↵	← Plotting $y = x$
>>subplot(222) ↵	← Subfigure selection for $y = e^{-x}$
>>ezplot('exp(-x)') ↵	← Plotting $y = e^{-x}$
>>subplot(212) ↵	← Subfigure selection for $y = (1 - e^{-x})$
>>ezplot('1-exp(-x)') ↵	← Plotting $y = (1 - e^{-x})$

same window whose implementation needs above attached text box commands and whose final output is the figure D.1(e). We are supposed to have four figures when the integer input argument of the **subplot** is 22 (two for rows and two for columns). The arguments 221, 222, 223, and 224 provide handle on the four figures consecutively. But the figures could have been plotted on 223 and 224 are absent so we ignore them. The argument 21 creates two subfigures (two rows and one column) handled by 211 and 212, but 211 is absent so we ignore that.

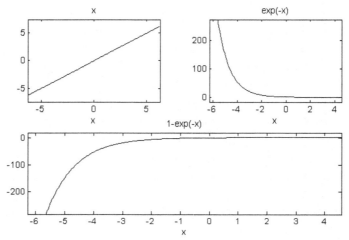

Figure D.1(e) Plots of $y = x$ and $y = e^{-x}$ in the upper row and $y = (1 - e^{-x})$ in the lower row in the same window

Let us see the input arguments of the **subplot** for different subfigures (each third brace set [ ] is one subfigure in the following tabular representation) as follows:

Subfigures needed	First two input integers of subplot	Third input integer of subplot	Commands we need
[ ] [ ]	22	[1] [2]	subplot(221) subplot(222)
[ ] [ ]		[3] [4]	subplot(223) subplot(224)
[ ] [ ]	22 for upper two (lower two remain empty)	[1] [2]	subplot(221) subplot(222)
[ ]	21 for the lower one (upper one remains empty)	[2]	subplot(212)
[ ]	21 for the upper one (lower one remains empty)	[1]	subplot(211) subplot(223) subplot(224)
[ ] [ ]	22 for the lower two (upper two remain empty)	[3] [4]	
[ ] ⎡ ⎤	22 for the left two (right two remain empty)	[1] ⎡ ⎤	subplot(221) subplot(223) subplot(122)
[ ] ⎣ ⎦	12 for the right one (left one remains empty)	[3] ⎣ 2 ⎦	
⎡ ⎤ [ ]	22 for right two (left two remain empty)	⎡ ⎤ [2]	subplot(222) subplot(224) subplot(121)
⎣ ⎦ [ ]	12 for the left one (right one remains empty)	⎣ 1 ⎦ [4]	

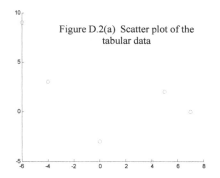

Figure D.2(a) Scatter plot of the tabular data

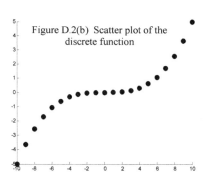

Figure D.2(b) Scatter plot of the discrete function

✦ ✦ Scatter data plot by using small circles for discrete signal

Instead of having a graph as continuous line, it is possible to have the graph in terms of bold dots or round circles like the figure D.2(a). The function **scatter** executes this kind of graph for which the common syntax is **scatter**(x data as a row matrix, y data as a row matrix, size of the circle, color of the circle). The command also accepts the first two input arguments. The size of the circle is any user-given integer number. The larger is the number, the bigger is the size for example 75, 100, etc.

Let us graph the table D.1 data as the scatter plot. The command we need is placed on the right side attached text box. Upon execution of the command, we see the figure D.2(a). The color of the circle is blue by default but any three element row matrix sets the user-defined color. The

Table D.1: Some x - y data						
x	-6	-4	0	4	5	7
y	9	3	-3	-5	2	0

Scatter plot for the table D.1 data:
```
>>x=[-6 -4 0 4 5 7]; ↵
>>y=[9 3 -3 -5 2 0]; ↵
>>scatter(x,y) ↵
```

three element row matrix refers to red, green, and blue components respectively each one within 0 and 1. Black color means all zero, white means all 1, red means other two components zero, and so on. The circle displayed in the figure D.2(a) is all empty but one fills the circle by using the reserve word **filled** under quote and including another input argument to the **scatter**. Let us say we intend to scatter graph with circle size 100 and the circles should be filled with black color. The necessary command is **scatter(x,y,100,[0 0 0],'filled')**, graph is not shown for space reason.

This type of graph is suitable for representation of the function which is discrete in nature. For instance the discrete function $y[n] = \dfrac{n^3}{200}$ over $-10 \le n \le 10$ is to be plotted with black circles of size 100 where n is integer. As a procedure, we form a row matrix **n** to generate (section 1.3) the interval with start value -10, increment 1, and end value 10 on writing **n=-10:10;**. The scalar code of appendix A computes the $y[n]$ values and assigns those to

-333-

workspace **y**. However the complete code is placed on the right side attached text box which brings about the graph of the figure D.2(b).

> Scatter plot for discrete function:
> \>\>n=-10:10; y=n.^3/200; ↵
> \>\>scatter(n,y,100,[0 0 0],'filled') ↵

❖ ❖ Discrete function or data plotting using vertical lines

Just now we discussed how one graphs the discrete functional data by using bold dots. There is another option which graphs any discrete data by using vertical lines proportionate to the discrete functional values. A discrete function may exist in two forms – data and expression based. The built-in function that graphs a discrete function is **stem**

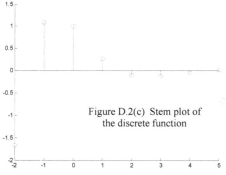

Figure D.2(c) Stem plot of the discrete function

which has a syntax **stem**(x data as a row matrix, y data as a row matrix). If we have some expression based discrete function, first the sample values of the discrete function need to be calculated through the scalar code (appendix A) and after that the graphing is performed.

Let us plot the discrete function $f[n]=$ $2^{-n}\cos n$ over the integer interval $-2 \le n \le 5$. Attached on the right side text box is the implementation of the command which results the figure D.2(c). In the implementation, the workspace **n** and **f** hold the eight integers from -2 to 5 and sample values of $f[n]$ both as a row matrix respectively.

> Discrete function plot using vertical lines:
> \>\>n=-2:5; ↵
> \>\>f=2.^(-n).*cos(n); ↵
> \>\>stem(n,f) ↵

By default the vertical line color of the stem plot is blue. User-defined color of the vertical lines is obtainable by adding one more input argument to **stem** mentioning the color type but under quote (**r** for red, **g** for green, **b** for blue, **c** for cyan, **m** for magenta, **y** for yellow, **k** for black, and **w** for white). If we wish to set the vertical line color as green for the graph of the figure D.2(c),

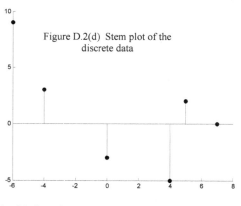

Figure D.2(d) Stem plot of the discrete data

we exercise the command **stem(n,f,'g')**. The vertical line head circles are filled by using the command **stem(n,f,'g','filled')** where **filled** is a reserve word placed under quote.

If the reader wants to plot the table D.1 data considering discrete signals $x[n]$ and $y[n]$ as stem plot, following needs to be exercised:

>>x=[-6 -4 0 4 5 7]; ⏎   ← x holds $x[n]$ data

>>y=[9 3 -3 -5 2 0]; ⏎   ← y holds $y[n]$ data

>>stem(x,y,'k','filled') ⏎   ← calling the stem for plot with black line and filled circles

Figure D.2(d) is output from above executions in which the horizontal and vertical axes of the figure correspond to $x[n]$ and $y[n]$ data respectively.

Appendix E

Algebraic equation solver

By virtue of the built-in master function **solve**, we find the solution of a single or multiple algebraic equations when the equations are its input arguments. The notion of the solution is symbolic and a substantial number of simultaneous linear, algebraic, or trigonometric equations can be solved by using the **solve**. The common syntax of the implementation is **solve**(equation-1, equation-2, ... so on in vector string form – appendix A, unknowns of the equations separated by a comma but put under quote).

The return from the function **solve** is in general a structure array which is beyond the discussion of the text (reference 47). Very briefly a structure array is composed of several members. In order to view the solution from the **solve**, one needs to call a member of the array. If **s** is a structure array and **u** is one of its members, we call the member by using the command **s.u**. One can assign the **s.u** to some other workspace variable if it is necessary. Following points must be considered while using the **solve**:

(a) We solve equations related to variables which have power 1 or more. One variable equation of power 1 for example $2x - 7 = 5x$ is written as **2*x-7=5*x** in code form, the two variable equations $7x + y - 7 = 0$ and $2y + 4 = 5x$ are written as **7*x+y-7=0** and **2*y+4=5*x** respectively, and so on.

(b) Since the solution approach is completely symbolic, we enter rational value of the equation coefficient in case of decimal data for example the equation $2.4x - 7.5 = 5x$ had better be written as **24*x/10-75/10=5*x**.

(c) The **i** or **j** is the unit imaginary number in MATLAB (appendix B.6). Such use sometimes turns the function **solve** non-executable. If any variable **i** is in the equations, we use a dummy variable for example **c**.

(d) The return from the **solve** is usually in rational form for instance **24/10** instead of 2.4. If we need the decimal value, we employ the command **double** on the return.

(e) The equations are assigned under quote while entering as input arguments to the **solve** or assigning to some variables.

Let us go through the solution finding for following three examples.

♦♦ Example 1

The solution of the equation $2x + 7 = 9x$ is $x = 1$ which we wish to find. We execute the following for the solution at the command prompt:

```
>>s=solve('2*x+7=9*x','x') ↵

s =
```

1

The **s** in last execution is any user-chosen variable. The **s=1** return indicates the $x=1$ solution. The independent variable in the given equation is x that is why the second input argument of **solve** is '**x**'.

♦ ♦ **Example 2**

The equation set $\begin{Bmatrix} 6x - y = -8 \\ 9x = 8y + 5 \end{Bmatrix}$ has the solution $x = -\dfrac{23}{13} = -1.7692$ and

$y = -\dfrac{34}{13} = -2.6154$ and our objective is to obtain the solution.

The given two equations have the codes **6*x-y=-8** and **9*x=8*y+5** respectively. The related variables in the two equations are x and y therefore we carry out the following at the command prompt:

>>s=solve('6*x-y=-8','9*x=8*y+5','x','y') ⏎

```
s =
        x: [1x1 sym]
        y: [1x1 sym]
```

The **s** in above execution is also any user-chosen variable. The **solve** returns the solution to the **s**. As we mentioned earlier the return from the **solve** is a structure array and its members are the **x** and **y** (related variables in the given equations). The return is an object (called symbolic object, indicated by the **sym**) rather than data. Should we pick up the solution of x and y from the **s**, we need to exercise the commands **s.x** and **s.y** respectively. Let us see what we obtain as the solution:

For the x value:

>>s.x ⏎

ans =

-23/13
>>double(s.x) ⏎

ans =
        -1.7692

For the y value:

>>s.y ⏎

ans =

-34/13
>>double(s.y) ⏎

ans =
        -2.6154

The result is as expected. We could have assigned the return to some variable for example **s.x** or **double(s.x)** to **a** by writing **a=s.x** or **a=double(s.x)**.

♦ ♦ **Example 3**

For multiple equations it is not feasible that we enter all equations as one line to the **solve**. Instead we first assign the given equations to some user-chosen variables and then call the **solve** with these variable names as the input arguments. The equation set $\{x - y - 3.2z + 2u = -8, 8.5y - 7z + u = 5, x - 4y + 2z = 76, -3.4x + 6z + 7u = -12\}$ has the solution $u = \dfrac{29598}{1387} = 21.3396$, $x = \dfrac{348525}{2774} = 125.6399$, $y = \dfrac{95869}{2774} = 34.5598$, and $z = \dfrac{245775}{5548}$

=44.2997 which we find by exercising ongoing function and symbology as follows:

```
>>e1='x-y-32*z/10+2*u=-8'; ↵        ← assigning the first equation to e1
>>e2='85*y/10-7*z+u=5'; ↵           ← assigning the second equation to e2
>>e3='x-4*y+2*z=76'; ↵              ← assigning the third equation to e3
>>e4='-34*x/10+6*z+7*u=-12'; ↵      ← assigning the fourth equation to e4
>>s=solve(e1,e2,e3,e4,'x','y','z','u') ↵  ← calling the solve on e1, e2, e3, and e4
```

```
s =                         ← s holds the solution as a structure array
    u: [1x1 sym]            ← u is a member of s
    x: [1x1 sym]            ← x is a member of s
    y: [1x1 sym]            ← y is a member of s
    z: [1x1 sym]            ← z is a member of s
```

The e1, e2, e3, and e4 are all user-chosen variables in above. The next step is to see the values returned by the solve:

for the rational value of x :

```
>>s.x ↵

ans =

348525/2774
```

for the rational value of y :

```
>>s.y ↵

ans =

95869/2774
```

for the rational value of z :

```
>>s.z ↵

ans =

245775/5548
```

for the rational value of u :

```
>>s.u ↵

ans =

29598/1387
```

for the decimal value of x :

```
>>double(s.x) ↵

ans =
        125.6399
```

for the decimal value of y :

```
>>double(s.y) ↵

ans =
        34.5598
```

for the decimal value of z :

```
>>double(s.z) ↵

ans =
    44.2997
```

for the decimal value of u :

```
>>double(s.u) ↵

ans =
        21.3396
```

When we assign the equations, we use the quote but inside the solve the assignees do not have quote for example e1 not 'e1'. All four values as a four element row matrix are seen by:

```
>>[s.x s.y s.z s.u] ↵

ans =

[ 348525/2774, 95869/2774, 245775/5548, 29598/1387]
```

All four values in decimal form as a row matrix are seen as follows:

```
>>double([s.x s.y s.z s.u]) ↵

ans =
    125.6399  34.5598  44.2997  21.3396
```

Obviously the return is in the order we typed in.

Appendix F
MATLAB functions exercised in the text

Function name	Purpose	Page
abs	returns the absolute value of real or complex numbers	294, 306
angle	extracts phase angle from a complex number or matrix in radian	306
axis	sets user defined axes on a drawn graph	161, 177
basspec	implements basic spectral subtraction algorithm, author written	186
bin2dec	converts binary string to its decimal equivalent	214
blackman	generates samples of a Blackman window	79
burg	built-in word for specifying the Burg power spectrum type	211
butter	designs Butterworth filter of various types	113
cart2pol	converts a number from rectangular to polar system	307
char	determines keyboard characters from integer ASCII codes	222
charfcn[0](n)	generates the unit sample $\delta[n]$ symbolically	92
cheby1	designs various Chebyshev filters of type I	114
cheby2	designs various Chebyshev filters of type II	115
cond	determines the condition number of a matrix	128
conj	finds complex conjugate of a number	307
conv	performs convolution of two discrete signals	50
corrcoef	determines the correlation coefficient of two or more random variables	157
cos	finds the trigonometric cosine of all elements in a matrix when the matrix is its input argument	294

Continuation of the last table:

Function name	Purpose	Page
cosd	finds trigonometric cosine when its input argument is in degree	294
cov	determines the covariance of two or more random variables	139
datastats	finds comprehensive statistical properties like mean, range, etc of some user-supplied data	26
dec2bin	converts decimal integer number to its binary equivalent	214
deg2rad	converts a number from degree to radian	306
det	determines the determinant of a square matrix	124
diag	extracts different diagonals from a rectangular matrix	319
dir	displays files that are present in the current path or directory of MATLAB	18
double	turns any symbolic data to double precision value or decimal	336
downsample	downsamples or decimates a discrete signal by integer factor when the signal is its input argument and as a row or column matrix	41
eig	determines the eigenvalue and eigenvector of a square matrix	126
end	terminates the execution of a for-loop or if-else checking	299, 302
eps	MATLAB synonym for small quantity ε	69
exp	takes exponent of a number	294
ezplot	draws y versus x type graph from y expression and x interval	330
factor	factorizes an integer	7
fft	computes the discrete Fourier transform of a discrete signal	56
fftshift	flips a discrete signal about its half index	62
figure	opens a new window for graphing	180
filt	forms a filter object or $H(z)$ system	117
filter	performs filtering operation on a discrete signal	98

Continuation of the last table:

Function name	Purpose	Page
find	determines the element position in a matrix subject to condition	311
fix	discards the fractional part of a decimal number regardless of the magnitude and returns the integer part	309
fliplr	flips a discrete signal from left to right	308
flipud	flips a discrete signal from up to down	308
for-end	beginning and ending statements of a for-loop	302
freqz	computes values or graphs frequency spectra based on supplied system function $H(z)$	101
function	reserve word for writing a function file	325
g	generates one cycle glottal pulse samples, author written	201
gn	generates multi cycle glottal pulse samples from single cycle samples, author written	202
hamming	generates samples of a Hamming window	79
hann	generates samples of a Hanning window	80
heaviside(n)	generates the unit sequence $u[n]$ symbolically	90
help	search function for getting assistance about an M-file function	13
hist	determines and graphs the frequency of a discrete signal which contains range type data	148
huffmandict	determines Huffman codes from probability vector	218
if-else	built-in words for conditional or logical checking	299
ifft	computes the inverse discrete Fourier transform from the forward counterpart	56
imag	extracts imaginary part from a complex number or matrix	307
impz	computes inverse Z transform numerically	99

Continuation of the last table:

Function name	Purpose	Page
interp1	performs one dimensional interpolation on a discrete signal	47
inv	determines the inverse of a square matrix	125
iztrans	computes inverse Z transform in expression form	93
legend	puts user-supplied differentiating texts on a drawn graphics mainly on y versus x type plots	137
length	determines the number of elements in a row or column matrix	192
load	imports data to workspace from a saved mat file	224
log	computes natural logarithm of a number	294
log10	takes logarithm on common base 10	294
log2	takes logarithm on the base 2	294
lookfor	search function for any M-file function or file existing in MATLAB in accordance with user-supplied words	13
lpc	computes linear prediction coefficients	169
mat2gray	maps user-provided data between 0 and 1	31
max	finds the maximum from numerical data supplied as a row or column matrix	304
mean	finds the statistical average of some user-supplied data	24
median	finds the statistical median of some user-supplied data	25
min	finds the minimum from numerical data supplied as a row or column matrix	304
mmax	finds multiple maxima from numerical data supplied as a row or column matrix, author written	318
moment	computes moments of user-defined order on a discrete signal	155
mtop	converts a row or column matrix to user defined Toeplitz matrix, author written	130
nnz	finds the number of nonzero elements in a matrix	316

Continuation of the last table:

Function name	Purpose	Page
nonzeros	extracts nonzero elements from a matrix	316
normrnd	generates Gaussian random number samples from user-supplied mean and variance	191
numel	determines the number of elements in a row or column matrix	33
ones	generates matrix of ones	312
padarray	pads array by user-defined elements in all directions	313
parallel	determines the equivalent of two parallel Z transform system functions	119
phone	displays telephone dial pad signal's time and frequency domain waveshapes	199
pi	π	294
plot	graphs the y trace (s) versus x type data	329
pol2cart	converts a number from polar to rectangular	307
poly	determines the characteristic polynomial of a square matrix	126
pretty	displays any coded expression in mathematical form	314
princomp	determines the principal components from a matrix	144
psd	built-in function for object based power spectral density computing and graphing	211
pwsd	computes the discrete power spectral density of a discrete signal, author written	210
quant	quantizes a discrete signal from the user-defined resolution	44
rad2deg	converts a number from radian to degree	306
range	finds the statistical range of some user-supplied data	24
rank	determines the rank of a matrix	123
real	extracts the real part from a complex number or matrix	307
rectwin	generates the samples of a rectangular window	79

Continuation of the last table:

Function name	Purpose	Page
rem	computes the remainder of an integer after integer division	310
resample	resamples a discrete signal of given sampling frequency to another sampling frequency	43
reshape	rearranges a row or column matrix to rectangular matrix	321
rle	determines run length encoding of integer discrete signal, author written	215
roots	determines the roots of a polynomial	316
round	rounds the fractional or decimal number to its nearest integer	310
save	saves workspace variables to a file	224
scatter	graphs x-y points in terms of scattered dots	333
series	determines the equivalent of two series Z transform system functions	118
signal	a directory containing signal processing functions	14
simple	simplifies any expression which is in coded form in MATLAB	91
simplify	simplifies any expression which is in coded form in MATLAB	93
sin	finds the trigonometric sine of all elements in a matrix when the matrix is its input argument	294
size	determines array or matrix dimensions	320
solve	solves algebraic equations	336
soundview	shows the sound wave in a tiny window	23
specgram	computes as well as graphs the short time Fourier transform	84
spectrum	built-in word for specifying the power spectrum type of discrete signal	211
sqrt	computes square root of its input argument elements whether scalar or matrix	294
std	finds the statistical standard deviation of some user-supplied data	24
stem	graphs discrete signals by vertical lines	334

Continuation of the last table:

Function name	Purpose	Page
subplot	divides a figure window into subwindows according to user-definition	331
subs	substitutes user-given value(s) to some mathematical expression	93
sum	sums all elements in a row or column matrix	313
svd	decomposes a rectangular matrix into three matrices based on singular values	128
svd_noise	computes the approximate signal on singular value discarding, author written	197
sym	declares independent variables of an expression but mainly applied on fractional numbers for conversion from decimal to symbol or rational form	314
syms	declares independent variables of an expression	314
tabulate	determines the amplitude frequency of a discrete signal which contains only integer samples	146
tfdata	extracts numerator and denominator polynomial coefficients from filter object	119
title	includes a title statement in a drawn graphics	9
toeplitz	forms Toeplitz matrices	315
top2m	determines the average of the diagonal elements of a rectangular matrix, author written	321
trapz	computes trapezoidal integration taking x and y data	152
unifrnd	generates uniform random number samples from user-supplied range	190
upsample	upsamples or interpolates a discrete signal by integer factor when the signal is its input argument and as a row or column matrix	41
wavfinfo	displays digital audio file information in MATLAB	19
wavplay	plays a softcopy digital audio contents in MATLAB	18

Continuation of the last table:

Function name	Purpose	Page
wavread	reads a softcopy digital audio in MATLAB	18
wavwrite	writes workspace audio data to a portable softcopy file, mainly to .**wav**	225
welch	built-in word for specifying the Welch power spectrum	211
who	displays variables present in the workspace	225
wiener2	implements Wiener filtering in integer time index domain	189
win	generates single sample of any user-defined window, author written	74
xcorr	computes autocorrelation of a discrete signal	159
zeros	generates matrix of zeroes	312
zplane	graphs poles and zeroes of a system function $H(z)$ in z plane	101
ztrans	computes forward Z transform in expression form	91

References

>> >> Audio Signal Fundamentals >> >>

[1] Bob McCarthy, *"Sound Systems: Design and Optimization - Modern Techniques and Tools for Sound System Design and Alignment"*, 2007, Focal Press, Elsevier, Jordan Hill.
[2] Stanley R. Alten, *"Audio in Media"*, 1981, Wadsworth Publishing Company, Belmont, California.
[3] Ken C. Pohlmann, *"Principles of Digital Audio"*, 1989, Second Edition, Howard W. Sams & Company, Indiana.
[4] Philipos C. Loizou, *"Speech Enhancement-Theory and Practice"*, 2007, CRC Press, Boca Raton, Florida.
[5] Alan V. Oppenheim and Ronald W. Schafer, *"Discrete-Time Signal Processing"*, 1989, Prentice Hall, New Jersey.
[6] Richard C. Dorf, *"The Electrical Engineering Handbook-Circuits, Signals, and Speech and Image Processing"*, Third Edition, 2006, CRC Press, Taylor & Francis Group, Florida.
[7] _____, *"X European Signal Processing Conference"*, Sept. 4-8, 2000, Tampere, Finland.
[8] Steven L. Gay and Jacob Benesty, *"Acoustic Signal Processing for Telecommunication"*, Kluwer Academic Publishers, 2000, Boston.
[9] Sunil Bharitkar and Chris Kyriakakis, *"Immersive Audio Signal Processing"*, Springer, 2006, New York.
[10] Ulrich Karrenberg, *"An Interactive Multimedia Introduction to Signal Processing"*, Springer-Verlag, 2002, Heidelberg.
[11] Saeed V. Vaseghi, *"Multimedia Signal Processing-Theory and Applications in Speech, Music and Communications"*, John Wiley & Sons, Ltd, 2007, Chichester.
[12] Udo Zölzer, *"Digital Audio Signal Processing"*, John Wiley & Sons, Ltd, 1998, Chichester.
[13] James H. McClellan, Ronald W. Schafer, and Mark A. Yoder, *"DSP First – A Multimedia Approach"*, Prentice-Hall, Inc., 1998, Upper Saddle River, New Jersey.
[14] Ashok Ambardar, *"Digital Signal Processing: A Modern Introduction"*, Thomson, 2007, Toronto.
[15] Mark Kahrs and Karlheinz Brandenburg, *"Applications of Digital Signal Processing to Audio and Acoustics"*, Kluwer Academic Publishers, 1998, Massachusetts.
[16] Robert L. Mott, *"Sound Effects – Radio, TV, and Film"*, Focal Press, 1990, Boston.
[17] Derek Cameron, *"Audio Technology Systems: Principles, Applications, and Troubleshooting"*, Reston Publishing Company, Inc., 1978, Virginia.
[18] Jay Rose, *"Producing Great Sound for Digital Video"*, Miller Freeman Books, 1999, California.
[19] John Garas, *"Adaptive 3D Sound Systems"*, Kluwer Academic Publishers, 2000, Massachusetts.

>> >> MATLAB/SIMULINK Texts >> >>

[20] Mohammad Nuruzzaman, *"Tutorials on Mathematics to MATLAB"*, 2003, AuthorHouse, Bloomington, Indiana.
[21] Mohammad Nuruzzaman, *"Modeling and Simulation in SIMULINK for Engineers and Scientists"*, 2005, AuthorHouse, Bloomington, Indiana.
[22] Mohammad Nuruzzaman, *"Digital Image Fundamentals in MATLAB"*, 2005, AuthorHouse, Bloomington, Indiana.
[23] Duffy, Dean G., *"Advanced Engineering Mathematics with MATLAB"*, Second Edition, 2003, Chapman & Hall, CRC, Boca Raton.
[24] Hanselman, Duane C. and Littlefield, Bruce R., *"Mastering MATLAB 5: A Comprehensive Tutorial"*, 1998, Prentice Hall, Upper Saddle River, New Jersey.
[25] Shampine, Lawrence F. and Reichelt, Mark W., *"The MATLAB ODE Suite"*, 1996, The Math-Works, Inc., Natick, MA.
[26] Marcus, Marvin, *"Matrices and MATLAB - A Tutorial"*, 1993, Prentice Hall, Englewood Cliffs, N. J.

[27] Ogata, Katsuhiko, *"Solving Control Engineering Problems with MATLAB"*, 1994, Englewood Cliffs, N. J. Prentice Hall.

[28] Part-Enander, Eva, *"The MATLAB Handbook"*, 1998, Harlow: Addisson Wesley.

[29] Prentice Hall, Inc., *"The Student Edition of MATLAB for MS-DOS Personal Computers"*, 1992, Prentice Hall, Englewood Cliffs, N. J.

[30] Saadat, Hadi., *"Computational Aids in Control Systems Using MATLAB"*, 1993, McGraw-Hill, New York.

[31] Gander, Walter. and Hrebicek, Jiri., *"Solving Problems in Scientific Computing Using MAPLE and MATLAB"*, 1997, Third Edition, Springer Verlag, New York.

[32] Biran, Adrian B and Breiner, Moshe, *"MATLAB for Engineers"*, 1997, Addison Wesley, Harlow, Eng.

[33] D. M. Etter, *"Engineering Problem Solving with MATLAB"*, 1993, Prentice Hall, Englewood Cliffs, N. J.

[34] Shahian, Bahram. and Hassul, Michael., *"Control System Design Using MATLAB"*, 1993, Prentice Hall, Englewood Cliffs, N. J.

[35] Prentice Hall, Inc., *"The Student Edition of MATLAB for Macintosh Computers"*, 1992, Prentice Hall, Englewood Cliffs, N. J.

[36] Ogata, Katshuiko, *"Designing Linear Control Systems with MATLAB"*, 1994, Prentice Hall, Englewood Cliffs, N. J.

[37] Bishop, Robert H., *"Modern Control Systems Analysis and Design Using MATLAB"*, 1993, Addsison Wesley, Reading, MA.

[38] Moscinski, Jerzy and Ogonowski, Zbigniew., *"Advanced Control with MATLAB and Simulink"*, 1995, E. Horwood, Chichester, Eng.

[39] Alberto Cavallo, Roberto Setola, and Francesco Vasca, *"Using MATLAB Simulink and Control Systems Toolbox - A Practical Approach"*, 1996, Prentice Hall, London.

[40] Jackson, Leland B., *"Digital Filters and Signal Processing with MATLAB Exercises"*, Third Edition, 1996, Kluwer Academic Publishers, Boston.

[41] Kuo, Benjamin C. and Hanselman, Duanec., *"MATLAB Tools for Control System Analysis and Design"*, 1994, Prentice Hall, Englewood Cliffs, N. J.

[42] Chipperfield, A. J. and Fleming, P. J., *"MATLAB Toolboxes and Applications for Control"*, 1993, London, New York: Peter Peregrinus on Behalf of the Institute of Electrical Engineers.

[43] Math Works Inc., *"MATLAB Reference Guide"*, Math Works Inc., 1993, Natick, Massachusetts.

[44] Cleve Moler and Peter J. Costa, *"MATLAB Symbolic Math Toolbox"*, User's Guide, Version 2.0, May 1997, Natick, Massachusetts.

[45] James W. Nilsson and Susan A. Riedel, *"Using Computer Tools for Electric Circuits"*, 1996, Fifth Edition, Addison Wesley Publishing Company, Natick, Massachusetts.

[46] Theodore F. Bogart, *"Computer Simulation of Linear Circuits and Systems"*, 1983, John Wiley and Sons, Inc., New York.

[47] Mohammad Nuruzzaman, *"Technical Computation and Visualization in MATLAB for Engineers and Scientists"*, February, 2007, AuthorHouse, Bloomington, Indiana.

[48] Mohammad Nuruzzaman, *"Electric Circuit Fundamentals in MATLAB and SIMULINK"*, October 2007, BookSurge Publishing, Charleston, South Carolina.

[49] Mohammad Nuruzzaman, *"Signal and System Fundamentals in MATLAB and SIMULINK"*, July 2008, BookSurge Publishing, Charleston, South Carolina.

[50] Mohammad Nuruzzaman, *"Modern Approach to Solving Electromagnetics in MATLAB"*, January 2009, BookSurge Publishing, Charleston, South Carolina.

Subject Index_____

rt>

Printed in the USA